Problem Solving in Environmental Engineering

Prof. Dr. Eng. Isam Mohammed Abdel-Magid Ahmed

Dr. Mohammed Isam Mohammed Abdel-Magid

Second version, improved, updated and revised, Dammam, 2015

First edition, published by University of Dammam Press, Dammam, 2013

ISBN-13: 978-1517007904
ISBN-10: 1517007909

Printed by: **CreateSpace**

Prof. Dr. Eng. Isam Mohammed Abdel-Magid Ahmed: Head Proofreading and revision department at the Centre of Scientific Publications and Dammam University, Professor of water resources and environmental engineering, Building 800, Room 240 Environmental Engineering Department, College of Engineering, University of Dammam, Box 1982, Dammam 31451, KSA, Fax: 96638584331+, Phone: 966530310018+, E-mail: iahmed@ud.edu.sa, isam_abdelmagid@yahoo.com, Web site: http://www/sites.google.com/site/isamabdelmagid/

Dr. Mohammed Isam Mohammed Abdel-Magid: Department of Internal Medicine, Khasab Hospital & Polyclinic, P. O. Box 306, Postal 811, Khasab, Musandam, Sultanate of OmanPhone: +97470445235, +971558215655, +249969263307, E-mail: mohammed_isam1984@yahoo.com, website: http://sites.google.com/site/mohammedisam2000

Preface to Second Edition

The first version of the book on "problem solving in environmental engineering" was printed by Dammam University Press as per recommendations of the dean of the college of engineering Dr. Dr. Abdul-Rahman ben Salih Hariri. It was decided to explore practically the expected benefits from the book as a course supplement and a practical-based text before its publication. The response to the previous edition of this book has been overwhelming. Used version of the book served a vital role in knowledge transfer to sanitary and environmental engineering undergraduate students. Constructive comments, positive observations, helpful notes, productive clarifications, useful interpretations, beneficial critiques, practical expansions and profitable remarks were received from a vast number of environmental personnel and authorities. Such a gather and a forum motivated enhancements, updates and revision of the book, a process that yielded this new version. Therefore, in this edition, certain portions have been updated, some new material has been added, and a good number of new problem solved issues have been included.

It is our understanding that computer modeling is a vital part of the practice of environmental engineering. To meet this goal, most of the examples in the book were constructed as computer software applications that demonstrate how to transform an environmental engineering problem into a working computer model that can be manipulated and edited. To avoid repetition, and to maintain the size of the print, some examples were left without accompanying models. In such cases, the omitted model will have some other model similar to it in the nearby text, to which there will be a pointer to the reader such as "*See Program .. Listing*".

All models were written using Microsoft Visual Basic.NET 10 programming language. The Integrated Development Environment (IDE) used was Microsoft Visual Studio 2010 Professional Edition. Examples were test-ran under Microsfot Windows XP and 7 workstations, although the pre-compiled code should run on other platforms with .NET 4.x support, such as Linux with MonoDevelop installed (more options are presented in Chapter 1).

To meet requirements set for this book, brevity constituted a vital consideration throughout. For those who deem that it has been too brief and concise, they are kindly requested to seek further information in standard references on the subject.

Special thanks and gratitude are due to Dammam Construction and Environmental engineering students, colleagues, anonymous reviewers, engineers scattered throughout the region and beyond for their specific suggestions to improve the text, generous help with scientific comments and for e-mail discussions of activity corrections and other weighty matters to provide recommendations for modification and amendment.

The authors hope that this new edition will continue to remain a popular book on environmental engineering field as it proved to be with the first edition whereby requests for copies are continuously received from readers worldwide. The authors would appreciate and welcome any meticulous suggestions and thoughtful detailed comments, towards improvement or detection of errors from students, readers, reviewers, faculty and others.

Prof. Dr. Eng. Isam Mohammed Abdel-Magid Sr
Dr. Mohammed Isam Mohammed Abdel-Magid Jr
Al-Khobar, Doha, 2014

Preface of First Edition

Books and text material in the field of environmental and sanitary engineering are dearth tackling principles, fundamentals, methods, models, concepts, designs and related scientific notions. Nonetheless, rare are the books that deal with practical problems and problem solving in environmental engineering.

This book has been written to deal with substance delineated in the 3-credit hour course of "Sanitary Engineering, ENVE 471" given for senior final year students of department of construction engineering of the College Of Engineering at Dammam University. The named course contents encompassed: Water sources and selection. Water demand. Water and wastewater characteristics, Quality and health impacts. Water and wastewater treatment unit operations. Water storage and distribution within buildings and municipality Wastewater reclamation and reuse. Solid and hazardous wastes engineering and management.

Aims and objectives of the book are proposed to accomplish the main learning outcomes for students enrolled in the course, which incorporated the following:

- Understanding and explaining the role of sanitation in the urban water cycle and its relation to public health and environment.
- Developing rational approaches towards sustainable wastewater management via pollution prevention, appropriate treatment, resource recovery and re-use at both centralized and decentralized levels.
- Understanding the relevant physical, chemical and biological processes and their mutual relationships within various sanitation components.
- Evaluating various urban drainage and sewerage schemes, wastewater, sludge and solid waste treatment technologies.
- Applying procedures to identify, evaluate and control hazards to the environment in general and human health hazards in

particular that may cause sickness or impaired health and well-being.

- Promoting environmental protection, improvement and sustainability, enhancing the overall quality of life.

Knowledge to be acquired from the course is expected to feature:
- Demonstrate effective problem-solving abilities
- Identify, monitor and evaluate environmental health hazards
- Utilize current technology of waste collection, handling, treatment, final disposal and management.
- Apply appropriate regulations and guidelines in safeguarding environmental conditions.
- Use effective oral and written communication skills
- Organize and maintain records and reports.

Cognitive skills to be developed are expected to contain the following:
- Capturing ability of reasonable scientific judgment and concepts of appropriate decision making.
- Students will be able to apply the knowledge of environmental pollution control that they have learnt in this course in practical construction engineering domain.
- Students should be able to design and apply necessary procedures and precautions to establish sound environmental systems in construction engineering field

The booklet has been built on the foundations of two previously published texts namely:
- Selected problems in water supply published and printed by Khartoum University Press (KUP), P. O. Box 321, Khartoum, Sudan, 1986, and
- Selected problems in wastewater engineering published by: Khartoum University Press (KUP), Khartoum and National Council for Research (NCR), Khartoum, Khartoum, Sudan, 1986.

The majority of incorporated problems within the book represent examination problems that were given by the author to students at university of Khartoum (U of K), University of United Arab

Emirates (AEU), Sultan Qaboos University (SQU), Omdurman Islamic University (OIU), Sudan University for Science and Technology (SUST), and Sudan Academy for Sciences (SAS). This is to expose students to different and changing design parameters and governing conditions.

"Problem solving in environmental engineering" is principally intended as a supplement and a complementary guide to basic principles of environmental and sanitary engineering. Nonetheless, it can be sourced as a standalone problem solving text in environmental engineering. The book targets university students and candidates taking first degree courses in construction, environmental, civil, mechanical, biomedical, and chemical engineering or related fields. The manuscript is estimated to have valuable benefits to postgraduate students and professional sanitary and environmental engineers. Equally, it is anticipated that the book will excite problem solving learning and accelerate self-teaching. By writing such a script it is hoped that the included worked examples and problems will ensure that the booklet is a treasured support to student-centered learning. To accomplish such objectives great attention was paid to offer solutions to selected problems in a well-defined, clear and discrete layout exercising step-by-step procedure and explanation of the related solution employing crucial procedures, methods, approaches, equations, data, figures and calculations.

The author acknowledges support and encouragement from different students, colleagues, fiends, institutions and publishers. The author acknowledges the inspiration, motivation and stimulus help offered to him by Dean Dr. Abdul-Rahman ben Salih Hariri, Dean College of Engineering of University of Dammam. Special and sincere vote of thanks would go to Mr. Mugbil ben Abdullah Al-Ruwais the director of Dammam University Press and his supreme staff for their patience, devotion and neat typing of the book.

Prof. Dr. Eng. Isam Mohammed Abdel-Magid Ahmed
Dammam University, Al-Khobar, 2013

Table of contents

List of Tables

List of Figures

List of computer programs

List of Appendices

Abbreviations, Notations and Symbols Used in the Book

Symbol	Description	Units
A	Cross-sectional area, area	m^2
a_s	Amount of sludge deposited	m
B	Width of channel	m
BOD	Biological oxygen demand	mg/l
b	Slope	s/m^6
C	Solids concentration	g/m^3
C_D	Newton's drag coefficient	g/m^3
C_e	Effluent concentration	g/m^3
C_g	Gas concentration in gas phase	g/m^3
CL	Gas concentration in liquid phase	g/m^3
C_i	Influent concentration	g/m^3
C_s	Oxygen saturation concentration	mg/l
C	Coefficient of runoff to rainfall	
c	Solids concentration	Kg/m^3
c_1	Mass of dry SS/unit volume of liquid sludge	Kg/m^3
C_s	Oxygen saturation concentration	mg/l
d	Filter depth	m
d	Diameter	m^2/s
D	Coefficient of diffusion	mg/l
d_L	Thickness of liquid film	m
D	Diameter	m
D_c	Critical oxygen deficit	mg/l
DR	Dilution rate	
DWF	Dry weather flow	m^3/d
E	Treatment efficiency	%
E_v	Amount of evaporation	m^3/d
F	Recirculation factor	
F_c	Cake correction factor	
F_f	Fraction of filter area	m^2
F/M	Food-microorganisms ratio	kgBOD/kg MLSS.d
Fr	Froude number	
g	Gravitational acceleration	m/s^2

H	Depth	m
h	Head loss	m
H	Depth	m
I_r	Average infiltration into sewer	m^3/d
I	Average rainfall intensity	mm/hr
k_s	Half velocity constant	mg/l
k	Empirical coefficient of Velz formula	Per m
k_1	First order reaction rate constant	Per day
k'_2	Re-aeration constant	Per day
k'	Bacterial die-away rate	Per day
k_d	Decay coefficient for volatile sludge solids	Per day
k_n	Kinetic coefficient	
k_p	Removal rate constant for pond	
K	Efficiency coefficient	Dimension less, %
k'	Rate constant	/s
kb	Bunsen absorption coefficient	
K_d	Distribution coefficient	
K_H	Henry's constant	g/J
k_2	Overall coefficient of gas transfer	/s
L_e	Concentration of effluent BOD	mg/l
L_i	Concentration of influent BOD	mg/l
L_o	Ultimate BOD	mg/l
L	Length	m
L_e	Depth of expanded bed	m
MLVSS	Mixed liquor volatile suspended solids	mg/l
m	Average rate of mass transfer	g/s
N_G	Gross power	
N	Number of viable microorganisms	
N	Number	
N_e	Effluent bacterial number	/100 ml
N_i	influent bacterial number	/100 ml
n	Frictional coefficient	
of	Oxygen transfer rate at field conditions	kgO_2/kWh
O_c	Oxygen transfer rate at standard conditions	kgO_2/kWh

OC	Oxygenation capacity	$g/m^3.s$
OD	Oxygen demand	gO_2/s
OE	Oxygenation efficiency	gO_2/J
P	Total pressure	kPa
p	Porosity of bed	
P_W	Water vapor pressure	kPa
P	Population number	
PE	Population equivalent	
p	Pressure	$Pa, N/m^2$
Q	Discharge of water flow	m^3/s
Q	Flow rate	m^3/d
Q_s	Stream discharge	m^3/s
Q_w	Wastewater flow	m^3/s
q	Discharge	m^3/s
R	Recirculation ratio, recycling ratio	m
Rh	Hydraulic radius	
Re	Reynolds number	
R	Universal gas constant	J/K.mole
Re	Reynolds number	
r	Specific resistance to filtration	m/kg
r1	Ratio of re-circulated flow to the raw incoming waste	
r'	Constant	
SI	Saturation index	
SS	Suspended solids	g/m^3
SA	Sludge age	d
SDI	Sludge density index (Donaldson)	Per day
SLR	Sludge loading ratio	Per day
SVI	Sludge volume index	ml/g
s	Slope, gradient	m/m, %
s.g.	Specific gravity	
s^-	Coefficient of compressibility	
S^*	Growth limiting substrate concentration	mg/l
T	Temperature	^0C
Tw	Average trade waste discharge	m^3/d
t	Time	s
T	Temperature	^0C

t	Time, or detention time	s
T_o	Standard temperature	271.3 K, or 0 °C
U	Specific utilization rate	Per day
V	Volume	m^3
V_g	Specific yield	m^3 gas/ m^3 digester volume d
VOL	Volumetric organic loading	Kg BOD/m^3 d
VS	Volatile solids	Kg /m^3
V_s	Settled volume of sludge in a 1 l cylinder in 30 minutes	ml
v	Velocity	m/s
v_f	Velocity for a pipe running full	m/s
v_p	Velocity for a pipe with partial flow	m/s
v_{st}	Settling velocity	m/s
V	Volume	m^3
v	Settling velocity	m/s
v_b	Face velocity of backwash water	m/s
v_H	Horizontal velocity	m/s
v_s	Scour velocity	m/s
W	Influent BOD load to filter	Kg/d
W	Width	m
W_p	Wetted perimeter	m
X_T	Total removal	%
x	Distance	m
X	Volumetric composition	%
X_T	Total removal	%
Y	Sludge yield coefficient (yield of dry suspended solids)	Kg /m^2.s
Y_t	Ultimate gas yield	m^3/ Kg VS
y	Throat width	m
α	Coefficient	
ε	Dynamic viscosity	N.s/m^2
ν	Kinematic viscosity	m^2 /s
θ	Time of revolution	s

θ_c	Mean cell residence time	d
ϕ	Angle	degrees
μ	Growth rate of microorganisms	Per day
μ_{max}	Maximum growth rate of microorganisms	Per day
η	Efficiency	%
μ	Dynamic viscosity	N.s/m2, Pa.s
ν	Kinematic viscosity	m2/s
ρ	Density	Kg/m3
ϕ	Shape factor	

Chapter One

Computer Modeling in Environmental Engineering

It goes without saying that computers are dominating the world of information technology. What started as simple ideas of enhancing manual computations and aiding in math solving, has become the most complex, inter-connected, multi-layered system of all times. Compters nowadays serve all sorts of functions: some are general every-day-purpose kind of things (writing your letters with a word processor, doing some tax calculations with an electronic calculator, playing video games, ...), while others are more specific to certain environments and specialized in purpose (a database server saving customers' data, a managerial program running a hospital management system, an autopilot system driving a combat airplane, an embedded Linux-derived kernel in your Android smartphone, ...), and those lists are far from exhaustive.

Using computers in engineering modeling is one of the recipes for being a successful engineer, traditional manual computations and hand-writing modular calculations are all being lost to the oblivion. There are multiple computer modular designs, with Differences in how they are being implemented. Those implementations vary according to more than one factor such as:

1. The underlying programming language used. Some modules are general in purpose, and as such are programmed using a general purpose language (such as Visual Basic, C/C++, Java, etc ...). Others are specialized in their working, necessitating a specialized programming language to be used

(such as Prolog for artificial intelligence software, PHP for website programming, SQL for database access, …).

2. The computer platform for target execution (most commonly used in market today being the Intel-compatible x86 '32-bit' or AMD64 '64-bit' architectures).

3. The operating system environment (again, common OSes in the market including – but not limited to – MS-Windows XP/7/8, Linux with all its flavors – Fedora, Mandriva, Gentoo, SuSE, among others; MAC OS, UNIX, Solaris, and much more)

4. And above all, the engineering principles that form the backbone for modular design.

This book is not intended to be an introduction to computer programming, nor is it going to explain the details of computer architecture or operating systems. It is assumed, however, that the reader has a basic working knowledge in the following areas to be able to read, follow, test, and implement the programming examples given through this text:

- Basic computer knowledge and operation (as how to open/setup/use a computer workstation, be it a PC or otherwise),

- Installing and running any operating system with which the user is familiar (note that the programs of the book were written using Visual Basic 10 under Visual Studio 2010 Professional Edition, and were tested and ran on a MS-Windows XP and 7 workstations. However, they can be run on other systems, but one needs to use some tricks and workarounds. For example, if running a Linux operating system – such as Fedora Linux – the user can download and install WINE – WINdows Emulator – from www.winehq.org, to be able to run the Windows-native executables). The source code files can be viewed and edited using any text editor if the user is running an OS other than Windows (or is using Windows without VS installed), but to be able manipulate and rebuild the program executables the user will need other workarounds, one option is to use MonoDevelop from   www.go-mono.com/mono-downloads/download.html, as an IDE that can run and compile .NET framework programs). Another option is setting up a Windows installation under your operating system by using a virtual machine (such as Oracle VM Virtual Box, which in fact was used by the authors to install Windows XP under Fedora Linux 20 to test those very programs!). That being said, the programs should run on any system that supports EXE file formats and a .NET framework JIT compiler (Just-In-Time compiler), or a similar software, to be able to run the CIL (Common-Intermediate Language) compiled executables.

- At least basic-to-intermediate programming knowledge with Visual Basic (preferably), although the example programs are straightforward and the code structure is maintained in a clean way to make it easy for programmers experienced with other programming languages (and even VB beginners) to follow program logic and run/test the programs.

- Microsoft Visual Studio 2010 Enterprise Edition was used to program, compile, and debug the examples presented in this book. It can be obtained from www.visualstudio.com/en-us/products/visual-studio-express-vs.aspx (for the express edition, a lightweight edition for separate programming languages), or go.microsoft.com/fwlink/?linkid=240162, or simply looking for "Microsoft Visual Studio Download" in Google. Again, other development systems (like MonoDevelop stated above) can be used to compile and debug the programs.

The time for flat programs (that run in a linear fashion from the first instruction to the last) has been gone long ago. Programs are being multilayered, event-driven in nature. That means the program will need to be segmented into parts, each part dealing with (or responding to) a certain event. That is usually (but not necessarily) a part of an object-oriented programming (OOP) design. Today's programs usually are written for GUI (Graphical User Interface) environments, but this is not a necessity. Indeed, sometimes having the overhead of calling graphical routines lifted off the program presents a huge advantage in program's speed and execution, that the GUI is simply omitted, and the program is written as a text-based (or console) program (If you ever went to book a flight and saw your travel agent working that ugly black screen, then you get it). Some programming languages are text-based by nature (such as Prolog), unless mixed with other languages, or being used with an extension (such as Visual Prolog extension for Prolog).

In this book, all programs were written as GUI programs for windowing systems. As such, the programs have two parts: the GUI design (what the user sees), and the code doing the work behind the scenes. The code for each program is included in the text along each example presented. The code provided goes into the main program window source file (usually named *Form1.vb* by default, if not indicated otherwise in the text), and as all the programs here are single-windowed (for simplicity's sake, some examples are complex enough that - in real-world professional design - they would be written as two- or three-windowed programs), this will be the case. As for the GUI design part, all the user interfaces are included as snapshots in Appendix (M) in the back of the book. The user can refer to them when reading/testing the programs to link the controls used in the design with the code (as the default names assigned to the controls are retained, that is, TextBox1, Label2, Form1, and so on, are left unchanged, to make it easy for the reader to follow what-goes-where).

The programming code was designed to be at the minimum level of complexity and length needed to perform the task at hand, at the same time not to be too short, scrambled, or ambiguous, so that new and beginner programmers can follow without stumbling at each line. For each task performed, especially if a function performs a lengthy operation, or a specific programming concept deemed complex or advanced, the reader will find prompt documentation through the source code in the relevant parts to make following the code an easy and productive task.

Visual Basic.NET was selected as a programming language for the book's examples for many reasons:

- MS-Windows is one of the most (if not *the* most) popular operating system in use today. Chances are, when buying a new PC or laptop (or even a mobile phone), the user will face this OS first and foremost.

- .NET Fx (pronounced "dot net framework") is becoming so popular as a dominant programming platform in computer industry, that it will be counterproductive not to learn and understand how to implement engineering models using this platform.

- Visual Basic.NET, which is a descendent of the old beginners' friend BASIC language, is holding its name as being an easy-to-learn language. It may be handicapped with its capabilities being less than other more sophisticated high-level languages, but the more important thing is that the language has so many layers of abstraction inserted between the programmer and the machine. That means the programmer doesn't need to bother about the low-level workings of the computer, he just focuses on programming. For this, BASIC, and it's descendent VB, which then evolved to VB.NET, is considered one of the most easy programming languages to learn, especially for beginners. And this is why it was the perfect choice for this text.

The basic idea behind the programs are to explain how to transform certain engineering problems into valid programming concepts and working computer modules. That means the programs are easy to read and transform into another programming language (if, for example, you are considering programming cross-platform modules using Java). They can also reprogrammed using another language under another platform other than Windows (for example using Qt/C++ to program under Linux).

Nevertheless, the reader is encouraged to use the modular design concepts introduced in this book's examples to construct working programming modules for the exercises presented throughout the text.

The source code presented in this book is considered open-source, and as such you are free to handle it however you care: manipulate it, use it, add/remove from it, upgrade it, and do whatever else you can think of, as long as it is an educational process. If the code is used in a commercial level program, please indicate the source of the code in your files, or kindly contact the authors. Again, the code is free, this is merely a common courtesy.

It's time to turn to Chapter 2 and start modular design.

Chapter Two

Selected Water and Wastewater Characteristics

Water and wastewater characteristics of importance are: physical, radiation, chemical and biological (bacteriological & microbiological).

2.1 Physical characteristics

Physical properties are subject to natural forces making it easier to measure & determine their values & effects. Physical properties of significance include: concentration of solids, turbidity, taste, odor, color, temperature, electrical conductivity, salinity, density, standard volume, viscosity, surface tension, moisture content, humidity, radiation and dissolved oxygen.

2.1.1 Temperature

Temperature of water & wastewater may change due to climatic effects, hot discharge (thermal pollution) & industrial discharges. Increase in temperature affect performance purification or treatment units, reduces concentration of dissolved oxygen, accelerating rates of chemical & biochemical reactions, reduce solubility of gases, increase rate of corrosion of materials, increase toxicity of dissolved elements, increase undesired growth, and increase problems of taste & odor.

Example 2.1

Convert 20, 160, & - 40°F to Celsius & Kelvin reading. Convert 30, 75 & - 10°C to Fahrenheit & Rankine degrees.

Solution

1. Given: T = 20, 160, & - 40°F; T = 30, 75 & -10°C.
2. Determine °C20 = (5/9)*(°F - 32) = (5/9)*(20 - 32) = - 6.7.
   Similarly °C160 = 71.1,°C-40 = - 40.
3. Find K-6.7 = °C+ 273.15 = - 6.7 + 273.15 = 266.45
   Similarly K71.1 = 344.25, K-40 = 233.15.
4. Find °F30 = (9/5)*C + 32 = (9/5)*30 + 32 = 86
   Similarly °F75 = 167, °F-10 = 14.
5. Compute °R86 = 86 + 459.67 = 545.67
   Similarly °R167 = 626.67, & °R14 = 473.67

Program 2.1 Listing:

Converting degrees to Celsius, Kelvin to Fahrenheit & Rankine

```
'*************************************************************
'Program 2.1
'Converts temperature to different scales
'*************************************************************

Public Class Form1
    'The members of the list of selections
    Dim ttl(3) As String
    'Variables used in calculations
    Dim f, c, k, r As Double

    Private Sub Form1_Load(ByVal sender As
        System.Object, ByVal e As
        System.EventArgs) Handles MyBase.Load

        ttl(0) = "Fahrenheit"
        ttl(1) = "Centigrade"
        ttl(2) = "Kelvin"
        ttl(3) = "Rankine"

    'Initialize all variables to '0'
        r = f = c = k = 0

        Me.MaximizeBox = False
    'Text shown on top
        Me.Text = "Program 2.1 "
    'Label control on left upper part of the form
        Label1.Text = "Convert temperature from:"

        ListBox1.Items.Clear()        'Initialize
        For i = 0 To ttl.Length - 1
```

```vb
    'Add members to the list
        ListBox1.Items.Add(ttl(i))
    Next

    'The labels on left and right lower part
    'of the form will be invisible initially,
    'together with the Textbox and the
    'NumericUpDown control.
    Label2.Visible = False
    TextBox1.Visible = False

    Label4.Text = "Decimal Places:"
    NumericUpDown1.Maximum = 10
    NumericUpDown1.Minimum = 0
    NumericUpDown1.Value = 2
    Label4.Visible = False
    NumericUpDown1.Visible = False

    'Set sizes for label at upper right side
    'of the form. AutoSize is set to false
    'to allow user resizing.
    Label3.AutoSize = False
    Label3.Height = 95
    Label3.Width = 150
    viewResults()
End Sub


'*********************************************************
'The following code will recalculate the results and
'show them when ever the user changes his selection.
'*********************************************************
    Private Sub ListBox1_SelectedIndexChanged(
        ByVal sender As System.Object,
        ByVal e As System.EventArgs) Handles
        ListBox1.SelectedIndexChanged

        'Make sure the selected unit is a valid one,
        'else exit the sub safely.
        If ListBox1.SelectedIndex < 0 Then
            Label2.Visible = False
            TextBox1.Visible = False
            Label4.Visible = False
            NumericUpDown1.Visible = False
            Exit Sub
        End If
        Label2.Text = "Enter temperature in " +
        ListBox1.SelectedItem.ToString
        Label2.Visible = True
```

26

```vbnet
            TextBox1.Visible = True
            Label4.Visible = True
            NumericUpDown1.Visible = True
            calculateResults()
            viewResults()
    End Sub

    '****************************************************
    'The following code will recalculate the results
    'and show them when ever
    'the user changes the input.
    '****************************************************
    Private Sub TextBox1_TextChanged(ByVal sender As
        System.Object, ByVal e As
        System.EventArgs) Handles
        TextBox1.TextChanged
        calculateResults()
        viewResults()
    End Sub

    Function FnCen(ByVal f)
     'Function to change Fahrenheit to Centigrade
        Return 5 / 9 * (f - 32)
    End Function

    Function FnFah(ByVal c)
     'Function to change Centigrade to Fahrenheit
        Return 9 / 5 * c + 32
    End Function

    Function FnKel(ByVal c)
     'Function to change Centigrade to Kelvin
        Return c + 273.15
    End Function

    Function FnRan(ByVal f)
     'Function to change Fahrenheit to Rankine
        Return f + 459.67
    End Function

    '****************************************************
    'The following function is used to format the
    'output of the calculations it uses the value
    'selected in the NumericUpDown control as a
    'guide to the number of decimal places in
    'output, then returns 'n' as a string formatted
    'according to the set number of decimals.
    '****************************************************
    Function formatN(ByVal n As Double) As String
     'Use the function as
```

27

```vb
        'FormatNumber(NumberToFormat, NumberOfDecimals)
        Return FormatNumber(n,
            NumericUpDown1.Value).ToString
    End Function

    '************************************************
    'The following code calculates the results of
    'changing the degrees.
    'It first checks for valid input in the TextBox,
    'then calculates based on the selected item in
    'the ListBox.
    '************************************************
    Sub calculateResults()
        Dim t = Val(TextBox1.Text)
        If Not IsNumeric(t) Or t = 0 Then
            r = f = c = k = 0
            Exit Sub
        End If

        Select Case ListBox1.SelectedIndex
            Case 0
                f = Val(TextBox1.Text)
                c = FnCen(f)
                k = FnKel(c)
                r = FnRan(f)
            Case 1
                c = Val(TextBox1.Text)
                f = FnFah(c)
                k = FnKel(c)
                r = FnRan(f)
            Case 2
                k = Val(TextBox1.Text)
                c = k - 273.15
                f = FnFah(c)
                r = FnRan(f)
            Case 3
                r = Val(TextBox1.Text)
                f = r - 459.67
                c = FnCen(f)
                k = FnKel(c)
            Case Else
                'Safeguard mechanism. If any
            'error in the selection, Zero all
            'the variables.
                r = f = c = k = 0
        End Select

    End Sub
```

```
'*****************************************************
'Show the calculations done in calculateResults()
'in the Label3 control.
'*****************************************************
Sub viewResults()
    Label3.Text = "C=" + formatN(c)
        + Chr(10) + Chr(13) _
        + "F=" + formatN(f)
        + Chr(10) + Chr(13) _
        + "K=" + formatN(k)
        + Chr(10) + Chr(13) _
              + "R=" + formatN(r)
End Sub


'*****************************************************
'Whenever a different number of decimals is
'selected, re-show the results.
'*****************************************************
Private Sub NumericUpDown1_ValueChanged(ByVal
    sender As System.Object, ByVal e As
    System.EventArgs) Handles
    NumericUpDown1.ValueChanged
     viewResults()
End Sub
End Class
```

2.1.2 Conductivity

Conductivity denotes intensity of an aqueous solution to carry an electric current. This ability is influenced by: concentration, type mobility, valence & relative concentration of ions; & temperature. Generally, solutions of most inorganic acids, bases & salts are relatively good conductors. Also conductivity is defined as: the electrical conductance of a conductor of unit length & unit cross-sectional area.

Example 2.2

The TDS of a certain sample is recorded as 890 mg/L & its electrical conductivity (EC) amounted to 1025 mmhos/cm. Determine the electrical conductivity of another sample which has a TDS of 1450 mg/L.

Solution

1. Find the constant "a" for the sample as: a = TDS/EC
   a = 890/1025 = 0.87
2. Determine the electrical conductivity for second sample as:
EC$_2$ = TDS$_2$/a = 1450/0.87 = 1667 mmohs/cm.

Program 2.2 Listing:
Determining electrical conductivity given TDS

```
'***********************************************************
'Program 2.2: Conductivity
'Computes electric conductivity of a sample of known
'TDS concentration
'***********************************************************
Public Class Form1
    'Variables used in calculations
    Dim TDS1, EC1, TDS2, EC2 As Double

    Private Sub Form1_Load(ByVal sender As
        System.Object, ByVal e As
        System.EventArgs) Handles MyBase.Load

        Me.MaximizeBox = False
        Me.Text = "Electric conductivity of a
        sample of known TDS concentration"
        Label1.Text = "Enter total Dissolved
        Solids (TDS) concentration(Sample1) mg/L:"
        Label2.Text = "Enter electrical conductivity
        of Sample1    umhos/cm:"
        Label3.Text = "Enter total Dissolved Solids
        (TDS) concentration(Sample2) mg/L:"
        Label4.Text = "Electrical conductivity for
        Sample2:"
        Label5.Text = "umhos/cm"
        Label6.Text = "Decimal Places:"
        NumericUpDown1.Maximum = 10
        NumericUpDown1.Minimum = 0
        NumericUpDown1.Value = 2

    End Sub

'***********************************************************
    'The following code calculates the result of EC2
    'depending on the input
    'values in textboxes 1-3.
    'The Val(x) function is used to convert from
    'String to Numerical.
```

```vb
'The FormatNumber is used to format the output
'to specific decimals.
'***************************************************
Sub calculateResults()
    Dim a As Double
    TDS1 = Val(TextBox1.Text)
    EC1 = Val(TextBox2.Text)
    TDS2 = Val(TextBox3.Text)
    a = TDS1 / EC1
    EC2 = TDS2 / a
    'Use the function as
  'FormatNumber(NumberToFormat, NumberOfDecimals).
    TextBox4.Text = FormatNumber(EC2,
        NumericUpDown1.Value).ToString
End Sub

'***************************************************
'Whenever a different number of decimals is
'selected, re-show the results.
'***************************************************
Private Sub NumericUpDown1_ValueChanged(ByVal
    sender As System.Object, ByVal e As
    System.EventArgs) Handles
    NumericUpDown1.ValueChanged

    calculateResults()
End Sub

Private Sub TextBox1_TextChanged(ByVal sender
    As System.Object, ByVal e
    As System.EventArgs) Handles
    TextBox1.TextChanged

    calculateResults()
End Sub

Private Sub TextBox2_TextChanged(ByVal sender As
    System.Object, ByVal e As
    System.EventArgs) Handles
    TextBox2.TextChanged

    calculateResults()
End Sub

Private Sub TextBox3_TextChanged(ByVal sender As
    System.Object, ByVal e As
    System.EventArgs) Handles
    TextBox3.TextChanged

    calculateResults()
```

```
        End Sub
End Class
```

2.1.3 Solids concentration

Solids designate matter that is suspended or dissolved in water & wastewater. Generally solids are: dissolved, suspended, volatile & fixed, & settleable solids.

Example 2.3

Solids analysis were conducted on the meat-processing wastewater. Samples were taken in 300 mL bottles. The tests were conducted in triplicate as shown in the table below: (SQU, 1991).

Total Solids Test Data			
Dish Number	1	2	3
Weight of empty dish, g	51.494	51.999	50.326
Weight of dish + dry solids, g	51.587	52.081	50.437
Weight of dish + ignited solids, g	51.541	52.042	50.383

i] Determine the concentration of total solids in mg/L.

ii] Find the total volatile solids content.

Solution

	Sample (1)	Sample (2)	Sample (3)
Weight of dish + solids (g)	51.587	52.081	50.437
Weight of dish (g)	51.494	51.999	50.326
Weight of solids (mg)	93	82	111
Volume used = 300×10⁻³ = 0.3L			
Concentration of solids	93 ÷ 0.3 = 310	293	370

$$concentration_{average} = \frac{310+293+370}{3} = \frac{973}{3} = 324\,mg/L$$

	Sample (1)	Sample (2)	Sample (3)
Weight of dish + solids (g)	51.587	52.081	50.437
Weight of ignited solids (mg)	51.541	52.042	50.383
Weight of volatile solids (mg)	46	39	54
Concentration of volatile solids	153	130	180

$$concentration_{average} = \frac{153 + 130 + 180}{3} = \frac{463}{3} = 154 \, mg/L$$

Program 2.3 Listing:

Concentration of total solids and total volatile solids content

```
'********************************************************
'Program 2.3: Salinity
'Computes solid contents
'********************************************************
Public Class Form1
    Dim m1, m2, m3, volume As Double
    Const T1 = 104        'Centigrade

    Private Sub Form1_Load(ByVal sender As
        System.Object, ByVal e As
        System.EventArgs) Handles MyBase.Load

        Me.MaximizeBox = False
        Me.Text = "Program 2.3: Determines
            the solids content"
        Label1.Text =
        "Enter weight of crucible m1 (gm):"
        Label2.Text =
        "Enter volume of sample V (mL):"
        Label3.Text = "Enter weight of dish+solids
        at 104 C (mg):"
        Label4.Text = "Enter weight of dish+ solids
        after heating to 550 C (mg):"

        Label5.Text = "Concentration of total
        solids (mg/L)="
        Label6.Text = "Concentration of volatile
        solids (mg/L)="
        Label7.Text = "Concentration of fixed
```

33

```vbnet
        solids (mg/L)="
        Label8.Text = "Decimal Places:"
        NumericUpDown1.Maximum = 10
        NumericUpDown1.Minimum = 0
        NumericUpDown1.Value = 2
    End Sub

    Sub calculateResults()
        m1 = Val(TextBox1.Text)
        volume = Val(TextBox2.Text)
        m2 = Val(TextBox3.Text)
        m3 = Val(TextBox4.Text)

        Dim msds, csds, mvols, cvols, cfixs As Double
        msds = m2 - m1                    'solids
        csds = msds * 10 ^ 6 / volume 'conc. of solids
        mvols = m3 - m1                   'volatiles
        cvols = mvols * 10 ^ 6 / volume
        cfixs = csds - cvols

        TextBox5.Text = FormatNumber(csds,
           NumericUpDown1.Value).ToString
        TextBox6.Text = FormatNumber(cvols,
           NumericUpDown1.Value).ToString
        TextBox7.Text = FormatNumber(cfixs,
           NumericUpDown1.Value).ToString
    End Sub

    '*************************************************
    'Whenever a different number of decimals is
    'selected, re-show the results.
    '*************************************************
    Private Sub NumericUpDown1_ValueChanged(ByVal
       sender As System.Object, ByVal e As
       System.EventArgs) Handles
       NumericUpDown1.ValueChanged

        calculateResults()
    End Sub

    Private Sub TextBox1_TextChanged(ByVal sender As
       System.Object, ByVal e As
       System.EventArgs) Handles
       TextBox1.TextChanged

        calculateResults()
    End Sub

    Private Sub TextBox2_TextChanged(ByVal sender As
       System.Object, ByVal e As
```

```
         System.EventArgs) Handles
         TextBox2.TextChanged

             calculateResults()
      End Sub

      Private Sub TextBox3_TextChanged(ByVal sender As
         System.Object, ByVal e As
         System.EventArgs) Handles
         TextBox3.TextChanged

             calculateResults()
      End Sub

      Private Sub TextBox4_TextChanged(ByVal sender As
         System.Object, ByVal e As
         System.EventArgs) Handles
         TextBox4.TextChanged

             calculateResults()
      End Sub
End Class
```

Density, Specific Volume, Specific Weight, & Specific Gravity

Density is the mass of a substance per unit volume or in a qualitative manner it is the measure of the relative "heaviness" of objects with a constant volume. Density is temperature dependent.

2.1.4 Density

$\rho = m/V$

where:

ρ = Density of the fluid, kg/m^3.

m = Mass, kg.

V = Volume, m^3.

2.1.5 Specific volume

Specific volume is the volume per unit mass, i.e. it is the reciprocal of the density.

$\kappa = 1/\rho$

where:

κ = Specific volume of the fluid, m^3/kg.
ρ = Density of the fluid, kg/m^3.

1.1.6 Specific weight
Specific weight is the weight of unit volume.
$\gamma = m^*g/V = \rho^*g$
where:
γ = Specific weight, N/m^3.
m = Mass, kg.
g = Gravitational acceleration, m/s^2.
V = Volume, m3.
ρ = Density of the fluid, kg/m^3.

2.1.7 Specific gravity
Specific gravity is the ratio of the density of fluid to the density of water at some specified temperature. Equation 2.10 indicates the specific gravity concept.

$\rho = \rho_{water}$ at 4°C

Example 2.4
The specific weight of water at ordinary temperature & pressure conditions is 9.806 kN/m^3. The specific gravity of mercury is 13.55. Find density and specific weight of mercury, & density of water.

Solution
a) Given: g = 9.806 kN/m^3, $s.g._{Hg}$ = 13.55. $s.g = \dfrac{\rho}{\rho_w}$
b) Determine the density of water as:
$\rho = \gamma_{water}/g$ = 9.806 kN/m^3/9.81 m/s^2 = 1 Mg/m^3 = 1 g/cm^3.
c) Find specific weight of mercury as:
  $\gamma_{Hg} = s.g._{Hg}^*\gamma_{water}$ = 13.55*9.806 = 133 kN/m^3
d) Compute density of mercury as:
  $\rho_{H}g = s.g._{Hg}^*\rho_{water}$ = 13.55*1 = 13.55 Mg/m^3

Program 2.4 Listing:
Density and specific weight of a substance

```
'*******************************************************
'Program 2.4: Density
'Computes Density and specific gravity
'*******************************************************
Public Class Form1
    Const gamma = 1000   'density of water KN/m3
    Const g = 9.801      'acceleration due to gravity

    Dim t(3) As String
    'Variables used in calculations
    Dim m, Ro, V, sv, sw, sg As Double

    Private Sub Form1_Load(ByVal sender As
        System.Object, ByVal e As
        System.EventArgs) Handles MyBase.Load

        t(0) = "Compute density"
        t(1) = "Compute specific volume"
        t(2) = "Compute specific weight"
        t(3) = "Compute specific gravity"
        ListBox1.Items.Clear()        'Initialize
        For i = 0 To t.Length - 1
        'Add members to the list
            ListBox1.Items.Add(t(i))
        Next

        Me.MaximizeBox = False
        Me.Text = "Program 2.4:Computes Density and
            specific gravity"

        Label5.Text = "Decimal Places:"
        NumericUpDown1.Maximum = 10
        NumericUpDown1.Minimum = 0
        NumericUpDown1.Value = 2
        Label2.Visible = False
        Label3.Visible = False
        Label4.Visible = False
        Label5.Visible = False
        TextBox1.Visible = False
        TextBox2.Visible = False
        TextBox3.Visible = False
        NumericUpDown1.Visible = False
    End Sub

    '*******************************************************
    'The following code shows/hides the labels and
```

```
'textboxes according to the selected item from
'the list. It also sets
'the text displayed in each label.
'************************************************
Private Sub ListBox1_SelectedIndexChanged(ByVal
    sender As System.Object, ByVal
    e As System.EventArgs) Handles
    ListBox1.SelectedIndexChanged

    If ListBox1.SelectedIndex < 0 Then Exit Sub

    Select Case ListBox1.SelectedIndex
        Case 0
            Label2.Text =
    "Enter the mass in Kg:"
            Label3.Text =
    "Enter the volume in m3:"
            Label4.Text =
    "The density 'Ro' (Kg/m3):"
        Case 1
            Label2.Text =
    "Enter the mass in Kg:"
            Label3.Text =
    "Enter the volume in m3:"
            Label4.Text =
    "Specific volume (m3/Kg):"
        Case 2
            Label2.Text =
    "Enter the mass in Kg:"
            Label3.Text =
    "Enter the volume in m3:"
            Label4.Text =
    "Specific volume (N/m2):"
        Case 3
            Label2.Text =
    "Enter density of material Kg/m3:"
            Label4.Text = "Specific gravity:"
    End Select

    TextBox1.Text = ""
    TextBox2.Text = ""
    TextBox3.Text = ""
    Label2.Visible = True
    Label4.Visible = True
    Label5.Visible = True
    TextBox1.Visible = True
    TextBox3.Visible = True
    NumericUpDown1.Visible = True
    If ListBox1.SelectedIndex = 3 Then
        Label3.Visible = False
```

```
            TextBox2.Visible = False
        Else
            Label3.Visible = True
            TextBox2.Visible = True
        End If
    End Sub

    '****************************************************
    'The following code calculates the results
    'according to the
    'selected item from the list box and displays
    'the result in TextBox3.
    '****************************************************
    Sub calculateResults()
        If ListBox1.SelectedIndex < 0 Then Exit Sub

        Select ListBox1.SelectedIndex
            Case 0
                m = Val(TextBox1.Text)
                v = Val(TextBox2.Text)
                Ro = m / v        'the density
                TextBox3.Text = formatN(Ro)
            Case 1
                m = Val(TextBox1.Text)
                V = Val(TextBox2.Text)
                sv = V / m        'specific volume
                TextBox3.Text = formatN(sv)
            Case 2  .
                m = Val(TextBox1.Text)
                V = Val(TextBox2.Text)
                sw = m * g / V  'specific weight
                TextBox3.Text = formatN(sw)
            Case 3
                Ro = Val(TextBox1.Text)
                sg = Ro / gamma  'specific gravity
                TextBox3.Text = formatN(sg)
        End Select
    End Sub

    '****************************************************
    'Whenever a different number of decimals is
    'selected, re-show the results.
    '****************************************************
    Private Sub NumericUpDown1_ValueChanged(ByVal
        sender As System.Object, ByVal
        e As System.EventArgs) Handles
        NumericUpDown1.ValueChanged
        calculateResults()
    End Sub
```

```
Private Sub TextBox1_TextChanged(ByVal sender
    As System.Object, ByVal e As
    System.EventArgs) Handles
    TextBox1.TextChanged
     calculateResults()
End Sub

Private Sub TextBox2_TextChanged(ByVal sender
    As System.Object, ByVal e As
    System.EventArgs) Handles
    TextBox2.TextChanged
     calculateResults()
End Sub

'*************************************************
'The following function is used to format the
'output of the calculations
'it uses the value selected in the NumericUpDown
'control as a guide to
'the number of decimal places in output, then
'returns 'n' as a string
'formatted according to the set number of decimals.
'*************************************************
Function formatN(ByVal n As Double) As String
     'Use the function as
     'FormatNumber(NumberToFormat, NumberOfDecimals).

     Return FormatNumber(n,
      NumericUpDown1.Value).ToString
   End Function
End Class
```

2.1.8 Viscosity

Viscosity of a fluid is a measure of its resistance to shear or angular deformation. Viscosity is the property that relates the applied forces to the rates of deformation of the fluid. Viscosity may be found from Newton's law.

$$F = \tau . A = \mu . A . \frac{dv}{dy}$$

Where:
τ = Shear stress, $N*m^{-2}$.
F = Shear force acting on a very wide plate free to move, N.
A = Area, m^2.
μ = Dynamic (absolute) viscosity, $N*s*m^{-2}$.

40

du/dy =   Velocity gradient (rate of angular deformation, rate of shear), s^{-1}.

Coefficient of dynamic viscosity, m, may be described as "The shear force per unit area required to drag one layer of fluid with unit velocity past another layer through unit distance in the fluid".

$$\mu = \rho^* v$$

Where:

μ = Dynamic (absolute) viscosity, N*s*m^{-2}.

v = Kinematic viscosity, m^{2}*s^{-1} (usually defined as the ratio of dynamic viscosity to mass density).

ρ = Density, kg*m^{-3}.

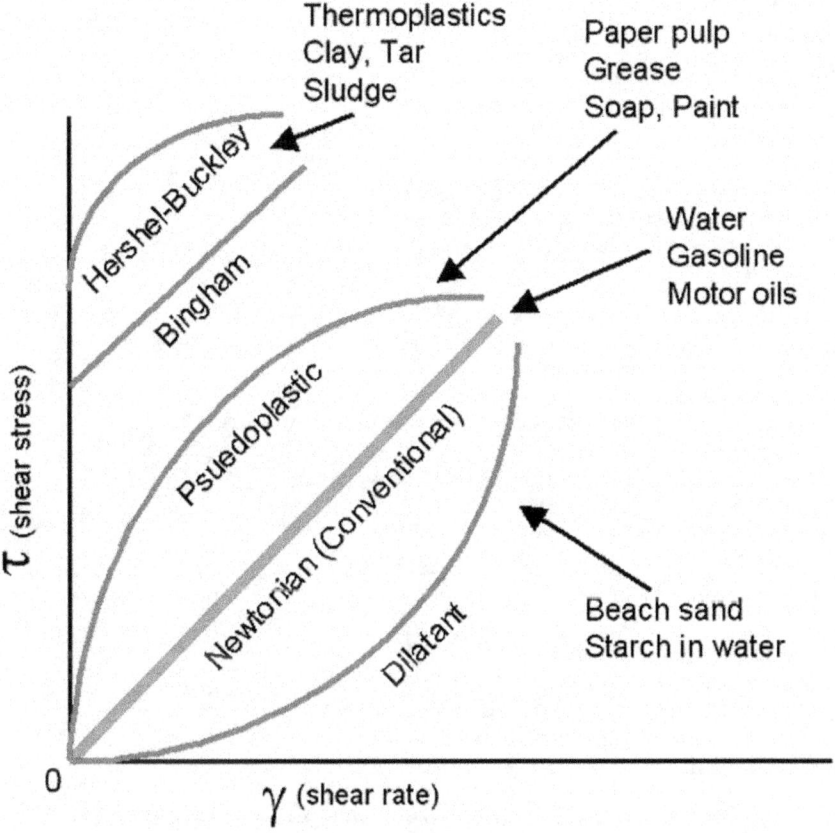

Fig. 2.1. Stress-velocity gradient relationship.

Example 2.5

For a liquid with a dynamic viscosity of $4.5*10^{-3}$ Pa*s & a density of 912 kg/m^3 determine its kinematic viscosity.

Solution

1. Given: $\mu = 4.5*10^{-3}$ Pa*s, $\rho = 912$ kg/m^3.
2. Determine the kinematic viscosity as:
$$\nu = \mu/\rho = 4.5*10^{-3} \text{ Pa*s}/912 \text{ kg/m}^3 = *10^{-6} \text{ m}^2/\text{s}.$$

Program 2.5 Listing:
Kinematic viscosity of a fluid

```
'*********************************************************
'Program 2.5: Viscosity
'Computes kinematic viscosity from dynamic viscosity
'*********************************************************
Public Class Form1
    'Variables used in calculations
    Dim mue, Row, FnKinem

    Private Sub Form1_Load(ByVal sender As
        System.Object, ByVal e As
        System.EventArgs) Handles MyBase.Load

        Me.MaximizeBox = False
        Me.Text = "Computes kinematic viscosity
            from dynamic viscosity"

        Label1.Text = "Enter dynamic viscosity u
            in  Ns/m2:"
        Label2.Text = "Enter density of liquid  Ro
            in Kg/m3:"
        Label3.Text =
            "The Kinematic viscosity equals:"
        Label4.Text = "m2/s"

        Label5.Text = "Decimal Places:"
        NumericUpDown1.Maximum = 10
        NumericUpDown1.Minimum = 0
        NumericUpDown1.Value = 2
    End Sub

    '************************************************
    'The following code will recalculate the results
    'and show them when ever
```

42

```
'the user changes the input.
'************************************************
Private Sub TextBox1_TextChanged(ByVal sender As
    System.Object, ByVal e As
    System.EventArgs) Handles
    TextBox1.TextChanged
    calculateResults()
End Sub

Private Sub TextBox2_TextChanged(ByVal sender As
    System.Object, ByVal e As
    System.EventArgs) Handles
    TextBox2.TextChanged
    calculateResults()
End Sub

'************************************************
'The following code calculates the result
'according to the inputs in
'the text boxes. It then shows result in TextBox3.
'************************************************
Sub calculateResults()
    mue = Val(TextBox1.Text)
    Row = Val(TextBox2.Text)
    FnKinem = mue / Row
    TextBox3.Text = FormatNumber(FnKinem,
        NumericUpDown1.Value).ToString
End Sub

Private Sub NumericUpDown1_ValueChanged(ByVal
    sender As System.Object, ByVal
    e As System.EventArgs) Handles
    NumericUpDown1.ValueChanged
    calculateResults()
End Sub
End Class
```

2.1.9 Surface Tension

Surface tension is a property of a liquid that permits the attraction between molecules to form an imaginary film that is able to resist tensile forces at the interface between two immiscible liquids or at the interface between a liquid & a gas.

$$\sigma = h(\rho_1 - \rho_2)\frac{g \cdot r}{2\cos\varphi} = h(\rho_1 - \rho_2)\frac{g \cdot D}{4\cos\varphi}$$

Where:

h =   Column height by which liquid rose along capillary tube, m.

s =   Surface tension, N/m.

ϕ =   Angle of contact subtended by the heavier fluid & tube (= 0° for most organic liquids & water against glass {provided the glass is wet with a film of the liquid}, = 130° for mercury against glass).

ρ_l =   Density of liquid, kg/m³.

ρ_g =   Density of gas (or light liquid), kg/m³.

g =   Local gravitational acceleration, m/s².

D =   Inside diameter of capillary tube, m

Example 2.6

Compute the diameter of a clean glass tube required so that the rise of water at 30°C in the tube due to capillary action is not to exceed 2 mm.50

Solution

1. Given: h = 2*10⁻³ m, T = 30°C.
2. For a temperature of 30°C & from table find σ = 0.0712 N/m & g = 9.765 kN/m³.
3. Compute the diameter as:

$$\sigma = h\left(\rho_1 - \rho_2\right)\frac{g \cdot r}{2\cos\varphi} = h\left(\rho_1 - \rho_2\right)\frac{g \cdot D}{4\cos\varphi}$$

d = 4*σ*cosφ/g*h = [4*0.0712*cos 0]/[9.765*10³*2*10⁻³]
= 14.6 mm.

Program 2.6 Listing:
Diameter of a tube given capillary action

```
'*****************************************************
'Program 2.6: Tension
'Computes capillary heights in surface tension problems
'*****************************************************
Imports System.Math

Public Class Form1
```

44

```vbnet
'Arrays used to store temperature values
Dim temp(40), sigma(40), density(40) As Double
Dim ttl(2) As String
'Variables used in calculations
Dim diam, T, Rog As Double
'Raw termperature data
Dim sigmaData, densityData As String

'constants
Const pi = 3.142857
Const g = 9.81      'acceleration due to gravity
Const phiw = 0      'angle of contact water & glass
Const phim = 130    'angle of contact mercury & glass
Const Hgd = 13.55

Private Sub Form1_Load(ByVal sender As
    System.Object, ByVal e As
    System.EventArgs) Handles MyBase.Load
    ttl(0) = "Determine capillary height h"
    ttl(1) = "Determine the diameter of the
        glass tube for a specific h"
    ttl(2) = "Determine surface tension"

    'Temperature-Surface tension table for
    'water- in dyne/cm
    sigmaData =
    "74.92,74.78,74.64,74.50,74.36,74.22,
    74.07,73.93,73.78,73.64,"
    sigmaData = sigmaData +
    "73.49,73.34,73.19,73.05,72.90,72.75,
    72.59,72.44,72.28,72.13,"
    sigmaData = sigmaData +
    "71.97,71.82,71.66,71.5,71.35,71.18,
    71.02,70.86,70.7,70.53,"
    sigmaData = sigmaData +
    "70.38,70.21,70.05,69.88"
    'Temperature-Density table- kg/m3
    densityData =
    "999.965,999.941,999.902,999,849,
    999.781,999.7,999.605,999.498,"
    densityData = densityData +
    "999.377,999.244,999.099,998.943,
    998.774,998.595,998.405,998.203,"
    densityData = densityData +
    "997.992,997.77,997.538,997.296,997.044,
    996.783,996.512,996.232,"
    densityData = densityData +
    "995.944,995.646,995.34,995.025,994.702,
    994.374,994.03,993.68,"
    densityData = densityData + "993.33,992.296"
```

```
    'Parse the above data, storing it from String
  'format into Array format..
    readData(sigmaData, densityData)

    Me.MaximizeBox = False
    Me.Text = "Program 2.6:Computes capillary
  heights in surface tension problems"

    'Add the options to the list box
    ListBox1.Items.Clear()
    For i = 0 To ttl.Length - 1
        ListBox1.Items.Add(ttl(i))
    Next

    'Options used when calculating the
  'capillary height
    RadioButton1.Text = "Water tube"
    RadioButton2.Text = "Mercurial tube"
    RadioButton1.Checked = True
    RadioButton1.Visible = False
    RadioButton2.Visible = False
    'Hide all labels and text boxes at the start
    Label2.Visible = False
    Label3.Visible = False
    Label4.Visible = False
    Label5.Visible = False
    Label6.Visible = False
    TextBox1.Visible = False
    TextBox2.Visible = False
    TextBox3.Visible = False
    TextBox4.Visible = False
    TextBox5.Visible = False

    Label7.Text = "Decimal Places:"
    NumericUpDown1.Maximum = 15
    NumericUpDown1.Minimum = 0
    NumericUpDown1.Value = 8
    Label7.Visible = False
    NumericUpDown1.Visible = False
End Sub

'**************************************************
'This function reads the data stored in two
'strings, the sigmaData and densityData strings.
'It uses call-back to itself to add elements to
'the sigma and destiny arrays,
'chopping the first value stored in each string
'every time it loops, until the last value in
'the string is reached.
'**************************************************
```

46

```vbnet
Sub readData(ByVal sigmaData As String,
    ByVal densityData As String)
    'Declared static so one instance of it is used
    Static count = 5
    temp(count) = count
    If sigmaData.IndexOf(",") < 0 Then
    'if last value in the string is reached..
        sigma(count) = Val(sigmaData)
    'convert to N/m
        sigma(count) = sigma(count) / 1000
        density(count) = Val(densityData)
    Else
        'if there are still values in the string..
        'Read from inside-out: first determine the
    'index of the first ',' in the string,
        'then take the substring from the beginning
    'to just before the ',' and then
        'convert this string to numerical value..
    'Simple!!

        sigma(count) = Val(sigmaData.Substring(0,
        sigmaData.IndexOf(",")))
    'convert to N/m
        sigma(count) = sigma(count) / 1000
        density(count) =
        Val(densityData.Substring(0,
        densityData.IndexOf(",")))

        'Remove the first value, which is already
    'stored in the array, from the string,
        'using the index of the ',' removing
    'everything before the comma (equals the
        'first value)..
        sigmaData = sigmaData.Remove(0,
        sigmaData.IndexOf(",") + 1)
        densityData = densityData.Remove(0,
        densityData.IndexOf(",") + 1)
        count += 1     'increase the count by one
    'call-back to the function, passing the
    'new strings
        readData(sigmaData, densityData)
    End If
End Sub

Private Sub ListBox1_SelectedIndexChanged(ByVal
    sender As System.Object, ByVal
    e As System.EventArgs) Handles
    ListBox1.SelectedIndexChanged

    If ListBox1.SelectedIndex < 0 Then
```

```vb
        Label7.Visible = False
        NumericUpDown1.Visible = False
        Exit Sub
End If

Select Case ListBox1.SelectedIndex
    Case 0
        'Determine capillary height h
        Label2.Text = "Enter diameter of
tube in (mm) (between 10 to 60mm):"
        Label3.Text = "Enter temperature
t (C) (5<t<38):"
        Label4.Text = "Enter the density of
the gas (kg/m3):"
        Label6.Text =
"The column height h (m):"
     Case 1
        'Determine the diameter of the glass
'tube for a specific h
        Label2.Text = "Enter the density of
the liquid or gas, kg/m3:"
        Label3.Text = "Enter surface tension
coefficient, N/m:"
        Label4.Text = "Enter angle of contact
in degrees:"
        Label5.Text = "Enter the exent of the
liquid rise in the tube, m:"
        Label6.Text = "the required diameter
of the tube (m):"
     Case 2
        'Determine sigma
        Label2.Text = "Enter the density of
the liquid or gas, kg/m3:"
        Label3.Text = "Enter the diameter of
the capillary tube,m:"
        Label4.Text = "Enter angle of contact
in degrees:"
        Label5.Text = "Enter the exent of the
liquid rise in the tube, m:"
        Label6.Text = "the surface tension
coefficient (N/m)="
End Select

 'Make the labels and text boxes visible after
'the user makes the selection
  Label2.Visible = True
  Label3.Visible = True
  Label4.Visible = True
  Label6.Visible = True
  Label7.Visible = True
```

```
          NumericUpDown1.Visible = True
          TextBox1.Text = ""
          TextBox2.Text = ""
          TextBox3.Text = ""
          TextBox4.Text = ""
          TextBox5.Text = ""
          TextBox1.Visible = True
          TextBox2.Visible = True
          TextBox3.Visible = True
          TextBox5.Visible = True

          If ListBox1.SelectedIndex = 0 Then
              Label5.Visible = False
              TextBox4.Visible = False
              RadioButton1.Visible = True
              RadioButton2.Visible = True
          Else
              Label5.Visible = True
              TextBox4.Visible = True
              RadioButton1.Visible = False
              RadioButton2.Visible = False
          End If

End Sub

'****************************************************
'The following code will recalculate the results
'and show them when ever
'the user changes the input.
'****************************************************
Private Sub TextBox1_TextChanged(ByVal sender As
    System.Object, ByVal e As
    System.EventArgs) Handles
    TextBox1.TextChanged
    calculateResults()
End Sub

Private Sub TextBox2_TextChanged(ByVal sender As
    System.Object, ByVal e As
    System.EventArgs) Handles
    TextBox2.TextChanged
    calculateResults()
End Sub

Private Sub TextBox3_TextChanged(ByVal sender As
    System.Object, ByVal e As
    System.EventArgs) Handles
    TextBox3.TextChanged
    calculateResults()
End Sub
```

49

```vbnet
Private Sub TextBox4_TextChanged(ByVal sender As
    System.Object, ByVal e As
    System.EventArgs) Handles
    TextBox4.TextChanged
     calculateResults()
End Sub

Private Sub NumericUpDown1_ValueChanged(ByVal
    sender As System.Object, ByVal
    e As System.EventArgs) Handles
    NumericUpDown1.ValueChanged
     calculateResults()
End Sub

'*****************************************************
'The following code calculates the result
'according to the inputs in
'the text boxes. It then shows result in TextBox5.
'*****************************************************
Sub calculateResults()
    If ListBox1.SelectedIndex < 0 Then Exit Sub
    Dim s, Roww, h, Ro, sig, phi, d As Double

    Select ListBox1.SelectedIndex
        Case 0
            diam = Val(TextBox1.Text)
            T = Val(TextBox2.Text)
            Rog = Val(TextBox3.Text)

            If diam < 10 Or diam > 60 Then
                TextBox5.Text = "The Diameter
            should be between 10 and
            60 mm."
                Exit Sub
            End If
            If T < 5 Or T > 38 Then
                TextBox5.Text = "The Temperature
        should be between 5 and 38 degrees."
                Exit Sub
            End If
            'get the value of sigma and density
        'of water corresponding to
        'temperature t
            For i = 5 To 38
                If T = temp(i) Then
                    s = sigma(i)
                    Roww = density(i)
                    Exit For
                End If
```

```
                Next i
                'compute height h
                If RadioButton1.Checked Then
            'Use water tube
                    h = 4 * s * Cos(phiw * pi / 180)
              / (g * diam * (Roww - Rog))
                Else
                    'Use mercurial tube
                    h = 4 * s * Cos(phim * pi / 180)
              / (g * diam * (Roww - Rog))
                End If
                TextBox5.Text = formatN(h)

        Case 1
            Ro = Val(TextBox1.Text)
            sig = Val(TextBox2.Text)
            phi = Val(TextBox3.Text)
            h = Val(TextBox4.Text)

            phi = phi * pi / 180     'in radians
            d = 4 * sig * Cos(phi) / (Ro * g * h)
            TextBox5.Text = formatN(d)

        Case 2
            Ro = Val(TextBox1.Text)
            d = Val(TextBox2.Text)
            phi = Val(TextBox3.Text)
            h = Val(TextBox4.Text)

            phi = phi * pi / 180     'in radians
            sig = h * Ro * g * d / (4 * Cos(phi))
            TextBox5.Text = formatN(sig)

    End Select
End Sub

'*****************************************************
'The following function is used to format the
'output of the calculations
'it uses the value selected in the NumericUpDown
'control as a guide to
'the number of decimal places in output, then
'returns 'n' as a string
'formatted according to the set number of decimals.
'*****************************************************

Function formatN(ByVal n As Double) As String
    'Use the function as
  'FormatNumber(NumberToFormat, NumberOfDecimals).

    Return FormatNumber(n,
```

51

```
        NumericUpDown1.Value).ToString
    End Function

    Private Sub RadioButton1_CheckedChanged(ByVal
        sender As System.Object, ByVal
        e As System.EventArgs) Handles
        RadioButton1.CheckedChanged
        calculateResults()
    End Sub

    Private Sub RadioButton2_CheckedChanged(ByVal
        sender As System.Object, ByVal
        e As System.EventArgs) Handles
        RadioButton2.CheckedChanged
        calculateResults()
    End Sub
End Class
```

2.1.10 Bulk Modulus (Bulk Modulus of Elasticity)

Bulk modulus is a property that is used to evaluate the degree of compressibility. Large values of bulk modulus indicates fluid is relatively incompressible (i.e. a large pressure change is needed to create a small change in volume).

Where:

E_v = Bulk modulus, N*m^{-2}.

dP = Differential change in pressure, Pa.
dV = Differential change in volume, m^3.
V = Volume, m^3.
ρ = Density of fluid, kg/m^3.

$$E_v = -\frac{dP}{\left(\dfrac{dV}{V}\right)} = -\frac{dP}{\left(\dfrac{\rho}{d\rho}\right)}$$

Example 2.7

A liquid compressed in a cylinder has a volume of 2 L at 1000 kPa & a volume of 1805 cm^3 at 2000 kPa. What is its bulk modulus of elasticity?

Solution

1. Given: $V_1 = 2000$ mL, $V_2 = 1805$ mL, $P_1 = 1000$ kPa, $P_2 = 2000$ kPa.
2. Determine the bulk modulus of the liquid as:
    $E_v = - dP/(dV/V) = - (2000 - 1000)kPa/[(1805 - 2000)/2000] = 10.3$ MPa.

Program 2.7 Listing:

Bulk modulus of elasticity of a liquid

```
'********************************************************
'Program 2.7: Bulk
'********************************************************
Public Class Form1
    Dim v1, v2, p1, p2, Eb As Double

    Private Sub Form1_Load(ByVal sender As
        System.Object, ByVal e As
        System.EventArgs) Handles MyBase.Load

        Me.MaximizeBox = False
        Me.Text = "Program 2.7: Computes Bulk
            modulus given dV and dP"

        label1.text = "Enter initial volume V1 in mL:"
        label2.text = "Enter final volume V2 in mL:"
        label3.text =
        "Enter initial pressure P1 in kPa:"
        label4.text =
        "Enter final pressure P2 in kPa:"
        Label5.Text = "The bulk modulus Eb in kPa:"
        Label6.Text = "Decimal Places:"
        NumericUpDown1.Maximum = 10
        NumericUpDown1.Minimum = 0
        NumericUpDown1.Value = 2

    End Sub

    '********************************************************
    'The following code will recalculate the results
    'and show them when ever
    'the user changes the input.
    '********************************************************
    Private Sub TextBox1_TextChanged(ByVal sender As
        System.Object, ByVal e As
        System.EventArgs) Handles
```

53

```vbnet
        TextBox1.TextChanged
          calculateResults()
    End Sub

    Private Sub TextBox2_TextChanged(ByVal sender As
        System.Object, ByVal e
        As System.EventArgs) Handles
        TextBox2.TextChanged
          calculateResults()
    End Sub

    Private Sub TextBox3_TextChanged(ByVal sender As
        System.Object, ByVal e
        As System.EventArgs) Handles
        TextBox3.TextChanged
          calculateResults()
    End Sub

    Private Sub TextBox4_TextChanged(ByVal sender As
        System.Object, ByVal e
        As System.EventArgs) Handles
        TextBox4.TextChanged
          calculateResults()
    End Sub

    Private Sub NumericUpDown1_ValueChanged(ByVal sender
        As System.Object, ByVal e
        As System.EventArgs) Handles
        NumericUpDown1.ValueChanged
          calculateResults()
    End Sub

    '************************************************
    'The following code calculates the result
    'according to the inputs in
    'the text boxes. It then shows result in TextBox5.
    '************************************************
    Sub calculateResults()
          v1 = Val(TextBox1.Text)
          v2 = Val(TextBox2.Text)
          p1 = Val(TextBox3.Text)
          p2 = Val(TextBox4.Text)

          Eb = -(p2 - p1) / ((v2 - v1) / v1)

          TextBox5.Text = FormatNumber(Eb,
        numericupdown1.value).ToString
    End Sub
End Class
```

54

2.1.11 Dissolved oxygen
Example 2.8
Compute the saturation value of dissolved oxygen in water exposed to water-saturated air containing 20.9 percent oxygen under a pressure of 732 mm of mercury. Assume temperature of 20°C & a chloride concentration in water of 8000 mg/l. (SQU, 1991).

Solution
1. Given: P = 732 mm Hg, for t = 20°C then from tables P_w = 18 mm Hg.
2. Find

$$C_{S_{20°C}} = 8.7 - \frac{3000}{100} x\, 0.009 = 8.43\, mg/L$$

3. Find

$$C_s' = C_s \frac{P - P_w}{760 - P_w}$$

$$C_s' = 8.43 \frac{732 - 18}{760 - 18} = \frac{8.43 \times 714}{742}$$
$$= 8.11\, mg$$

Example 2.9
At what depth of the water is the partial pressure of the oxygen 31 kPa? (U of K, 1988).

Solution
1. Given: partial pressure = 31 kPa
2. Determine total pressure = 31 ÷ 0.21 = 147.6 kpa
3. Then, $147.6 \times 10^3 = (9.81h + 101.3) \times 10^3$
   9.81h = 46.32 $\Longrightarrow$ h = 4.7 m

Program 2.9 Listing:
Depth of water given partial pressure of oxygen

```
'***********************
'EXAMPLE 2.9
'***********************
```

```
Public Class Form1

    Private Sub Button1_Click(ByVal sender As
        System.Object, ByVal e As
        System.EventArgs) Handles Button1.Click
        Dim O2, Height As Double
        O2 = Val(TextBox1.Text)
        Height = (O2 * 100 / 21) - 101.3
        Height /= 9.81

        Label2.Text =
        "Depth of water at which O2 is at " + _
            O2.ToString + " Pressure = " +
        Height.ToString + "m"
    End Sub

    Private Sub Form1_Load(ByVal sender As
        System.Object, ByVal e As
        System.EventArgs) Handles MyBase.Load
        Me.Text = "Example 2.9"
        Me.FormBorderStyle =
        Windows.Forms.FormBorderStyle.FixedSingle
        Me.MaximizeBox = False
        Label1.Text = "Enter partial pressure of O2:"
        Label2.Text = ""
        Button1.Text = "&Calculate"
    End Sub
End Class
```

Example 2.10

What is the saturation value of dissolved oxygen in a water containing 1000 mg/L of chloride ion with a temperature of 22°C at a pressure of 720 mm Hg? (SQU, 1991).

Solution

1) Given: $Cl^- = 1000$ mg/L, T = 22°C, P = 720 mm Hg
2) $(C_s)_{22°C} = 8.8 - 10 \times 0.008 = 8.72$ mg/L
3) From tables $P_w = 20$ mm Hg
4)
$$C_{s'22°C} = 8.72 \, x \, \frac{(720-20)}{(760-20)} = \frac{8.72 \, x \, 700}{740} = 8.25 \, mg/L$$

56

2.2 Radioactivity

Radioactivity is property of unstable atoms. It arises from the spontaneous breaking up of certain heavy atoms into atoms of other kinds which might themselves be radioactive. Radioactivity continues forming a transformation series. The disintegration results in emission of alpha particles, beta particles, & gamma rays. Each radioactive substance is characterized by a certain time known as its half-life period. The half-life period is the time taken for half the atoms in any given sample of the substance to decay. Half-life periods are different for each element. Amounts of radiation can be measured in curies, rems, millirems & rads. Curie is intensity of a sample of radioactive material in terms of atoms of the material that decay each second, or is the number of disintegrations occurring per second in one gram of pure radium. This rate (37 billion atoms per second for one gram of radium) is the basis of this measurement. Rem[1] is a measurement of effects of radiation on the body. Rad is the unit of measure for physical absorption of radiation. The international units of absorbed dose are sieverts & grays. The rate of decay of any nuclide is directly proportional to the number of atoms present & follows a first order reaction

$Ln (N_t/No) = - k_t$
Where:-
N_o =   Number of atoms found initially.
N_t =Number of atoms found at time t.
k_t = Decay constant.

Example 2.11

Radioactive radium is having a half-life of 1600 years. Find the mass remaining unchanged of 50 units of the substance after 3200, 4800, & 6400 years. Compute the time needed for the substance to lose 10 percent of its mass.

---

[1]   Unit used most often to measure radiation exposure for a person is millirem (mrem). It is one-thousandth of a rem

Solution

1. Determine the decay constant as: k_t = 0.693/$t_{\frac{1}{2}}$ = 0.693/1600 = 4.33*10⁻⁴ year-1.
2. Find mass remaining after time t as: N = No*e⁻ᵏ*t for t = 1600 N = 50*e⁻⁰·⁰⁰⁰⁴³³*1600 = 25; for t = 3200, N = 12.5; for t = 4800, N = 6.25; for t = 6400, N = 3.13.
3. Plot graph of mass remaining versus time, & find from graph for a remaining mass of, (= 50 - (10*50/100)) 45, the required time = 243 years, or weight lost = 10*50/100 = 5
4. weight remaining = 50 - 5 = 45
5. Or, use Ln 45/50 = - 4.33*10⁻⁴*t to find t = 243 years.

Example 2.12

Radioactive radium is having a half-life[2] of 1600 years. Find the mass remaining unchanged of 240 units of the substance after 2500, 5200, and 6400 years. Compute the time needed for the substance to lose 30 percent of its mass. Plot the decay curve of the radioactive substance.

Solution

1. Determine the decay constant as: kt = Ln2/$t\frac{1}{2}$= 0.693/$t\frac{1}{2}$ = 0.693/1600 = 0.000433 year⁻¹.
2. Find mass remaining after time t as: N = N₀*e⁻ᵏ*ᵗ for t = 2500, N = 81.3; for t = 5200, N = 25.27; for t = 6400, N = 15.
3. Plot graph of mass remaining versus time, and find from graph for a remaining mass of, (= 240 - (30*240/100)), the required time = 824 years, or
4. Or, use the equation for weight lost = 30*240/100 = 72, then weight remaining = 240 - 72 = 168 as:
5. Ln 168/240 = - 0.000433*t
6. to find t = 824 years

---

[2] Decay constant = Ln2 divided by half-life.

2.3 Chemical characteristics
2.3.1 General
Example 2.13

Arrange the following solutions in order of decreasing acidity (i.e., highest [H+] first, lowest last): Solution A, pH=8; Solution B, pOH=4; Solution C, $[H+]=10^{-6}$; Solution D, $[OH-]=10^{-5}$. (SQU, 1991).

Solution

For A: $[H^+] = 10^{-8}$, (pH = 8 = -log $[H^+]$)
For B: $[H^+] = 10^{-10}$, (pH = 14 – 4 =10)
For C: $[H^+] = 10^{-6}$

For D: $[H^+] = \dfrac{10^{-14}}{[OH^-]} = \dfrac{10^{-14}}{10^{-5}} = 10^{-9}$

In increasing order (remembering that exponents are negative)
C, A, D, B, $10^{-6} > 10^{-8} > 10^{-9} > 10^{-10}$

Program 2.13 Listing:
Arranging solutions in order of decreasing acidity

```
'*******************
'EXAMPLE 2.13
'*******************
Imports System.Math

Public Class Form1

    Private Sub Form1_Load(ByVal sender As
        System.Object, ByVal e As
        System.EventArgs) Handles MyBase.Load

        Me.Text = "Example 2.13"
        Me.FormBorderStyle =
        Windows.Forms.FormBorderStyle.FixedSingle
        Me.MaximizeBox = False
        'setup user interface
        Label1.Text = "Determine [H+] by using:"
        Label2.Text = "which equals:"
        Label3.Text = ""
        Label4.Text = ""
        Label5.Text = ""
```

```
        Label6.Text = ""
        Label7.Text = ""
        Button1.Text = "&Order solutions"

        Dim lst(3) As String
        lst(0) = "[H+]"
        lst(1) = "[OH-]"
        lst(2) = "pH"
        lst(3) = "pOH"
        ComboBox1.Items.AddRange(lst)
        ComboBox1.SelectedIndex = -1
        ComboBox2.Items.AddRange(lst)
        ComboBox2.SelectedIndex = -1
        ComboBox3.Items.AddRange(lst)
        ComboBox3.SelectedIndex = -1
        ComboBox4.Items.AddRange(lst)
        ComboBox4.SelectedIndex = -1
End Sub

Private Sub Button1_Click(ByVal sender As
    System.Object, ByVal e As
    System.EventArgs) Handles Button1.Click
    Dim solution(3) As Double
    Dim sstr(3) As Char
    sstr(0) = "A"
    sstr(1) = "B"
    sstr(2) = "C"
    sstr(3) = "D"
    '***set the value of each solution's [H+]
    '***according to the
    '***selected item in the relevant combobox
    If ComboBox1.SelectedIndex >= 0 And
    TextBox1.Text <> "" Then
        Select Case ComboBox1.SelectedIndex
            Case 0  'Calculate by [H+]
                solution(0) = Val(TextBox1.Text)
            Case 1  'Calculate by [OH-]
                solution(0) =
            -14 - Val(TextBox1.Text)
            Case 2  'Calculate by pH
                solution(0) = -Val(TextBox1.Text)
            Case 3  'Calculate by pOH
                solution(0) =
            -(14 - Val(TextBox1.Text))
        End Select
    End If

    If ComboBox2.SelectedIndex >= 0 And
        TextBox2.Text <> "" Then
        Select Case ComboBox2.SelectedIndex
```

```
            Case 0  'Calculate by [H+]
                solution(1) = Val(TextBox2.Text)
            Case 1  'Calculate by [OH-]
                solution(1) =
           -14 - Val(TextBox2.Text)
            Case 2  'Calculate by pH
                solution(1) = -Val(TextBox2.Text)
            Case 3  'Calculate by pOH
                solution(1) =
           -(14 - Val(TextBox2.Text))
        End Select
    End If

    If ComboBox3.SelectedIndex >= 0 And
        TextBox3.Text <> "" Then
        Select Case ComboBox3.SelectedIndex
            Case 0  'Calculate by [H+]
                solution(2) = Val(TextBox3.Text)
            Case 1  'Calculate by [OH-]
                solution(2) =
           -14 - Val(TextBox3.Text)
            Case 2  'Calculate by pH
                solution(2) = -Val(TextBox3.Text)
            Case 3  'Calculate by pOH
                solution(2) =
           -(14 - Val(TextBox3.Text))
        End Select
    End If

    If ComboBox4.SelectedIndex >= 0 And
        TextBox4.Text <> "" Then
        Select Case ComboBox4.SelectedIndex
            Case 0  'Calculate by [H+]
                solution(3) = Val(TextBox4.Text)
            Case 1  'Calculate by [OH-]
                solution(3) =
           -14 - Val(TextBox4.Text)
            Case 2  'Calculate by pH
                solution(3) = -Val(TextBox4.Text)
            Case 3  'Calculate by pOH
                solution(3) =
           -(14 - Val(TextBox4.Text))
        End Select
    End If

    Dim tmp As Double
    Dim tmp2 As char
    Dim repeat As Boolean = True
    'This loop will order the items in an
'ascending order.
```

```
    'to do so, it uses a bubble search algorithm
'which goes like
    'this: compare items (1)&(2), if (1) < (2),
'switch them.
    'then go to (2)&(3) and do the same, upto
'the last one.
    'then repeat from the start until no
'switches are done, then exit

    While repeat
        repeat = False
        For i = 0 To 2
            If solution(i) < solution(i + 1) Then
                tmp = solution(i)
                solution(i) = solution(i + 1)
                solution(i + 1) = tmp
                tmp2 = sstr(i)
                sstr(i) = sstr(i + 1)
                sstr(i + 1) = tmp2
                repeat = True
            End If
        Next
    End While

    Label3.Text = "Solutions in increasing order:"
    For i = 0 To 3
        Label3.Text += vbCrLf + "(" + sstr(i) +
    "): [H+] = 10^" + solution(i).ToString
    Next
End Sub

Private Sub ComboBox1_SelectedIndexChanged(ByVal
    sender As System.Object, ByVal
    e As System.EventArgs) Handles
    ComboBox1.SelectedIndexChanged
    'show "10^" before the textbox if user
  'selected the first or second items
    If ComboBox1.SelectedIndex = 0 Or
    ComboBox1.SelectedIndex = 1 Then
        Label4.Text = "10^"
    Else : Label4.Text = ""
    End If
End Sub

Private Sub ComboBox2_SelectedIndexChanged(ByVal
    sender As System.Object, ByVal
    e As System.EventArgs) Handles
    ComboBox2.SelectedIndexChanged
    'show "10^" before the textbox if user
  'selected the first or second items
```

```
        If ComboBox2.SelectedIndex = 0 Or
     ComboBox2.SelectedIndex = 1 Then
          Label5.Text = "10^"
     Else : Label5.Text = ""
     End If
End Sub

Private Sub ComboBox3_SelectedIndexChanged(ByVal
    sender As System.Object, ByVal
    e As System.EventArgs) Handles
    ComboBox3.SelectedIndexChanged
    'show "10^" before the textbox if user
  'selected the first or second items
    If ComboBox3.SelectedIndex = 0 Or
    ComboBox3.SelectedIndex = 1 Then
          Label6.Text = "10^"
    Else : Label6.Text = ""
    End If
End Sub

Private Sub ComboBox4_SelectedIndexChanged(ByVal
    sender As System.Object, ByVal
    e As System.EventArgs) Handles
    ComboBox4.SelectedIndexChanged
    'show "10^" before the textbox if user
  'selected the first or second items
    If ComboBox4.SelectedIndex = 0 Or
    ComboBox4.SelectedIndex = 1 Then
          Label7.Text = "10^"
    Else : Label7.Text = ""
    End If
End Sub
End Class
```

Example 2.14

Calculate the OH- concentration in grams of OH- per litre, of a solution containing $1 \times 10\text{-}10$ mole of H+ per litre. (SQU, 1992).

Solution

$[OH^-][H^+] = 10^{-14}$

$[OH^-] = 10^{-14} \div 10^{-10} = 10^{-4}$ mole/L

$MW_{[OH-]} = 16 + 1 = 17$

Concentration of $OH^- = 10^{-4} \times 17 = 1.7 \times 10^{-3}$ g/L = 1.7 mg/L

Example 2.15

Compare the degree of acidity of the following solutions: First solution of pH = 3.5; Second solution of [H+] = 3. 2×10-4 mg/Li and third solution of Hydrogen ion concentration, [H+] = 2.5×10-3 mole/L (SQU, 1993).

Solution

First solution: $pH_1 = 3.5$
Second solution: $pH_2 = -\log 3.2 \times 10^{-4} \times 10^{-3} = -\log 3.2 \times 10^{-7} = 6.5$
Third solution: $pH_3 = -\log 2.5 \times 10^{-3} = 2.6$
Comparison: Solution 3, solution 1, solution 2

Program 2.15 Listing:
Comparing degree of acidity of solutions

```
'*******************
'EXAMPLE 2.15
'*******************
Imports System.Math

Public Class Form1

    Private Sub Form1_Load(ByVal sender As
        System.Object, ByVal e As
        System.EventArgs) Handles MyBase.Load
        Me.Text = "Example 2.15"
        Me.FormBorderStyle =
        Windows.Forms.FormBorderStyle.FixedSingle
        Me.MaximizeBox = False
        'setup user interface
        Label1.Text = "Determine pH by using:"
        Label2.Text = "which equals:"
        Label3.Text = ""
        Label4.Text = "x10^"
        Label5.Text = "x10^"
        Label6.Text = "x10^"
        Button1.Text = "&Order solutions"
        Label4.Visible = False :
    TextBox4.Visible = False
        Label5.Visible = False :
    TextBox5.Visible = False
        Label6.Visible = False :
    TextBox6.Visible = False
```

64

```vbnet
        Dim lst(2) As String
        lst(0) = "[H+] in mg/L"
        lst(1) = "[H+] in mol/L"
        lst(2) = "pH"
        ComboBox1.Items.AddRange(lst)
        ComboBox1.SelectedIndex = -1
        ComboBox2.Items.AddRange(lst)
        ComboBox2.SelectedIndex = -1
        ComboBox3.Items.AddRange(lst)
        ComboBox3.SelectedIndex = -1
End Sub

Private Sub Button1_Click(ByVal sender As
    System.Object, ByVal e As
    System.EventArgs) Handles Button1.Click
    Dim solution(2) As Double
    Dim sstr(2) As Char
    sstr(0) = "1"
    sstr(1) = "2"
    sstr(2) = "3"
    '***set the value of each solution's pH
    '***according to the
    '***selected item in the relevant combobox
    If ComboBox1.SelectedIndex >= 0 And
    TextBox1.Text <> "" Then
        Select Case ComboBox1.SelectedIndex
            Case 0  'Calculate by [H+] in mg/L
                solution(0) =
        -Log10(Val(TextBox1.Text) *
        Pow(10, Val(TextBox4.Text) - 3))
            Case 1  'Calculate by [H+] in mol/L
                solution(0) =
        -Log10(Val(TextBox1.Text) *
        Pow(10, Val(TextBox4.Text)))
            Case 2  'Calculate by pH
                solution(0) = Val(TextBox1.Text)
        End Select
    End If

    If ComboBox2.SelectedIndex >= 0 And
    TextBox2.Text <> "" Then
        Select Case ComboBox2.SelectedIndex
            Case 0  'Calculate by [H+] in mg/L
                solution(1) =
        -Log10(Val(TextBox2.Text) *
        Pow(10, Val(TextBox5.Text) - 3))
            Case 1  'Calculate by [H+] in mol/L
                solution(1) =
        -Log10(Val(TextBox2.Text) *
        Pow(10, Val(TextBox5.Text)))
```

```
            Case 2   'Calculate by pH
                solution(1) = Val(TextBox2.Text)
        End Select
    End If

    If ComboBox3.SelectedIndex >= 0 And
        TextBox3.Text <> "" Then
        Select Case ComboBox3.SelectedIndex
            Case 0   'Calculate by [H+] in mg/L
                solution(2) =
        -Log10(Val(TextBox3.Text) *
        Pow(10, Val(TextBox6.Text) - 3))
            Case 1   'Calculate by [H+] in mol/L
                solution(2) =
        -Log10(Val(TextBox3.Text) *
        Pow(10, Val(TextBox6.Text)))
            Case 2   'Calculate by pH
                solution(2) = Val(TextBox3.Text)
        End Select
    End If

    Dim tmp As Double
    Dim tmp2 As Char
    Dim repeat As Boolean = True
    'This loop will order the items in an
'ascending order.
    'to do so, it uses a bubble search algorithm
'which goes like
    'this: compare items (1)&(2), if (1) < (2),
'switch them.
    'then go to (2)&(3) and do the same, upto
'the last one.
    'then repeat from the start until no switches
'are done, then exit

    While repeat
        repeat = False
        For i = 0 To 1
            If solution(i) > solution(i + 1) Then
                tmp = solution(i)
                solution(i) = solution(i + 1)
                solution(i + 1) = tmp
                tmp2 = sstr(i)
                sstr(i) = sstr(i + 1)
                sstr(i + 1) = tmp2
                repeat = True
            End If
        Next
    End While
```

```vb
        Label3.Text = "Solutions in increasing order:"
        For i = 0 To 2
            Label3.Text += vbCrLf + "(" + sstr(i) + "):
    pH = " + FormatNumber(solution(i), 2)
        Next
End Sub

Private Sub ComboBox1_SelectedIndexChanged(ByVal
    sender As System.Object, ByVal
    e As System.EventArgs) Handles
    ComboBox1.SelectedIndexChanged
    'show "10^" before the textbox if user
  'selected the first or second items
    If ComboBox1.SelectedIndex = 0 Or
    ComboBox1.SelectedIndex = 1 Then
        Label4.Visible = True :
    TextBox4.Visible = True
    Else : Label4.Visible = False :
    TextBox4.Visible = False
    End If
End Sub

Private Sub ComboBox2_SelectedIndexChanged(ByVal
    sender As System.Object, ByVal
    e As System.EventArgs) Handles
    ComboBox2.SelectedIndexChanged
    'show "10^" before the textbox if user
  'selected the first or second items
    If ComboBox2.SelectedIndex = 0 Or
    ComboBox2.SelectedIndex = 1 Then
        Label5.Visible = True :
    TextBox5.Visible = True
    Else : Label5.Visible = False :
    TextBox5.Visible = False
    End If
End Sub

Private Sub ComboBox3_SelectedIndexChanged(ByVal
    sender As System.Object, ByVal
    e As System.EventArgs) Handles
    ComboBox3.SelectedIndexChanged
    'show "10^" before the textbox if user
  'selected the first or second items
    If ComboBox3.SelectedIndex = 0 Or
    ComboBox3.SelectedIndex = 1 Then
        Label6.Visible = True :
    TextBox6.Visible = True
    Else : Label6.Visible = False :
    TextBox6.Visible = False
    End If
```

```
      End Sub
End Class
```

Example 2.16

Determine the weight in grams of $Na_2CO_3.10H_2O$ required to prepare half a litre of a o.1 N solution.

Solution

MW_{soda} = $MWNa_2CO_3.10H_2O$ = $23\times2 + 12 + 3\times16 +10(1\times2+16)$ = 286.

EW = $286 \div 2 = 143$ g.

$$N = \frac{wt/EW}{V} = 0.1 = \frac{wt/143}{0.5}$$

Weight = 0.1x0.5x143 = 7.15g

2.3.2 Hardness

Example 2.17

Differentiate between permanent and temporary hardness.

Solution

Hardness:
  a) Temporary: $Ca(HCO_3)_2$, $Mg(HCO_3)_2$, $Mg\ CO_3$
  b) Permanent: $CaCl_2$, $CaSO_4$, $MgCl_2$, $MgSO_4$

1)   The following water analysis is submitted for your review:

K^+	6 mg/L	Cl^-	120 mg/L
Na^+	72 mg/L	SO_4^{--}	Missing
Ca^{++}	90 mg/L	alkalinity	135 mg/L as $CaCO_3$
Mg^{++}	30 mg/L	CO_2	3 mg/L
		Temperature	25°C

  a)   Assuming that all of the constituents with the exception of the sulfate ion SO_4, have been analyzed correctly and

68

that the alkalinity is due solely to bicarbonate, determine the sulfate concentration.

b)   Calculate total and non-carbonate hardness of the water as mg/L $CaCO_3$.

c)   What conclusions can you draw from your computations? (UAE, 1989).

Solution

1. Construct table for determining concentration in milli-equivalent per liter (by dividing by the equivalent weight) as follows:

Substance (1)	Concentration (Given value) (2)	EW = MW/Z (3)	C, meq/L (4) = (2)/ (3)	C, mg/L $CaCO_3$ (5) = (4)*50
Ca^{++}	90	20	4.5	225
Mg^{++}	30	12.15	2.47	123.5
Na^+	72	23	3.13	156.5
K^+	6	39.1	0.153	7.7
			10.253	
Cl^-	120	35.5	3.38	169
Alkalinity = HCO_3^-, CO_3^-		61	2.7	135
SO_4^{--}	x	48	x/48	
CO_2	3	22		
			6.08+ (x/48)	

i) Since concentration of cations = that of anions

6.08+ (x/48) – 10.253 $\Rightarrow$ X = 200.3 mg/L SO_4^-

ii) Total hardness = Ca^{++} + Mg^{++} = 348.5 mg/L $CaCO_3$

Non-carbonate hardness = total – carbonate hardness = 348.5 – 135 = 213.5 mg/L $CaCO_3$

iii) Water is very hard (TH > 300)

2)   Write a computer program to calculate total and non-carbonate hardness of a water sample given cations and anions concentrations.

Program 2.17 Listing:

Total and non-carbonate hardness of a water sample for given concentrations of cations and anions

```
'********************
'PROGRAM 2.17
'********************
Public Class Form1

    Private Sub Form1_Load(ByVal sender As
        System.Object, ByVal e As
        System.EventArgs) Handles MyBase.Load

        Me.Text = "Example 2.17"
        Me.FormBorderStyle =
        Windows.Forms.FormBorderStyle.FixedSingle
        Me.MaximizeBox = False
        'setup user interface
        Label1.Text = "Ca++"
        Label2.Text = "Mg++"
        Label3.Text = "Na+"
        Label4.Text = "K+"
        Label5.Text = "Cl-"
        Label6.Text = "Alkalinity (CaCO3)"
        Label7.Text = "SO4-"
        Label8.Text = "CO2"
        Label9.Text = "Enter the concentrations
                in mg/L" + vbCrLf
        Label9.Text += "Leave the missing value empty."
        Button1.Text = "&Calculate"
    End Sub

    Private Sub Button1_Click(ByVal sender As
        System.Object, ByVal e As
        System.EventArgs) Handles Button1.Click
        Dim anConc(3), catConc(3) As Double
        Dim anEW(3), catEW(3) As Double
        Dim anConcMeqL(3), catConcMeqL(3) As Double
        Dim anConcMgL(3), catConcMgL(3) As Double
    'mark the missing anion/cation
        Dim missing As Integer
        Dim totalAnions, totalCations As Double
        Dim totalHard, nonCarbHard As Double

        'Retrieve values for column (2) of the table
        catConc(0) = Val(TextBox1.Text)
        catConc(1) = Val(TextBox2.Text)
        catConc(2) = Val(TextBox3.Text)
```

```
   catConc(3) = Val(TextBox4.Text)
   anConc(0) = Val(TextBox5.Text)
   anConc(1) = Val(TextBox6.Text)
   anConc(2) = Val(TextBox7.Text)
   anConc(3) = Val(TextBox8.Text)
   'Calculate column (3) of the table
   'using predetermined values
   catEW(0) = 20 : catEW(1) = 12.15 :
catEW(2) = 23 : catEW(3) = 39.1
   anEW(0) = 35.5 : anEW(1) = 61 :
anEW(2) = 48 : anEW(3) = 22
   'Calculate column (4) of the table
   totalCations = 0
   For i = 0 To 3
      If catConc(i) = 0 Then :
    missing = i 'mark the missing cation
      Else : catConcMeqL(i) =
         catConc(i) / catEW(i)
            totalCations += catConcMeqL(i)
      End If
   Next
   totalAnions = 0
   For i = 0 To 2
      If anConc(i) = 0 Then :
    missing = i + 4 'mark the missing anion
      Else : anConcMeqL(i) = anConc(i) / anEW(i)
            totalAnions += anConcMeqL(i)
      End If
   Next
   'Calculate column (5) of the table
   For i = 0 To 3
      If i <> missing Then
      catConcMgL(i) = catConcMeqL(i) * 50
   Next
   For i = 0 To 2
      If i <> missing - 4 Then
      anConcMgL(i) = anConcMeqL(i) * 50
   Next
   'Find the missing ion concentration
   If missing < 4 Then
   'missing ion is a cation
      catConcMeqL(missing) =
   totalAnions - totalCations
      catConc(missing) =
   catConcMeqL(missing) * catEW(missing)
      Label9.Text = "Missing concentration = "
         + catConc(missing).ToString
   Else    'missing ion is an anion
      anConcMeqL(missing - 4) =
   totalCations - totalAnions
```

```
          anConc(missing - 4) =
       anConcMeqL(missing - 4) * anEW(missing-4)
          Label9.Text = "Missing concentration = "
          + anConc(missing - 4).ToString
       End If
       'calculate total hardness from Ca++ and Mg++
       totalHard = catConcMgL(0) + catConcMgL(1)
       nonCarbHard = totalHard - anConcMgL(1)
       Label9.Text += vbCrLf + "Total hardness = "
          + totalHard.ToString
       Label9.Text += vbCrLf +
          "Non-carbonate hardness = " +
          nonCarbHard.ToString
   End Sub
End Class
```

2.4 Biochemical oxygen demand (BOD)

The BOD test measures the amount of oxygen consumed by bacteria whilst oxidizing organic matter under aerobic conditions. The test determines the approximate quantity of oxygen that will be required to biologically stabilize the organic matter present. The test advantages include:

- Determination of the size of waste treatment facilities
- Measurement of the efficiency of some treatment processes.

It is to be noted that complete stabilization may demand a period too long for practical purposes. Therefore, incubation for five days at $20°$ C approximates an average temperature for slow moving streams during summer conditions. The test requires exclusion of light during the incubation period to prevent oxygen formation by algae in the sample.

Example 2.18

1) A BOD test is carried out on a sample of sewage which has a rate constant (k_1) value of 0.15 /day. Compute the value of the BOD_5 as compared to the ultimate BOD.
2) Write a computer program to compute the value of the BOD_5 as compared to the ultimate BOD in a BOD carried out on a sample of sewage given its rate const.
3) Verify your program by solving example 2.18a.

Solution

1. The first order re-aeration equation may be used, i.e.

$BOD_5^{20} = L_o(1 - 10^{-k_1 * t})$ where:

BOD_5^{20} = Amount of biochemical oxygen demand removed over 5 days at 20°C.

L_o = Initial BOD at zero time = ultimate BOD.

k_1 = Rate constant (per day).

t = Time, days.

2. Substituting given values in the above mentioned equation, it gives:

$BOD_5^{20} = L_o(1 - 10^{-k_1 * t}) = L_o(1 - 10^{-0.15 \times 5}) = 0.82$

3. Thus, the 5-day BOD amounts to about 82 percent of the ultimate BOD.

Table (2.1) Strength of sewage

Strength	BOD_5 (mg/L)
Weak	Less than 200
Medium	200 - 350
Strong	351 - 500
Very strong	Greater than 750

Program 2.18 Listing:

Value of BOD_5 as compared to the ultimate BOD

```
'********************
'PROGRAM 2.18
'********************
Imports System.Math

Public Class Form1

    Private Sub Form1_Load(ByVal sender As
        System.Object, ByVal e As
        System.EventArgs) Handles MyBase.Load

        Me.Text = "Example 2.18"
        Me.FormBorderStyle =
        Windows.Forms.FormBorderStyle.FixedSingle
        Me.MaximizeBox = False
        Label1.Text = "Select an option:"
        'Label2.Text = ""
        'Label3.Text = ""
```

```vbnet
        'Label4.Text = ""
        'Label5.Text = ""
        Button1.Text = "&Calculate"
        ComboBox1.Items.Clear()
        ComboBox1.Items.Add("Calculate BOD5")
        ComboBox1.Items.Add("Calculate BOD5 ratio")
        ComboBox1.Items.Add("Calculate k1")
        ComboBox1.Items.Add("Calculate Lo")
        ComboBox1.SelectedIndex = 0
End Sub

Private Sub Button1_Click(ByVal sender As
    System.Object, ByVal e As
    System.EventArgs) Handles Button1.Click
    Dim BOD5, Lo, k1, t, BOD5perc As Double

    'calculate the requested component
    'the base equation is: BOD5 = Lo(1 - 10^-k1*t)
    Select Case ComboBox1.SelectedIndex
        Case 0  'selected BOD5
            Lo = Val(TextBox1.Text)
            k1 = Val(TextBox2.Text)
            t = Val(TextBox3.Text)
            BOD5 = Lo * (1 - (Pow(10, (-k1 * t))))
            Label5.Text = "BOD5 = " +
        FormatNumber(BOD5, 2)
        Case 1  'selected BOD5 ratio
            k1 = Val(TextBox1.Text)
            t = Val(TextBox2.Text)
            BOD5perc =
        (1 - (Pow(10, (-k1 * t)))) * 100
            Label5.Text = "BOD5 as percentage of
        ultimate BOD = " +
        FormatNumber(BOD5perc, 2) + "%"
        Case 2  'selected k1
            Lo = Val(TextBox1.Text)
            BOD5 = Val(TextBox2.Text)
            t = Val(TextBox3.Text)
            k1 = -Log(1 - BOD5 / Lo) / (t * Log(10))
            Label5.Text = "k1 = " +
        FormatNumber(k1, 2)
        Case 3  'selected Lo
            BOD5 = Val(TextBox1.Text)
            k1 = Val(TextBox2.Text)
            t = Val(TextBox3.Text)
            Lo = BOD5 / (1 - (Pow(10, (-k1 * t))))
            Label5.Text = "Lo = " +
        FormatNumber(Lo, 2)
    End Select
End Sub
```

```vbnet
Private Sub ComboBox1_SelectedIndexChanged(ByVal
    sender As System.Object, ByVal
    e As System.EventArgs) Handles
    ComboBox1.SelectedIndexChanged
    'change the user interface according to the
    'selected item
    Select Case ComboBox1.SelectedIndex
        Case 0
            Label2.Text = "Lo="
            Label3.Text = "k1="
            Label4.Text = "t="
            TextBox3.Visible = True
            TextBox1.Text = ""
            TextBox2.Text = ""
            TextBox3.Text = ""
        Case 1
            Label2.Text = "k1="
            Label3.Text = "t="
            Label4.Text = ""
            TextBox3.Visible = False
            TextBox1.Text = ""
            TextBox2.Text = ""
        Case 2
            Label2.Text = "Lo="
            Label3.Text = "BOD5"
            Label4.Text = "t="
            TextBox3.Visible = True
            TextBox1.Text = ""
            TextBox2.Text = ""
            TextBox3.Text = ""
        Case 3
            Label2.Text = "BOD5="
            Label3.Text = "k1="
            Label4.Text = "t="
            TextBox3.Visible = True
            TextBox1.Text = ""
            TextBox2.Text = ""
            TextBox3.Text = ""
    End Select
    Label5.Text = ""
End Sub
End Class
```

75

Example 2.19 (see Program 2.18 Listing)

The 5-day BOD of a sample of sewage is 270 mg/l. The ultimate BOD is reported to be 390 mg/l. Determine the rate at which the sewage is being oxidized.

Solution

$BOD_5^{20} = L_o(1 - 10^{-k1*t})$

$270 = 390(1 - 10^{-5k1})$

This yields a value for k1 equal to 0.102 /day.

Example 2.20 (see Program 2.18 Listing)

A sample of sewage was incubated for 2 days and its BOD was found to be 200 ppm at a temperature of 20°C. Compute its 5-day BOD, assuming that the rate constant is 0.1 /day.

Solution

$BOD = L_o(1 - 10^{-k1t})$

Substituting given values, then ultimate BOD will be:

$L_o = 200/(1 - 10^{-0.1 \times 2}) = 541.9$ ppm

Thus,

$BOD_5 = 541.9(1 - 10^{-0.1 \times 5}) = 271$ ppm

Example 2.21 (see Program 2.18 Listing)

a)   A BOD test is carried out on domestic sewage which has a rate constant (k_1) value at 20°c of 0.16 day^{-1}. Calculate the BOD_5 (the BOD exerted) as a fraction of the ultimate BOD.

b)   A sewer carrying 0.3 m³/s foul sewage with a BOD of 250 mg/L joins two other sewers carrying flows of 0.1 m³/s and 0.2 m³/s with BOD_5 of 379 mg/L and 157 mg/L respectively. Calculate the BOD of the mixture.

Solution

a) $BOD_5 = Y = L(1 - 10^{-k1t})$

$K_1 = 0.16$ /d   t = 20°c

$Y/L = 1 - 10^{-0.16 \times 5} = 0.84$

i.e. the 5-day BOD is 84% of the ultimate BOD

b)

Sewer	Flow (m³/s)	BOD (mg/L)
I	0.3	250
II	0.1	379
III	0.2	157

$$BOD_{mixture} = \frac{BOD_1 Q_1 + BOD_2 Q_2 + BOD_3 Q_3}{Q_1 + Q_2 + Q_3}$$

$$= \frac{0.3 \times 250 + 0.1 \times 379 + 0.2 \times 157}{0.3 + 0.1 + 0.2} = 240.5 \, mg/L$$

Example 2.22 (see Program 2.18 Listing)

a) What does BOD stand for? What does this parameter indicate about water or wastewater quality?

b) Given that the ultimate BOD of a sample of sewage is 300 mg/L, find its 5-day BOD. Assume the rate constant to be 0.15 per day.

c) A treatment plant that has an influent BOD of 240 mg/L and an effluent BOD of 15 mg/L. For a flow rate of 15 ML/d:

    i)   Determine the efficiency of treatment.

    ii)   Find the BOD load the plant is disposing to the neighbouring surface water course. (SQU, 1991).

Solution

1. BOD = Biochemical oxygen demand (The amount of oxygen needed by bacteria for their metabolic activities).The more organic material there is in the water, the higher the BOD exerted by microbes, will be. It measures organic pollution in streams or lakes, measures strength of sewage and is used in design and operation of water pollution control plants.

2. $BOD_t = BOD_L(1-10^{-kt}) = 300(1-10^{-5 \times 0.15}) = 300(1 - 0.178) = 247$ mg/L

3.

$$BOD_i = 240 \longrightarrow \boxed{} \longrightarrow BOD_o = 15$$

$$Q = 15 \times 10^6 \times 10^{-3} \, m^3/d$$

i] $E = \dfrac{BOD_i - BOD_o}{BOD_i} = \dfrac{240 - 15}{240} x\ 100 = 94$

ii] BOD_{load} = BOD×Q = $15 \times 10^{-3}(kg/m^3) \times 15 \times 10^3(m^3/d)$ = 225 kg/d

2.5 Standards
Example 2.23
1) The lethal dose LD_{50} of cyanide for man is 70 mg. the ADI of cyanide for man is 0.05 mg per kg body weight. Determine the maximum permissible concentration in drinking water for the compound in mg/l. (Assumptions used: a daily in take of 3 litres of water per person, an average weight of 60 kg a person and a contribution of 10% to the daily in take of the compound by water consumption).
2) Write a computer program to find the maximum permissible concentration in drinking water for a compound in mg/l, given the lethal dose LD_{50} of the compoud for man. the ADI of for man per kg body weight.
3) Verify your program by solving example 2.23a.

Solution
1) Given: LD_{50} = 70 mg, ADI = 0.05 mg/kg body weight, MPC = 0.1 mg/l, Q = 3 L/d/c, w = 60 kg, contribution of daily in take = 10%.

2) Maximum permissible concentration will be:

$$ADIx\ \dfrac{percentageofcontribution}{consumption} x weight =$$

$$0.05\dfrac{mg}{kg.d} x 0.1\% x \dfrac{60\,kg}{3\,L/d} = 0.1mg/L$$

Program 2.19 Listing:

Maximum permissible concentration in drinking water for a compound

```
'****************************
'EXAMPLE 2.19: calculate MPC
'****************************
Public Class Form1

    Private Sub Form1_Load(ByVal sender As
        System.Object, ByVal e As
        System.EventArgs) Handles MyBase.Load
        Me.Text = "Example 2.19"
        Me.FormBorderStyle =
        Windows.Forms.FormBorderStyle.FixedSingle
        Me.MaximizeBox = False
        'set up user interface
        Label1.Text = "ADI (mg/kg)"
        Label2.Text = "% of contribution (%)"
        Label3.Text = "Daily consumption, Q (L/d)"
        Label4.Text = "Weight (kg)"
        Label5.Text = ""
        Button1.Text = "&Calculate MPC"
    End Sub

    Private Sub Button1_Click(ByVal sender As
        System.Object, ByVal e As
        System.EventArgs) Handles Button1.Click
        Dim ADI, perc, Q, Wt, MPC As Double
        ADI = Val(TextBox1.Text)
        perc = Val(TextBox2.Text) / 100
        Q = Val(TextBox3.Text)
        Wt = Val(TextBox4.Text)
        MPC = ADI * perc * Wt / Q
        Label5.Text = "Max. permissible conc. = " +
                FormatNumber(MPC, 2) + "mg/L"
    End Sub
End Class
```

Theoretical exercises 2.1

1) What are the most important **quality** parameters for supplying drinking water to a rural village?
2) What are the most significant parameters that affect water **quality**? State reasons.
3) Why are **non-pathogenic** microorganisms counted when they pose little hazard to human health? (UAE, 1989).
4) Outline your plan for establishing an accredited **reference laboratory** for testing quality of wastes? (SAS, 2007).
5) Write briefly about a water supply **source selection**.
6) Radioactive radium has a half-life of 1600 years. For 50 units of this material find the time required for it to lose 10 percent of its mass. Compute and plot its **decay curve**. (U of K, 2001).
7) What are the major hazards of **radioactive** wastes?
8) The rate of decay of any nuclide is directly proportional to the number of atoms present and follows a first order reaction
$N_t = N_o e^{-k \cdot t}$. Define symbols presented in the decay equation and outline how they can be determined.
9) Radioactive strontium, Sr^{90}, is having a half-life of 28 years. Find the **mass remaining** unchanged of 100 units of the substance after 56, 84, 112 and 144 years. Compute the time needed for the substance to lose 30 percent of its mass. Plot the radioactive **decay curve** of the substance (SAS, 2007).
10) What objections could there be if: **Turbidity, Chlorides, Nitrates, or Hardness**, are present in excess in water? (UAE, 1989).
11) Discuss the importance of **harness** in water and the chemical reactions associated with lime-soda softening.
12) What are the significances of the **BOD** test?
13) Why is the **BOD** test adopted as a standard one used to determine the pollution strength of sewage?
14) What is the difference between "**hard**" and "soft" water? Is it advisable to remove all the hardness from drinking water? State your reasons.
15) What is **DO** ? Why is it a significant parameter of water quality?
16) A wastewater sample was collected in a tightly closed container. The sample contains sulfate ions, dissolved oxygen and nitrate ions. Illustrate the order of biological degradation of these

compounds. Give your reasons. When do offensive **odors** appear? (SQU, 1992).

17) "Water may act as a vehicle for certain **disease**". Discuss this statement. (SQU, 1992).

18) Why can **Coliform bacteria** be used as indicators of quality for drinking water? (UAE, 1989), (SQU, 1992), (SQU, 1991).

19) Discuss the major objectives of **sedimentation**. (SQU, 1992).

20) A fresh wastewater containing nitrate ions, sulfate ions, and dissolved oxygen is placed in a sealed jar without air. In what sequence are these oxidized compounds reduced by **bacteria**? State your reasons. When do obnoxious odors appear? (SQU, 1991).

21) Name the most common classes of water related **diseases**. Give two examples for each disease. (SQU, 1991).

Practical exercises 2.2

Solids content

32) The following data are from total solids and total volatile solids tests on a wastewater. Calculate the **total and volatile** solids concentrations in mg/L.
   - Weight of empty dish = 68.942 g
   - Weight of dish + dry solids = 69.049 g
   - Weight of dish + ignited solids = 69.003 g
   - Volume of wastewater sample = 100 mL. (SQU, 1993).

Dissolved oxygen

33) What is the **dissolved** oxygen concentration in water at 20°C, 900 mg/L chloride concentration, and barometric pressure of 600 mm Hg? (UAE, 1989). (Ans.7.1 mg/L).

34) Compute the **saturation** value of dissolved oxygen in water exposed to water-saturated air containing 20.9 percent oxygen under a pressure 0f 732 mm of mercury. Assume temperature of 20 ☐ and a chloride concentration in water of 8000 mg/L. (SQU, 1991).

Radioactivity

35) A radioactive isotope has a half-life of 30 years. Starting by a mass of 20 kg, find the amount remaining after 15 and 42 years, respectively. What is the **time** required for 60 percent of the mass to disintegrate.

($N_t = N_o e^{-k \cdot t}$ $k = \dfrac{\ln 2}{t_{1/2}}$ of K, 2001).

36) A radioactive Phosphorous nuclide has a half-life of 14.3 days.

     i. How **long** would a sample containing 25 grams of this nuclide have to be stored in order to reduce the concentration to 0.39 grams?

     ii. Plot a graph of amount of sample remaining versus time and determine from your graph the amount of sample that remains after 100 days of storage. (SQU, 1993). (Ans. 0.195 g).

37) One of the most hazardous radioactive isotopes in the fallout of atomic bombs is strontium-90, $^{90}_{38}Sr$ for which $t_{0.5} = 28$ years. If 600 grams of Sr-90 descended on a family farm on the day a child was born in 1964, how many grams will still be on that farm when his granddaughter is born in 2014? (SQU, 1991). (Ans. 172 g).

ADI

38) The **ADI** of a certain toxic compound for man is 0.03 mg/kg body weight per day. Estimate the value of a standard in drinking water for this compound in mg/L. (Base your calculations on:

    - A daily intake of 3 liter water per capita,
    - An average weight of 65 kg per person,
    - And a contribution of 10 per cent to the daily intake of this compound by consumption of water) (UAE, 1989). (Ans. 0.065 mg/L).

39) The lethal dose LD 50 of a compound for man is 70 mg. The ADI of the compound for man is 0.05 mg/kg body weight. Assuming a daily intake of 3.5 litres of water per person, an average weight of 65 kg for a person & a contribution of 10% to the daily intake of the compound by water consumption, estimate the **maximum permissible** concentration of the compound in drinking water. (SQU, 1991).

40) The ADI of a certain toxic compound for man is 0.02 mg/kg body weight per day. Estimate the value of a **standard** in drinking water for this compound. Base your calculations on:
   - a daily intake of four litre water per capita per day,
   - a weight of 60 kg per person,
   - a contribution of ten percent to the daily intake of this compound by consumption of water. (Ans.0.03 mg/L).

Chemistry revision

41) One form of soap has the formula $C_{17}H_{35}COONa$. It is prepared by the reaction:

$$C_3H_5(C_{17}H_{35}COO)_3 + 3NaOH \longrightarrow 3C_{17}H_{35}COONa + C_3H_5(OH)_3$$

Calculate the number of grams of sodium hydroxide (lye) required to prepare 175 grams of soap. (SQU, 1991). (Ans.22.9 gNaOH).

42) Penicillin, the first of a now large number of antibiotics (antibacterial agents), was discovered accidentally by the Scottish bacteriologist Alexander Fleming in 19928, but he was never able to isolate it as a pure compound. This and similar antibiotics have saved millions of lives that would otherwise have been lost to infections. Penicillin F has the formula $C_{14}H_{20}N_2SO_4$. Compute the mass percent of each element. (SQU, 1991). (Ans. 54, 6.5, 9, 10, 20.5).

43) When a water treatment plant operator buys $Al_2(SO_4)_3.H_2O$ to be used in the coagulation process. How much is the percent aluminum he is buying? (Ans. 8 %).

44) What is the ratio of the molar concentration of OH- ions in a solution at pH = 11 to that at pH 9? (U of K, 1985).

45) From which one of the following reactions is light an important requirement? (U of K, 1985).
   a) $H_2S + 2H_2O + 2CO_2 \rightarrow 2CH_2O + H_2SO_4$
   b) $2H_2S + O_2 \rightarrow 2S + 2H_2O$
   c) $2S + 3O_2 + 2H_2O \rightarrow 2H_2SO_4$
   d) $2NH_3 + 3O_2 \rightarrow 2HNO_2 + 2H_2O$
   e) $2NO_2^- + O_2 \rightarrow 2NO_3^-$

46) Compare the ratios CH_4/CO_2 for the gas produced in the anaerobic digestion of:

i.   Ethanol O_2H_5OH
ii. Acetic acid CH_3COOH &
iii. oxalic acid COOH.COOH. List them in order of INCREASING values. (U of K, 1985). (Ans.3:1, 1:1, 1:7).

pH

16)   The pH of a solution changes from 6.5 to 2.5. By what factor did the strength of its **acidic** condition increase? (UAE, 1989).

17)   Arrange the following solutions in order of decreasing **acidity** (i.e. highest $[H^+]$ first, lowest last): Solution A, pH = 8; Solution B, pOH = 4; Solution C, $[H^+]$ = 10^{-6}; Solution A, $[OH^-]$ = 10^{-5}. (SQU, 1991).

18)   By how much units will the **pH** of water increase if 0.001 mole of Sodium Hydroxide, NaOH, is added to one liter? (SQU, 1993).

19)   Which is more **acidic**: a solution with a pH of 5 or a solution containing 50 grams of $\{OH^-\}$ per litre? (pH = - Log[H^+]

Hardness

20)   A surface water has the following analysis:

Calcium	72 mg/L	Chloride	**Missing**
Magnesium	48.8 mg/L	Alkalinity	250 mg/L $CaCO^3$
Sodium	9.2 mg/L	Sulphate	134.4 mg/L

1. Assuming that the analysis of the sample has been done correctly, with the exception of chloride ion and that the alkalinity is due to bicarbonate, compute the **chloride concentration**.

2. Calculate **total** and **non-carbonate** hardness of the water expressed as mg/L CaCO3.

3. What conclusions can you draw from your calculations? (UAE, 1989). (Ans. 10, 380, 130 mg/L $CaCO_3$).

32) Outline principal cations and anions causing **hardness**. In a harness determination test laboratory analysis of a water sample yielded the following data:

Cation (mg/L)	Anion (mg/L)
$Na^+ = 11.5$	$HCO_3^- = 91.5$
$Ca^{++} = y$	$SO_4^{--} = 24$
$Mg^{++} = 12.2$	$Cl^- = 35.5$
$Sr^{++} = 11$	

i) Express concentrations of ions in **milliequivalent**/liter.

ii) For an acceptable experimental **error** of 10 percent determine the concentration of Ca^{++} cation (y) in mg/L $CaCO_3$.

ii) Determine the **total, carbonate,** and **non-carbonate** hardness of the sample expressed in mg/L $CaCO_3$.

iv) Draw a **bar graph** of the sample.

v) Comment on your results. (SAS, 2008).

33) Outline main advantages and disadvantages of water hardness. How do you classify waters according to their degree of hardness? In a harness determination test laboratory analysis of a water sample yielded the following data:

Cation (mg/L)	Anion (mg/L)
$Na^+ = 11.5$	$HCO_3^- = 91.5$
$Ca^{++} = 20$	$SO_4^{--} = 24$
$Mg^{++} = 12$	$Cl^- = 10.7$
$Sr^{++} = 4.4$	

Express concentrations of ions in **milliequivalent**/liter.

1. Comment about the experimental **error**, assuming that an error of 10 percent is acceptable.

2. Draw a **bar graph** of the sample.

3. Determine the **total, carbonate,** and **non-carbonate** hardness of the sample.

4. Comment on your results.

34) Outline your plan for establishing an accredited reference laboratory for testing quality of wastes? (SAS, 2007).

35) A water sample contains the following:

Analysis	Concentration (mg/L)
Calcium	120
Magnesium	48.6
Sodium	92

Bicarbonate	244
Sulfate	96
Chloride	71

[a] Determine the **total** hardness.

[b) Find the **carbonate** hardness.

[c] Compute the **non-carbonate** hardness. (SQU, 1992). (Ans. 10, 4, 6 meq/L).

36) Tests for common ions are run on a sample of water and the results are as shown below.

Constituents [meq/L]			
Ca^{++}	6	HCO_3^-	7
Mg^{++}	4	SO_4^{--}	4
Na^+	4	Cl^-	3

Carbonate, **non-carbonate** and **total hardness**. (SQU, 1993). (Ans. 10, 7, 3 meq/L).

37) Outline advantages and disadvantages of **hardness** in water.

The ionic character of a groundwater is defined by the following hypothetical combinations:

Na_2SO_4	0.2 meq/L
$Mg(HCO_3)_2$	0.4 meq/L
NaCl	0.6 meq/L
$Ca(HCO_3)_2$	3.0 meq/L
$MgSO_4$	0.8 meq/L

i. Sketch a **milliequivalent**-per-litre bar graph.

ii. Determine the **total, carbonate** and **non-carbonate** hardiness of the water sample.

iii. Is the hardness of this water considered excessive? (Ans. 210, 170, 40 mg/L $CaCO_3$).

* In your opinion what is the most significant source of **groundwater pollution** in Oman? Why is groundwater contamination so difficult to detect and clean up? (SQU, 1992).

27) Outline reasons for measuring **hardness** in water. Analysis on a water sample revealed the following results: (SQU, 1992)

Item	Concentration
Calcium hardness	150 mg $CaCO_3$/L
Magnesium hardness	70 mg $CaCO_3$/L
Sodium ion	8.3 mg Na^+/L
Potassium ion	3.9 mg K^+/L
Alkalinity	190 mg $CaCO_3$/L
Sulfates ion	28.8 mg SO_4^-/L
Chloride ion	7.8 mg Cl^-/L

    i. If an experimental **error** of 10 % is acceptable, should the analysis be considered complete?

    ii. Draw a **milliequivalent**-per-litre bar graph.

    iii. Determine the **total, carbonate** and **non-carbonate** hardiness of the water sample.

    iv. List the **hypothetical combinations** of compounds for the sample. (Ans. Yes, 220, 190, 30 mg/L $CaCO_3$).

    v. Comment on your results. (SQU, 1991).

28)

a] Would hardness be acceptable in most drinking water supplies? Why or why not?

b] Differentiate between permanent and temporary hardness.

c] An analysis of hard water revealed the following data:

The Total hardness	250 mg CaO/l
Calcium hardness	130 mg Ca/l
Alkalinity	200 mg $CaCO_3$/1
Acidity	0.3 meq/l

    i] Draw a **bar diagram** in milliequivalent per litre of this water

    ii]Compute **total, carbonate** & **non-carbonate** hardness.

    iii] List the **hypothetical combinations** of positive & negative ions from the bar graph.

    iv] What conclusions can you draw from your computations? (SQU, 1991). (Ans. 250, 200, 50 mg/L $CaCO_3$).

29)

a) Define hardness of water, note two broad classifications of hardness, and discuss the sources and impacts of hardness.

b) An analysis of a water sample revealed the following data:

Constituents [mg/l]	
Ca++	60
Mg++	10
Na+	7
K+	20
HC_3^-	115 as $CaCO_3$
SO_4^{--}	96
Cl-	11

a] What is the percent **error** in cation-anion balance?

b] Calculate **carbonate** and **non-carbonate hardness** of the water expressed as mg $CaCO_3$/l.

c] Draw a **bar diagram** in milliequlvalent per litre of this water.

d] List the **hypothetical** chemical **combinations** of positive and negative ions from the bar graph.

e] What conclusions can you draw from your calculations? (SQU, 1991).

30)

a] discuss the merits & demerits of **hard** water.

b] The following water analysis is submitted for your review:

Mg^{++} as Mg	30.1 mg/l
Alkalinity as $CaCO_3$	135 mg/l
Ca^{++} as Ca	90 mg/l
Sodium as Na	72.2 mg/l
K^+ as K	6.3 mg/l
Sulfate as SO_4^{--}	missing, [Y;mg/l]
Chloride as Cl^-	120 mg/l
Carbon Dioxide as CO_2	3 mg/l
Temperature	25°C

i] Assuming that all of the constituents with the exception of the sulfate ion (SO_4--) have been analyzed correctly and that the alkalinity is due solely to bicarbonate, determine the **sulfate concentration**, y.

ii] Determine the **total** and **non-carbonate** hardness [express results in mg/l $CaCO_3$].

iii] Draw a **bar diagram** in milliequivalants per litre of the water.

iv] List the **hypothetical** chemical **combinations** of positive & negative ions from the bar graph.

v] Comment on your results. (SQU, 1991). (Ans. 210, 349, 135, 214 mg/L $CaCO_3$).

31)   Discuss the merits and demerits of **hard** water. An analysis of a water sample yields the following results:

Total hardness as $CaCO_3$	215 mg/L
Mg as Mg	185 mg/L
Sodium as Na	15.8 mg/L
Sulfate as SO_4	28.6 mg/L
Chloride as Cl	10 mg/L
Nitrate as N	1 mg/L
Carbon dioxide as CO_2	25.8 mg/L
pH	7.07

i. Indicate whether there is any **error** in the cation-anion balance

ii. Determine the **non-carbonate** hardness.

iii. Draw a **bar diagram** for the water.

iv. List the **hypothetical** chemical **combination**s of positive and negative ions from the bar diagram.

v. What conclusions can you draw from your calculations? (SQU, 1991).

32) Discuss the importance of **harness** in water and the chemical reactions associated with lime-soda softening. Surface water has the following analysis:

Calcium 71 mg/L, magnesium 48.8 mg/L,   sodium 9.2 mg/L, Bicarbonate 305 mg/L, sulfate 134.4 mg/L,   chloride 7.1 mg/L

- Calculate the number of **millequivalents** per litre of each substance.
- Calculate the **total** hardness, **carbonate** & **non-carbonate** hardness expresses as mg/L $CaCO_3$.
- Draw a **bar diagram** of the water. (U of K, 1985). (Ans. 380, 250, 130 mg/L $CaCO_3$).

33) Discuss the merits & demerits of hard water. An analysis of a water sample yields the following results:

89

Total hardness as $CaCO_3$	215 mg/l
Alkalinity as $CaCO_3$	185 mg/l
Mg as Mg	15.8 mg/l
Sodium as Na	8 mg/l
Sulfate as SO_4	28.6 mg/l
Chloride as Cl	10 mg/l
Nitrate as N	1 mg/l
Carbon Dioxide as CO_2	25.8 mg/l
pH	7.07

i] Indicate whether there is any **error** in the cation-anion balance.

ii] Determine the **non-carbonate** hardness.

iii] Draw a **bar diagram** for the water.

iv] List the **hypothetical** chemical **combinations** of positive & negative ions from the bar graph.

v] What conclusions can you draw from your calculations? (SQU, 1991). (Ans. 0%, 30 mg/L $CaCO_3$).

34) Surface water has the following analysis:

Calcium 71 mg/L,   magnesium 48.8 mg/L,   sodium 9.2 mg/L,
Bicarbonate 305 mg/L,   sulfate 134.4 mg/L,   chloride 7.1 mg/L

i.   Calculate the number of **millequivalents** per litre of each substance.

ii.   Calculate the **total** hardness, **carbonate** & **non-carbonate** hardness expresses as mg/L $CaCO_3$.

iii.   Draw a **bar diagram** of the water. (B.Sc., UK, 1985) (Ans. 380, 250, 130 mg/L $CaCO_3$).

35) A sample of water from a surface water stream is analyzed for the common ions with the following results:

Cations	Concentration (mg/L)	Anions	Concentration (mg/L)
Ca^{++}	60	Cl^-	11
Mg^{++}	10	HCO_3^-	115
Na^+	7	SO_4^{--}	96
K^+	20	NO_3^-	10

i)   What is the percent **error** in cation-anion balance? If an error of 10 percent is acceptable, should the analysis be considered complete?

ii)   Draw a **bar diagram** in milli-equivalent per liter of this water.

iii) List the **hypothetical** chemical **combinations** of positive and negative ions from the bar graph.

iv) Compute **total**, **carbonate** and **non-carbonate** hardness. (Ans. 7.5%, 191, 94, 97 mg/L CaCO₃).

v)   Comment on your results. (UAE, 1990).

36) What are the major merits and demerits of hardness? A useful test of the accuracy of a water analysis is to compare the reported cation and anion concentration in meq/L. In a perfect analysis, all cations and anions contained in the water would balance. The reported results of a water sample are as follows:

Cations (mg/L)	Anions (mg/L)
$Na^+ = 90$	$HCO_3^- = 220$
$Ca^{+2} = 60$	$SO_4^{--} = 64$
$Mg^{+2} = 20$	$Cl^- = 102$
$Fe^{+2} = 2$	$NO_3^- = 1$

i)   Does the analysis fall within a maximum acceptable **error** of 10 %?

ii)   Determine **total**, **carbonate** and **non-carbonate** hardness of the sample.

iii) Plot the **bar diagram** of the water.

iv) Indicate hypothetical chemical combinations of anions and cations of the sample. (OIU, UNESCOC, 2004). (Ans. Yes, 237, 180, 57 mg/L CaCO₃).

37) What is the significance of determining **hardness** in water analysis? The following mineral analysis was reported for a well water: (All units are in mg/L unless otherwise stated)

Fluoride = 1.1	Silica (SiO_2) = 13.4
Chloride = 4.0	**Bicarbonate = 318.0**
Nitrate = 0.0	Sulfate = 52.0
Sodium = 14.0	Iron = 0.5
Potassium = 1.6	Manganese = 0.07
Calcium = 96.8	Zinc = 0.27
Magnesium = 30.4	Barium = 0.2

    i)   Determine **total, carbonate** and **non-carbonate** hardness in mg/L $CaCO_3$ using the "common" polyvalent cation definition of hardness.

    ii)   How do you **classify** this water? State your reasons.

    iii) Comment about concentration of substances from a **health** point of view. (OIU, UNESCOC, 2004).

38) Would **hardness** be acceptable in most drinking water supplies? Why or why not? A surface water sample is chemically analyzed with the following analysis:

Cations	Concentration (mg/L)	Anions	Concentration (mg/L)
Ca^{++}	80	Cl^-	21.3
Mg^{++}	12.2	HCO_3^-	122
Na^+	11.5	SO_4^{--}	122.5
K^+	19.5	NO_3^-	24.8

    i)   What is the percent **error** in cation-anion balance?

    ii)   Draw a **bar diagram** in milliequivalent per liter of this water.

    iii) List the **hypothetical combinations** of positive and negative ions from the bar graph.

    iv) Compute **total carbonate** and **non-carbonate** hardness.

    v)   Comment on your results. (SQU, 1993).

39) What are the advantages and disadvantages of water **hardness**? Assuming an experimental error of 3 percent, find the missing **sulfate content** for the following data of a sample of water (all values in mg/L):

$Na^+ = 46$	$HCO_3^- = 366$
$Ca^{++} = 60$	$SO_4^{--} = y$
$Mg^{++} = 45$	$Cl^- = 71$

i) Draw a **bar graph** of the sample.

ii) Determine the **total, carbonate,** and **non-carbonate** hardness of the sample. (U of K, 2001).

40) What are the effects of an increase in the dose of each of the following, beyond accepted limits, in drinking water: **Fluoride, Nitrates, hardness** and **Gross Beta** particles. Assuming an experimental error of around 5 percent, find the missing **chloride content** (x in mg $CaCO_3$/L)for the following data of a sample of water (all values are in mg/L):

$Ca^{++} = 60$	$HCO_3^- = 183$
$Na^+ = 23$	$SO_4^- = x$
$Mg^{++} = 24.3$	$Cl^- = 35.5$

i) Determine **total** hardness of the sample.

ii) Indicate **hypothetical** chemical **combinations** of anions and cations of the sample. (U of K, 2001).

41) What are the effects of an increase in the dose of each of the following, beyond accepted limits, in drinking water: **E. coli, Lead, Nitrates, and hardness**. Assuming an experimental error of around 10 percent, find the missing sulfate content (y in mg $CaCO_3$/L)for the following data of a sample of water (all values are in mg/L):

$Ca^{++} = 60$	$HCO_3^- = 183$
$Na^+ = 23$	$SO_4^- = y$
$Mg^{++} = 24.3$	$Cl^- = 35.5$

i) Determine **total, carbonate** and **non-carbonate** hardness of the sample.

ii) Indicate **hypothetical** chemical **combinations** of anions and cations of the sample. (U of K, 2005).

42) Define **hardness** of water. Note the main classes of hardness, and discuss cons and pros.

    1.   Assuming an experimental error of 5 percent, find the missing **chloride concentration** for the following data of a sample of water (all values are in mg/L):

Cations		Anions	
Ca^{++}	36.0	Cl^-	y
Mg^{++}	40.1	HCO_3^-	183.0
Sr^{++}	8.8	SO_4^-	38.4
Fe^{++}	25.2	NO_3^-	24.8
		SiO_3^-	15.2

    a)   Determine the **total, carbonate** and **noncarbonated** hardness of the water sample.

    b)   Draw a **bar graph** of the sample

    c)   List the **hypothetical** chemical **combinations** of calcium and magnesium ions from the bar graph.

    d)   Using the following table **classify** this water sample.

    e)   Comment on your results. (B.Sc. UoD, 2013) (Ans. 75, 310, 150, 160 mg/L $CaCO_3$).

Classification of water according to degree of hardness

Degree of hardness	Hardness as mg CaCO₃/L
Soft	$0-75$
Moderately soft	$75-150$
Moderately hard	$150-175$
Hard	$175-300$
Very hard	Greater than 300

43) Define **hardness** of water, note two broad classifications of hardness, and discuss the sources and impacts of hardness. Would harness be acceptable in most drinking water supplies? Why or why not? A sample of water from a surface water stream is analyzed for the common ions with the following results:

Cations	Concentration (mg/L)	Anions	Concentration (mg/L)
Ca^{++}	40	Cl^-	10.65
Mg^{++}	24.3	HCO_3^-	122
Na^+	6.9	SO_4^-	96
K^+	19.5	NO_3^-	18.6

   i)   What is the percent **error** in cation-anion balance? If an error of 10 percent is acceptable, should the analysis be considered complete?

   ii)   Draw a **bar diagram** in milli-equivalent per liter of this water.

   iii) List the **hypothetical** chemical **combinations** of positive and negative ions from the bar graph.

   iv)   Compute **total, carbonate** and **non-carbonate** hardness.

   v)   Using the following table **classify** this water sample.

   vi)   Comment on your results. (Ans. 4%, 200, 100, 100 mg/L CaCO₃).

Classification of water according to degree of hardness

Degree of hardness	Hardness as mg CaCO₃/L
Soft	$0-75$
Moderately soft	$75-150$
Moderately hard	$150-175$
Hard	$175-300$
Very hard	Greater than 300

44) Raw water has the following constituents expressed in meq/L: Ca^{2+} 4.6; Mg^{2+} 1.0; Na^+ 2.1; HCO_3^- 2.4; SO_4^{-2} 2.9; Cl^- 2.4 and CO_2

0.6. What is the total **hardness** expressed as $CaCO_3$. (B.Sc., UoD, 2013)

BOD

45) a) State the merits and demerits of combined and separate sewers systems. b) What system do you recommend for the city of Khartoum and why? c) Differentiate, in a plumbing contest, between one-pipe and two-pipe drainage systems for a multiple dwelling. d) What is the total BOD (in mg/l) of a 0.1 molar solution of ethanol (CH_3CH_2OH) and a 0.l molar solution of oxalic acid (COOH.COOH)? (U of K, 1984). (Ans. 11 g/L).

46) **Define** the term BOD. What are the **objectives** of BOD determination? Determine the 5-day BOD for: (UAE, 1991).
     i)   1 g of Alanine $CH_3CH(NH_2)COOH$
     ii)   200 ppm Glycine $CH_2(NH_2)COOH$

47) What are the advantages and disadvantages of the BOD **test**? Determine the theoretical oxygen demand for the complete oxidation of the following compounds: (UAE, 1989).
     i)   1 g Glucose $C_6H_{12}O_6$.
     ii)   1 g Oleic acid $C_{17}H_{33}COOH$.
     iii)   150 mg/L acetic acid CH_3COOH.

48) Find the total BOD (in mg/L) for 0.1 molar solution of oxalic acid COOHCOOH and a 0.2 molar solution of ethanol CH_3CH_2OH._(U of K, 2001).

49) The five-day BOD at $20°C$ of a sewage is 300 ppm. Assuming $k_1 = 0.18$ /d, compute the value of the **ultimate** BOD of this sewage.

50) An industrial waste has a 5-day BOD of 600 ppm and the k_1 value at $20°$ C is 0.2 per day. Find the **ultimate** BOD of the waste. What would be the 5-day BOD if the value of k_1 dropped to 0.1 per day? (Ans. 456 ppm).

51) The 5-day BOD at $20°C$ of a sewage is 200 mg/L. Assuming k' to be 0.17, find the **ultimate** BOD of the sample in mg/L. (UAE, 1989) (Ans. 233 mg/L).

52) The 5-day BOD of a waste is 240 mg/L. assuming a rate constant (to base e) at $10°C$ of 0.16/day, find: The **ultimate oxygen demand** and the 1 & 3 day BOD. (UAE, 1991). (Ans. 336, 74, 178 mg/L).

53) a) Discuss the significance of BOD test in characterization of

95

waste waters. b) Derive an equation for first stage BOD. c) Why are samples incubated at 20°C in the dark for 5 days in the BOD test? d) A BOD test is carried out on domestic waste water. 6ml of the waste in a 300ml bottle were analyzed. e) If the initial dissolved oxygen is 8 mg/l while the 5-day dissolved oxygen is equal to 4 mg/l, compute the BOD of the sample and the **ultimate** BOD, assuming k_1-rate of 0.1 day^{-1}. (U of K, 1985). (Ans. 200, 292 mg/L).

54) A small sewage works serves a population of 10000 people whose per capita BOD_5 production is 0.055 kg/day. Water consumption is metered at 200 1/cap./day. A small industry discharges an effluent of 15 l/sec. over a period of 3 hours per day. The BOD_5 of this industry was found experimentally after diluting the waste water at a dilution of 2%. At the end of 5 days incubation period 3 mg/l dissolved oxygen has been depleted in the bottle. The first stage ultimate oxygen demand for the sewage works is 400 mg/l. a) At what **rate** is the waste being oxidized (assume temperature of 20°C)? At this rate 60% of the oxygen demand will be exerted in how many hours? b) What is the oxidation **rate** at 25°C, given that:

$$k_{1T} = k_{1_{20}} (1.047)^{T-20} \ where \ k_{1_{20}} = \ reaction \ rate \ at \ 20 \ °C \ (U \ of \ K,$$
1984). (Ans. 101 hr, 0.12 /d).

55) Calculate the **5-day BOD** of a sample of domestic sewage as a fraction of the ultimate BOD, given that the rate constant value at 20° C is 0.11 /day. (Ans. 72%).

56) A sample of sewage was incubated for 2 days and the BOD of the sample was observed to be 165 ppm at 20° C. Determine the **5-day BOD** (Assume $k_1 = 0.1$). (Ans. 306 ppm).

57) A BOD test is carried out on domestic waste. 6 ml of the waste in a 300 ml bottle were tested. If the initial dissolved oxygen is 8 mg/l while the 5-day dissolved oxygen is equal to 4 mg/l, compute: **BOD$_5$** of the waste, and ultimate BOD, assuming k_1-rate constant of 0.1 /day. (Ans. 292 mg/L).

58) A waste was found to have an ultimate BOD of 600 ppm and was found to obey a first order reaction with a k_1 value of 0.2 per day. Calculate the **5-day BOD** of the waste. (Ans. 540 mg/L).

59) a] Why is the BOD test adopted as standard one used to determine the pollution strength of sewage? b] The 1 day BOD

of a waste is 60 mg/L while the 2 day BOD of the same waste is 98 mg/L. Find the **5-day BOD** of the waste. c] A treatment plant that has an influent BOD of 300 mg/L and an effluent BOD of 30 mg/L. For a flow rate of 20 ML/d:

   i] Determine treatment efficiency.

   ii] Compute the plant effluent BOD load. (SQU, 1993).

60) The 1 day BOD of a waste is 75 mg/L while the 2-day BOD of the same waste is 112 mg/L. Determine the **BOD$_5$** of the waste. (SQU, 1993).

61) Given that the ultimate BOD of a sewage is 300 mg/l, calculate the one day and five day BOD for rate constants 0.1, 0.3 and 1 /day. (Ans. 62, 150, 270, 205, 291, 300 mg/L).

62) The 1-day BOD of a waste is 60 mg/l while the 2-day BOD of the same waste is 98 mg/l. Determine the **BOD$_5$** of the waste.

63) For a BOD$_5^{20}$ of 200 mg/l and a rate constant of 0.11 per day, determine the 2-day BOD. (Ans. 111 mg/L).

64) A BOD test for a sample, at a temperature of 20°C, showed that the five-day BOD is 180 mg/L. Assume k' = 0.225 day^{-1} (to base e) determine one-day BOD and 1st-stage BOD of the wastewater. (OIU, UNESCOC, 2004). (Ans. 255 mg/L).

65) Assuming that each person contributes a BOD load of 0.06 kg/d, and a suspended solids load of 0.08 kg/d, find the daily domestic pollutional load of a town of 20000 inhabitants. If the DWF is 120 l/capita.d, compute the concentration of BOD and suspended solids in the sewage. (Ans. 500, 667 mg/L).

66) A treatment plant that has an influent BOD of 300 mg/L and an effluent BOD of 30 mg/L. For a flow rate of 20 ML/d.i. Determine treatment efficiency. Compute the plant effluent **BOD load**. (SQU, 1992). (Ans. 90%, 600 kg).

67) The treated sewage of 20000 people is discharged to a nearby stream which is flowing at a rate of 0.15 m^3/s with a BOD of 2 mg/l. The water consumption by the inhabitants' amount to 150 l/capita.d and the BOD contribution is 0.065 kg/capita.d. If the BOD in the stream below the outfall is not to surpass 4 mg/l, find the required efficiency of the treatment plant needed to cope with this condition. (Ans. 97%).

68) The 1-day BOD of a waste is 50 mg/L while the 2-day BOD of the same waste is 82 mg/L. Find the **5-day BOD** of the waste. (Ans. 124 mg/L).

69) What are the advantages of determining the value of BOD in a sample? Show how the following equation may be used to compute the BOD at a certain time and temperature:

$$BOD_t^T = L_o \left(1 - e^{-tk}\right)$$

The one-day BOD of a sample of wastewater was found to be 50 mg/L and the two-day BOD of the same sample is 90 mg/L. Determine the **five-day BOD** for the sample, assuming first order reaction at 20°C. (U of K, 2001). (Ans. 168 mg/L).

70) a) Why is the BOD test adopted as standard one used to determine the pollution strength of sewage? b) The 1 day BOD of a waste is 60 mg/L while the 2 day BOD of the same waste is 98 mg/L. Find the **5-day BOD** of the waste. (SQU, 1992). (Ans. 146 mg/L).

71) Define Biochemical Oxygen Demand (BOD). The 1 day BOD of a waste is 80 mg/L while the 2-day BOD of the same waste id 120 mg/L. Determine the **BOD$_5$** of the waste. (SQU, 1993). (Ans. 155 mg/L).

72) The BOD value of a wastewater was measured at 2 and 8 days and found to be 125 and 225 mg/L respectively. Determine the **5-day value** using the first-order rate model. (UAE, 1990). (Ans. 201 mg/L)

73) Define the BOD. The 1-day BOD of a waste is 60 mg/l while the 2-day BOD of the same waste is 98 mg/l. Determine the 5-**day BOD** of the waste. (U of K, 1986). (Ans. 146 mg/L).

74) Write briefly about main sources of groundwater pollution in Khartoum province. What are the significant parameters that govern the selection of water treatment units to treat surface water for drinking purposes? A BOD test for a sample, at a temperature of 20°C, showed that the one-day BOD is 81 mg/L and the two-day BOD amounts to 150 mg/L. Find the **5-day BOD**, and plot the BOD variation curve with time. (U of K, 2001).

75) Write briefly about parameters that affect quality of groundwater. What water treatment units would you recommend for treating a surface highly turbid water? Indicate reasons. A BOD test for a sample, at a temperature of 20°C, showed that the one-day BOD is 100 mg/L and the two-day BOD amounts to 150 mg/L. Find the **5-day BOD**, and plot the BOD variation curve with time. (U of K, 2005).

76) Which water characteristic you would consider most important when supplying water for drinking purposes? State your reasons. The 1-day BOD of a waste is 60 mg/L while the 2-day BOD of the same waste is 98 mg/L. Using the first order BOD reaction equation $BOD_t^T = L_o\left(1 - 10^{k'\cdot t}\right)$

determine the **5-day and** 30-day BOD of the waste. (Ans. 147 mg/L, 164 mg/L).

77) Define Biochemical Oxygen Demand. The BOD_5 of a wastewater is determined to be 150 mg/L at 20°C. The k value (to base e) is known to be 0.23 per day. What would the BOD_8 be if the test were run at 15°C? (UAE, 1990). (Ans. 168 mg/L).

78) A wastewater treatment plant receives 4000 m³/d with a COD of 400 mg/L. What is the plant **loading** in kg/d COD? (UAE, 1990). (Ans. 1600 kgCOD/d).

79) **Thomas's** method for determining the BOD exerted at time t is based on the following equation. Define the parameters shown in the equation.

$$\left(\frac{t}{y}\right)^{\frac{1}{3}} = \left(k_1 L\right)^{-\frac{1}{3}} + \frac{k_1^{2/3}}{6L^{1/3}} \cdot t$$

The BOD versus time data for the first 5-days of a BOD test obtained as follows:

Time (days)	BOD (mg/L)
2	10
4	16
6	20

Draw the BOD against time curve and compute the reaction **rate constant** and **ultimate** first-stage BOD. (UAE, 1989). (ans. 0.23 /d, 27 mg/L).

80) Distinguish between point and non-point sources of pollution, giving appropriate examples of each. What methods of water treatment would you select for a rural area? State your reasons. Variation of flow and BOD with time for a wastewater treatment unit are as presented in the following table:

Time, hour	BOD, mg/L	Flow, m³/s
0	155	18.9
2	110	18.4

4	40	17.9
6	20	9.3
8	135	9.8
10	350	16.8
12	275	16.4
14	325	17.0
16	370	29.1
18	690	26.4
20	570	23.9
22	280	32.1
24	230	35.9

Determine average flow, average BOD (**BOD$_{av}$**), and weighted average BOD. (U of K, 2001).

81) Outline advantages and disadvantages of the BOD test. How can a sample of municipal wastewater have a $BOD_3{}^{20} = 200$ mg/L and $BOD_{30}{}^{20} = 400$ mg/L? (OIU, UNESCOC, 2004).

82) The following results were obtained for the variation of BOD with time for a sample of wastewater:

BOD (mg/L)	Time (d)
0	0
160	1
250	2
300	3
330	4
380	5

Plot BOD-time curve for the data. Compute the **average** BOD. (OIU, UNESCOC, 2004).

(Hint: $BOD_t^T = L_o(1 - 10 - t \cdot k_1) = L_o(1 - e^{-t \cdot k'})$)

83) Indicate reasons for adopting the BOD as a standard test used to evaluate the pollution strength of sewage. **Thomas's** method for determining the BOD exerted at time t is based on the following equation:

$$\left(\frac{t}{y}\right)^{\frac{1}{3}} = (k_1 L)^{-\frac{1}{3}} + \frac{k_1^{2/3}}{6L^{1/3}} \cdot t$$

    i)   Define the parameters presented in the equation.

    ii)   Indicate advantages of this method.

84) The following BOD results were obtained on a sample of untreated wastewater at temperature of 20°C:

t (days)	BOD (mg/L)
0.5	36
1.0	65
2.0	109
3.0	138
4.0	158
5.0	172

1) Draw BOD versus time.
2) Compute the **reaction** rate constant.
3) Find **ultimate** 1st-stage BOD using Thomas method.
4) Determine the value if k_1 (to base 10). (UAE, 1989). (ans. 0.37 /d, 207 mg/L).

Population equivalent

85) Impacts of sewage quality parameters to population equivalent, **PE** parameter. The following equation represents the population equivalent evaluation equation. What does each parameter stands for?
$$PE = \frac{BOD_5 \, Q}{BOD_s}$$

86) Given reasons for referring population equivalent, **PE**, of sewage to a certain quality parameter. What are the main merits of the population equivalent parameter? Why is domestic sewage being adopted to be the standard sewage in PE evaluation equations. The daily wastewater production from a certain industry amounts to 5×10^6 L, with a 5-day BOD of 300 mg/L. Determine the population equivalent for this industry. (SAS, 2007).

87) The daily wastewater production and BOD_5^{20} from two industrial factories is as shown in the following table. (SAS, 2008).

Factory	Wastewater flow (m^3)	5-day BOD
Factory (A)	6000	250
Factory (B)	3000	400

Determine the **population equivalent** for the factories.

Chapter Three

Water Treatment

3.1 Sedimenation and Flotation

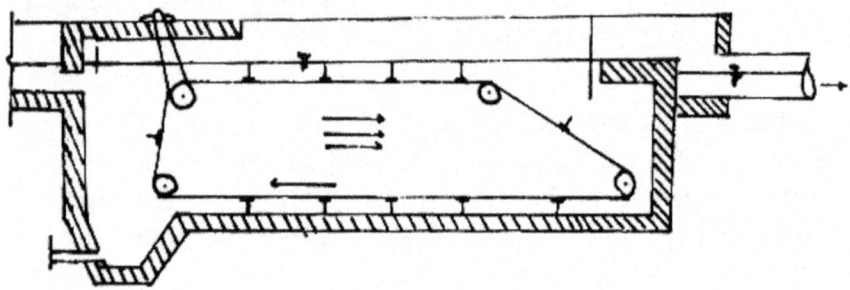

Introduction

In sedimentation particulate matter of mass density greater than that of the surrounding fluid will settle down under the effect of gravity; while in flotation particulate matter of lesser mass density will move upward and float.

In designing a sedimentation tank it is essential to determine its volume, plan area, shape, inlet and outlet devices and facilities for scum withdrawal.

Example 3.1

1. Find the settling velocity of spherical discrete particles 0.05 mm in diameter, given that their specific gravity is 2.6 settling in water at a temperature of 20° C.
2. Write a computer program to determine settling velocity of spherical discrete particles given its diameter, specific gravity, fluid in which particles are settling and fluids' temperature.

3. Verify your program by solving example 3.1a.

Solution

Stoke's law states that:

$$v = \frac{gx\, d^2(s.g-1)}{18v}$$

Where:

v =   Settling velocity, m/s.
g =   Gravitational acceleration, m/s^2.
d =   Diameter of spherical particle, m.
s.g.=   Specific gravity .
v =   Kinematic viscosity, m^2/s

Given in the problem: d = 0.05 mm = 5x10^{-5} m, s.g. = 2.6, and from tables of viscosity of water at 20° C, the kinematic viscosity is 1.003x10^{-6} m2/s.

Substituting in the above equation, then:

$$v = \frac{9.81\, x\,(5x\, 10^{-5})^2(2.6-1)}{18\, x\, 1.003\, x\, 10^{-6}} = 2.17\, x\, 10^{-3}\, m/s$$

Stoke's law is valid for laminar flow, then Reynolds number is less than 0.5. Therefore, Re should be checked:

Re = v.D/v = 2.17x10^{-3}x5x10^{-5}/1.003x10^{-6} = 0.118 (less than 0.5, OK.).

Program 3.1 Listing:
Settling velocity of spherical discrete particles in a fluid

```
'************************************************
'EXAMPLE 3.1: Calculates settling velocity
'************************************************
Public Class Form1
    Const g = 9.81      'gravitational constant

    Private Sub Button1_Click(ByVal sender As
        System.Object, ByVal e As
        System.EventArgs) Handles Button1.Click
        Dim Re, v, d, sg, rho As Double
        d = Val(TextBox1.Text) / 1000      'convert to m
        sg = Val(TextBox2.Text)
        rho = Val(TextBox3.Text)
```

```
            v = (g * (d ^ 2) * (sg - 1)) / (18 * rho)
            'check Re number
            Re = v * d / rho
            'output results
            Label4.Text = "Settling velocity = "
                    + v.ToString
            Label4.Text += vbCrLf + "Re number = "
                    + Re.ToString
            If Re < 0.5 Then Label4.Text += " (OK)"
    End Sub

    Private Sub Form1_Load(ByVal sender As
        System.Object, ByVal e As
        System.EventArgs) Handles MyBase.Load
        Me.Text = "Example 3.1: Settling velocity"
        Me.MaximizeBox = False
        Me.FormBorderStyle =
    Windows.Forms.FormBorderStyle.FixedSingle
        'set up user interface
        Label1.Text = "diameter (mm)"
        Label2.Text = "specific gravity"
        Label3.Text = "viscosity (m2/s)"
        Label4.Text = ""
        Button1.Text = "&calculate v"
    End Sub
End Class
```

Example 3.2 (see Program 3.1 Listing)

A sedimentation basin is designed to remove spherical discrete particles with relative density of 2.65 and diameter of 0.02 mm in water at $15°$ C. Given that the dynamic viscosity is equal to 1.1×10^{-3} Ns/m^2, calculate the settling velocity of the particles.

Solution

Using Stoke's law, then:

$$v = \frac{\rho \, gx \, d^2 (s.g-1)}{18\mu}$$

Where:

$\rho =$ density of fluid

$v =$ dynamic viscosity, $N.s/m^2$

Thus:

104

$$v = \frac{10^3 \times 9.81 \times \left(2 \times 10^{-5}\right)^2 (2.65 - 1)}{18 \times 1.1 \times 10^{-3}} = 3.217 \times 10^{-4} \, m/s$$

Check Re:

Re = ρv.D/μ = $10^3 \times 3.27 \times 10^{-4} \times 2 \times 10^{-5}/1.1 \times 10^{-3}$ = 5.9×10^{-3} < 0.5 (OK.).

Example 3.3

A sample from a river was analyzed for suspended solids. The analysis revealed that it contains 160 g/m^3 suspended solids. The sanitary engineer responsible suggested clarifying this water by the sedimentation process. When the clarification rate was determined experimentally, the cumulative frequency distribution for the settling velocities of these particles showed a straight line relationship:

10 % of the particles have a settling velocity greater than 1.5 m/hr.

10 % of the particles have a settling velocity less than 0.5 m/hr.

The amount of water to be purified is 1500 m^3/hour. Thus, the sanitary engineer proposed designing a sedimentation basin of length 50 m, width 25 m and depth of 3 m. assuming that this tank will work as an ideal sedimentation basin, find the suspended solids content of the effluent water.

Solution

From the data given above, plot the cumulative frequency distribution curve for the settling velocities of these particles as shown in figure 1.1 below:

Fig. (1.1) Cumulative frequency distribution curve

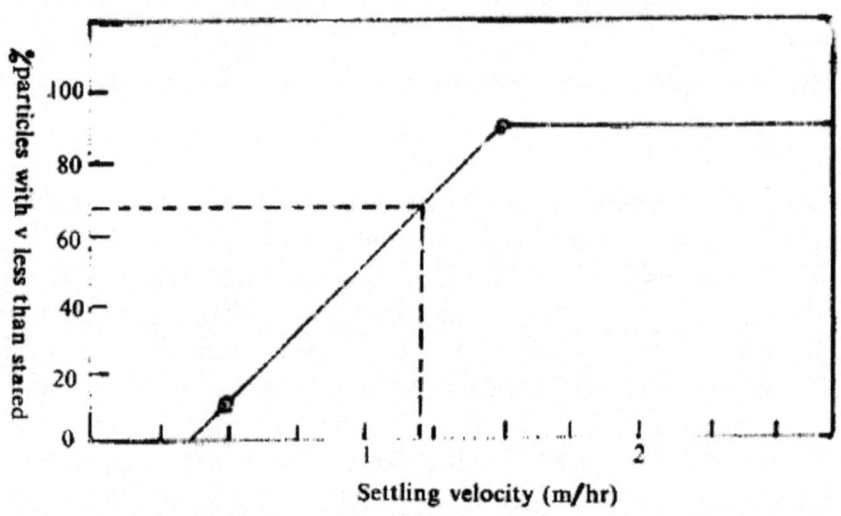

Settling velocity (m/hr)

The overall removal X_T is given by:

$$X_T = 100 - x_o + \frac{1}{v_o} \int_0^{x_o} v_t . dx$$

Where:

X^T = overall removal

v^0 = design settling velocity

x^0 = removal corresponding to v^0.

Given that the discharge Q is equal to 1500 m3/hour and since settling velocity

v^0 = discahge/area = Q/A and A is equal to width (W) times length (L), then:

106

v^0 = Q/WL = 1500/(25x50) = 1.2 m/hour

From figure 1.1 for vo = 1.2 m/hr, then X^0 = 66%

Therefore, the overall removal:

X^T = 100 – X^0 + removal = 100 – 66 +(1/2)(0.37 + 1.2)x66/1.2 =

77%

Given that the suspended soldids concentaruion entering the basin C^i

= 160 g/m^3 , then the concentration of suspended solids in the effluent

C^e could be found as:

C^e = (1 – XT)C^i = (1 – 0.77)x160 = 36.8 g/m^3

Example 3.4

Settling column tests performed on discrete particle suspension gave the following tabulated results:

Sampling depth (cm)	Sampling time (min)	% suspended solids remaining in sample
100	60	60
	180	43
	360	29
200	60	63
	180	55
	480	36
300	60	65
	120	63
	360	50

By using this data plot the cumulative batch settlement curve and calculate e the total removal for a horizontal settlement basin with surface area of one hundred m^2 when incoming flow rate is 4.8 m^3/min. Calculate also the removal to be expected for a similar basin when the flow is 0.02 m^3/sec.

Solution

From the given data the settling velocities could be found for each depth and sampling time knowing that:

Settling velocity = sampling depth/sampling time

Sampling depth (m)	Sampling time (sec)	Settling velocity (mm/s x 10^4)	% SS with settling velocity less than that stated
1	3600	2.78	60
1	10800	0.93	43
1	21600	0.46	29
2	3600	5.56	63
2	10800	1.85	55
2	28800	0.69	36
3	3600	8.33	65
3	7200	4.17	63
3	21600	1.39	50

Using this data achieved above, then the cumulative batch settlement curve could be plotted as in figure 1.2.

For a flow of 4.8 m³/min, the settling velocity

$V_1 = Q/A = 4.8/60 \times 100 = 8 \times 10^{-4}$ m/s

From the graph for a settling velocity of 8×10^{-4} m/s

$x_0 = 66$ %

To find the total removal X_T then the integral v.dx should be found. This could be done as follows:

Interval	dx	v_{st}	$v_{st}.dx$
0 - 10	10	0.05	0.5
10 - 20	10	0.15	1.5
20 - 30	10	0. 35	3.5
30 - 40	10	0. 65	6.5
40 - 50	10	1.05	10.5
50 - 60	10	1.75	17.5
60 - 66	6	4	24
Σv.dx			64

$$X_T = 100 - x_0 + \frac{1}{v_{st}} \int_0^{x_0} v_{st}.dx$$

Thus, $X_T = 100 - 66 + 64 \times 10^{-4}/8 \times 10^{-4} = 42$ %

Using a similar technique for water flow of 0.02 m³/sec, total removal could be found to be approximately 60%.

Fig (1.2) cumulative batch settlement curve

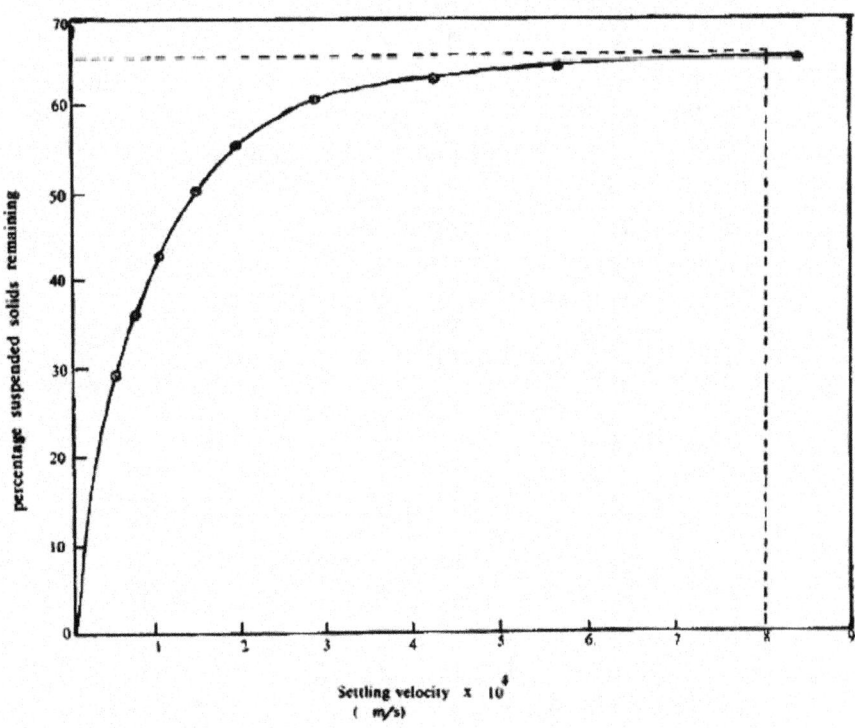

Settling velocity x 10^4
(m/s)

Example 3.5

Using the data obtained by the sanitary engineer in problem 1.3 calculate for ideal settling conditions:

- The amount of sludge deposited on the bottom of the tank after one year of operation and the average thickness of deposits in meters when the moisture content of the sludge is about 98 percent, given that the mass density of the dry sludge is 2500 kg/m^3.
- The length of the effluent weir to be applied.
- The Froude number.
- The Reynold's number in the horizontal direction given that the water temperature is 25° C.
- Find the magnitude of the critical scour velocity. Has this critical scour velocity been surpassed in the design of the sanitary engineer?

Solution

a) The amount of sludge deposited is given by:

As = Q C X_T = 1500x160x0.77 = 18.48x104 g/hr

After one year of operation, amount of sludge = 18.48x10⁴x24x365/1000 = 161.885x10⁴ kg/year

Thus, amount of sludge deposited per unit area = 161.885x10⁴ /(25x50) = 1295 kg/m².year

Given that sludge moisture content is 98%, therefore, the average thickness of deposits will equal to 64 m³/m².year.

b) Length of effluent weir should be greater than

$$\frac{Q}{5\,Hv_o} = \frac{1500}{5 \times 3 \times 1.2} = 83\,m$$

c) Froude number,

$$Fr = \frac{V_H^2}{g\,r_H} = \frac{Q^2(B+2h)}{g\,B^3\,h^3} = \frac{v_s^2 L^2\left(1+\frac{2h}{B}\right)}{g\,h^3}$$

Where:

r_H = Hydraulic radius = area/wetted perimeter = WH/(W + 2H) = (25x3)/(25 + 2x3) = 2.42 m

Fr = Froude number, dimensionless.

v_H = Average displacement velocity = Q/WH = 1500/(25x3) = 20 m/hour = 5.56x10⁻³ m/s.

Thus,

$$Fr = \frac{V_H^2}{g\,r_H} = \frac{5.57 \times 10^{-32}}{9.81 \times 2.42} = 0.13 \times 10^{-5}$$

d) Reynold's number

$$Re = \frac{v_H R_H}{v}$$

Where v = kinematic viscosity, m²/s

From tables, kinematic viscosity for a temperature of 25° C equals 0.893x10⁻⁶ m²/s

Therefore,

$$Re = \frac{5.56 \times 10^{-3} \times 2.42}{0.893 \times 10^{-6}} = 15067$$

Stability could be promoted and tank erratic behavior could be avoided if higher Froude numbers are applied but not so high as to endanger basin efficiency by turbulence or bottom scour. This leaves as a requirement

$Fr > 10^{-5}$ and $Re < 2000$

And in the case under study these requirements are not met. Therefore, the sanitary engineer's preliminary design could be improved without changing the surface loading by using, for instance, longitudinal baffles. These baffles only guide the water and get rid of irregularities in the flow while water pressure is kept the same at both sides.

c) Scour velocity, v_s is given by:

$$v_s = \sqrt{\left[\frac{40}{3}\left(\frac{\rho_s - \rho_w}{\rho_w}\right) \times gxd\right]}$$

For laminar settling

$$v = \frac{gx\,d^2\left[\frac{\rho_s - \rho_w}{\rho_w}\right]}{18\,v}$$

Combining these two equations then:

$$v_s^2 = 177.18\, v \cdot v\sqrt{\left[\frac{\rho_s - \rho_w}{\rho_w}\right]}$$

Assuming that the sludge contains 95% water, then $\rho_s = 1030$

Thus,

$$v_s^2 = 177.18\left(0.893 \times 10^{-6}\right) \cdot \frac{1.2}{60 \times 60}\sqrt{\left[\frac{1030 - 1000}{1030}\right]}$$

Which gives, $v_s = 23 \times 10^{-3}$ m/s

Exercise 3.1

Stoke's law

1. A sedimentation basin is designed to remove spherical discrete particles of diameter of 0.6 mm with a specific gravity of 1.01 from water at 25° C. Assuming ideal settling conditions in the basin, compute the **removal** of spherical discrete particles 0.3 mm in diameter with specific gravity of 1.03 by this basin keeping other conditions the same. (Ans. 75%).

2. A sedimentation tank 3 m deep is used to settle discrete solids from a wastewater sample of a temperature of 20°C. The suspended solids have a diameter 0.01 mm and a specific gravity 0f 2.6. Determine the tanks **detention time**. Compute the overflow rate for this tank and indicate the validity of **Stoke's** law. Take kinematic viscosity = 1.007×10^{-6} m^2/s for the given temperature. (SQU, 1993). (Ans.).

3. Given that the drag on a moving sphere at a low Reynolds number is a function of the dynamic viscosity, the settling velocity of the sphere and its radius, deduce an expression for the terminal velocity of discrete particles in water. Calculate the settling **velocity** of spherical discrete particles in water at 15° C if their relative density is 1.001 and their diameter is 0.5 mm. Compute also the settling **velocity** of particles with relative density of 1.42 and diameter of 0.1 mm. (Ans. 13, 2 mm/s).

4. What discrete particles will be better **removed** by plain sedimentation in a given horizontal sedimentation basin with a surface loading of 1 m/hr:
   - spherical particles with d = 0.1 mm and specific mass density 1.005,
   or
   - spherical particles with d = 0.05 mm and specific mass density of 2.65?
   (Take coefficient of viscosity as equal to 1.31×10^{-6} m^2/s.)
   (U of K, 1988). (Ans. Second particle).

$$S = \frac{g \times d^2 (s.g - 1)}{18 v}$$

Ideal, discrete settling

5. a) Differentiate between **discrete** and **flocculent** settling.

b) A sedimentation tank was designed to serve a population of 30000 whose average per capita domestic water consumption is 250 L/day. The water carries 120 mg/L suspended solids. A settling column test showed that the cumulative frequency distribution for the settling velocities of the particulate matter have a straight line relationship with characteristics as follows:

10% of the particles have a settling velocity larger than 0.5 mm/s

10% of the particles have a settling velocity smaller than 0.1 mm/s

If the designed basin has dimensions of 2x10x50 m for depth, width and length respectively, find:

1) The **detention time** of water in the tank.
2) The effluent suspended **solids concentration**.
3) The values of **Froude and Reynolds** numbers for the horizontal water movement (take coefficient of kinematic viscosity of water to be 1.31x10-6 m2/s).
4) The percentage of particles in the effluent having a settling velocity smaller than 0.2 mm/s. (Ans. 3.2 hr, 10 mg/L, 11%).

6. River water carries discrete particulate matter in an amount of 120 mg/l subdivided in three equal parts with settling velocities of 0.2 mm/sec, 0.4 mm/sec, and 0.6 mm/sec. respectively. For the plain sedimentation of this water in an amount of 0.5 m^3/sec. a settling tank is available with a length of 50 m,. a width of 20 m and a depth of 3 m. Even when working under ideal conditions, however, the suspended solids content of the effluent from this tank is too high. It is therefore decided to build a second tank of the same dimensions.

a) What are the suspended **solids content** of the effluent when both tanks are used in? i) Series. ii) Parallel. Assume idealized settling conditions in the tanks.

b) What is the **Reynolds** number in the horizontal direction for each tank given that the temperature of the water is 10°C? (Kinematic viscosity at this temperature equals 1.31 x 10^{-6} m^2/sec.)

c) What is the **Froude** number for the horizontal water movement at the temperature given in (b) above? (U of K, 1984). (Ans. 8.5, 7.9 g/m^3).

7. A settlement analysis was performed on a dilute suspension of a non-flocculating particulate matter. The tabulated data were found for samples collected at a depth of 1.4 m:

Sampling time(min)	0.5	1	2	6	8
Weight fraction remaining	0.66	0.58	0.47	0.15	0.12

For a clarification rate equal to 24.5 l/m^2/sec, what is the total **removal**? (Ans. 60%).

8. A sanitary engineer is asked to design a sedimentation tank for a town in the Gezira area with a population of 20000 and with an average per capita domestic water consumption of 200 L per day. The analysis of the batch settlement test revealed the following data:

Sampling time(sec)	% removed at			
	0.5 m	1m	1.5 m	2 m
0	0	0	0	0
3600	72	55	43	40
7200	83	69	62	56
10800	90	88	71	70

Based on these results the sanitary engineer decided to design a rectangular basin of depth 2.m with a detention time of 2 hours. Compute the required **area** needed for the basin and the minimum **removal** that could be achieved by the basin. Write any criticism that would be appropriate of the sanitary engineer's accessibility to the design. (Ans. 133 m^2, 70%).

9. Sand particles were sieved to analyzed their particle size distribution. For each size fraction obtained, an average settling velocity has been computed as shown in the table below:

Settling velocity (mm/s)	180	90	36	18	13	9
Weight fraction removed	0.45	0.54	0.65	0.79	0.89	0.97

Calculate the overall **removal** for an overflow rate of 6000 m^3 per unit area per day. (Ans. 72%).

10. In a batch settlement test of a flocculating material the results tabulated below were recorded:

Time (min)	Depth (m)	% removed
10	1	30
	2	25
	2.5	20
25	1	53
	2	45
	2.5	40
40	1	65
	2	55
	2.5	54
50	1	79
	2	68
	2.5	65

Calculate the **percentage of solids** that could be removed in 30 and 45 minutes at a depth of 2.5 m assuming that the flow is horizontal.

11. River water was analyzed for suspended solids and it was found to be 140 g/m^3. The settling column test revealed that the cumulative frequency distribution for the settling velocities of these particles shows a straight line relationship with the following characteristics:

10 % of the particles have a settling velocity larger than 0.6 m/hr.

10 % of the particles have a settling velocity smaller than 0.2 m/hr.

For the clarification of this water at amount of 350 m^3/hour a horizontal flow settling tank has been constructed with of width 10 m, length 30 m, and depth of 3 m. If the settling is occurring under ideal conditions, find the:

1) effluent suspended **solids concentration**.
2) rate of **sludge** accumulation at a distance of 25 m from the tank inlet.
3) value of the **Froude** number.
4) Value of the **Reynold** number for the horizontal water movement given that the kinematic viscosity of the water is 1.31x10^{-6} m^2/sec.
5) **Length** of the settling tank that would be required for 75 percent removal of the suspended solids.

6) Percentage of particles in the effluent having a settling velocity smaller than 0.3 mm/sec.
7) Detention time of tank.
a. How could the tank design be improved? (Ans. 13 g/m^3, 573 kg/m2.year, 19m, 99%, 2.6 hr).

12. What are the purposes for using sedimentation in water treatment? b) The results of a batch settlement test on a discrete particle suspension revealed the following

Sampling Depth (m)	% suspended Solids				
	1 hour	2 hours	3 hours	6 hours	8 hours
1	39	-	57	70	-
2	36	-	44	-	64
3	35	37	-	49	-

A horizontal flow sedimentation tank, of area 1000 m^2, is required for treating 0.25 m^3/s water.

I) Determine the **efficiency** of solids removal in the tank.
II) Find the effluent **solids concentration**.
III) Compute the **retention time** of solids within the tank.

(Take initial SS content to be 100 mg/L) (K of U, 1989).
(Ans. 56%, 44 mg/L, 3.3 hr).

13. a) Outline main advantages of sedimentation as a unit operation process in water treatment. b) What are the factors that govern the settling velocity of particles in water? C) A particle size distribution has been obtained from a sieve analysis of sand particles. For each weight fraction, an average settling velocity has been calculated. The data as follows:

Settling velocity (m/min)	3.0	1.5	0.6	0.3	0.22	0.15
Weight fraction remaining	0.55	0.46	0.35	0.21	0.11	0.03

i) What is the overall **removal** for an overflow rate of 4000 m^3/m^2/day?
ii) Compute the **solids concentration** of effluent from a sedimentation basin treating raw water of a solids content of 150 mg/L? what assumptions did you make? (UAE, 1990). (Ans. 57%, 61 mg/L).

14. Differentiate between discrete and flocculent settling of particles. A settling basin is designed to have a surface overflow rate of

34.56 m/day. Determine the overall **removal** for a suspension with the size distribution given below:

Particle size (mm)	0.1	0.08	0.07	0.06	0.04	0.02	0.01
Weight fraction greater than size (%)	10	15	40	70	83	99	100

(Assume specific gravity of particles to be 1.2 and take water temperature to be 20°C)

If the raw water, in (b) above, contains solids in an amount of 200 g/m^3, compute the **solids content** of the effluent. (UAE, 1989). (Ans. 32 g/m^3).

15. a) What are the main factors which determine the settling velocity of discrete particles? b) A settling column test was used to investigate the settling characteristics of a discrete particle suspension. The following results were recorded for a depth of 1.5 m:

Time [sec.]	480	600	1200	2400	3600	4800
Percent SS remaining in the sample	49	47	37	19	5	2

Find the theoretical **removal** of solids from this suspension in a horizontal flow tank with a surface overflow rate of 4.5 m^3/m^2.hour. (SQU, 1993). (Ans. 83 %).

16. A consultant Engineer is faced with the problem of designing a primary sedimentation tank for a town with a population of 20000 and with the average per capita domestic water consumption of 200 L/d. An analysis performed on sewage using a batch settlement test gave rise to the data shown in the following table:-

Sampling time (Sec.)	% removed at			
	0.5 m	1 m	1.5m	2 m
0	0	0	0	0
3600	62	50	43	40
7200	83	69	62	58
10800	-	88	76	70

It was decided to use a rectangular tank of depth 2 meters with design retention time of 2 hours. Estimate the required **area** of the tank and the minimum removal that the consultant Engineer expects on the basis of the data. Write a brief criticism of the consultants approach to the design. (U of K, 1983). (Ans. 167 m^2, 75%).

17. a) Define briefly the following processes: i) Sedimentation   ii) Flocculent settling iii) Plain sedimentation iv) Discrete settling. Denote applications and/or examples of each of these processes. b) Delineate major factors that influence the settling characteristics of discrete particulate matter in a fluid. c) The following data for suspended solids remaining in the sample (mg/l) were obtained from a batch settlement test on a discrete suspension:

Sampling depth	Sampling time (hr)					
(cm)	0	1	1.5	2	4	4.5
100	120	43.2	-	7.2	2.4	-
200	120	82.8	-	42	-	-
300	120	86.4	-	64.8	-	-
400	120	-	85.2	81.6	-	34.8

i)   Plot the removal of suspended solids versus overflow rate curve for a horizontal flow tank for the theoretical range of determined values of settling velocities.
ii) Compute the overall **removal** for a surface loading of 0.03 m³/100m² plan area.second.
iii) Evaluate tank **efficiency** based on given data.
iv) Compute effluent solids content, Comment on your result. (U of K, 1987). (Ans. 87%, 15 mg/L).

18. a) How can you improve the sedimentation of small particles? What are the main objectives of the process of sedimentation in water treatment? b) River water containing suspended particles in an amount of 150ppm, of which 60% has a settling velocity of 1.1 m/hr and 40% a settling velocity of 0.6 m/hr. For the purification of this water in an amount of 1600 m³/hr a settling tank need to be installed with dimensions of 3 and 20m for depth and width respectively. The length is required to achieve a suspended load reduction of 90%. Calculate for ideal settling conditions:

   i) The required **length** of the tank,
   ii) The **detention time** of the water in the tank,
   iii) The values of the **Reynold's** and **Froude's** numbers for the horizontal water movement and discuss their validity and justification. (Kinematic viscosity = 1.31×10^{-6} m²/s)

  iv) What conclusions can you draw from your
      computations? What would have been a better design
      for the same surface area of the basin? (U of K, 1988).
      (Ans. 100 m, 3.75 hr).
a) Define the following terms: sedimentation, flotation, plain
   sedimentation. Outline objectives sedimentation unit process.
   Certain particles are allowed to settle in a settling column
   test. The particles have a specific gravity of 1.25, & an
   average diameter of 0.12 mm. Using Stoke's law, determine
   the settling **velocity** of the particles in water. (The
   temperature of the water is 20°C. At this temperature take
   coefficient of dynamic viscosity to be $1.0087*10^{-3}$ N*s/m², &
   density = 998.2 kg/m³.). Validate your answer. (B.Sc. UoD,
   2013) (Ans. 1.9 mm/s).

$$E = \left[\frac{L_i - L_e}{L_i} \right] \quad \prime\prime = \prime \frac{100}{1 + 0.44\sqrt{\frac{W}{VF}}} \qquad Re = \frac{\rho\,vd}{\mu}$$

b) A suspension enters a horizontal sedimentation tank at a flow
   rate of 9 m³/minute. The tank is 4 m wide, 50 m long & 1.2 m
   deep. Compute, for ideal settling conditions, the value of the
   Froude & the Reynolds numbers for the horizontal water
   movement given that the water is at a temperature of 20°C.
   Indicate simple means to improve the operation of this tank
   and any remedy erratic behavior. (B.Sc. UoD, 2013) (Ans.
   $13X10^{-5}$, 231936).

$$Re = \frac{v_H R_H}{v} \qquad\qquad Fr = \frac{v_H^2}{g\,r_H}$$

19. River water contains suspended particles in an amount of 200
    g/m³. The cumulative frequency distribution for settling
    velocities of these particles shows a straight line with the
    following characteristics:
    10% of the particles have a settling velocity larger than 1.4
    m/hr,
    10% of the particles have a settling velocity smaller than 0.6
    m/hr.

For the clarification of this water in an amount of 1500 m^3/hr a settling tank has been built with the dimensions:
Length 75m, width 20m and depth 3m.
Calculate for ideal settling conditions:
  i)   The suspended solids content of the effluent.
  ii) The **frequency** distribution for the settling velocities of the suspended particles in the effluent.
  iii) The **sludge** deposit in kg/day over the first 50m length of the basin. Explain the results obtained.
  iv) The maximum allowable interval between **cleanings** when the depth of the sludge deposits – containing 98% - is limited to 1.2m.
  v)   The **length** of the effluent weir to be applied,
  vi) the values of the **Reynold's** and **Froude's** numbers for the horizontal water movement (at 10°C, v = 1.31x10^{-6} m^2/s. Which conclusions must be drawn from these values? What would have been a better design for the same surface area? (U of K, 1983). (Ans. 25 g/m^3, 4800 kg/d, 5 d, 100 m).

20. A sanitary engineer was asked to design a sedimentation tank for a town with a population of 30000 with an average per capita domestic water consumption of 250 L/day. The laboratory analysis revealed that the amount of suspended solids the water carries is 150 mg/L. The settling column test indicated that the cumulative frequency distribution for settling velocities of the particulate matter shows a straight line with characteristics as follows:

    10% of the particles have a settling velocity larger than 0.5 mm/s,

    20% of the particles have a settling velocity less than 0.1 mm/s.

    For the purification of this water the sanitary engineer proposed constructing a settling tank has of dimensions 1x15x60 m for depth, width and length respectively.
    Compute for discrete settling conditions:
      i)   The effluent suspended **solids concentration**.
      ii) The rate of **sludge** accumulation at a distance of 30m from tank inlet.

    iii) The amount of **sludge** deposited over the first 40m length of the basin. Explain the results obtained

    iv) The maximum allowable interval between **cleanings** when the depth of the sludge deposits is restricted to 1.2m given that sludge contains about 98% water.

    v) The values of the **Reynold's** and **Froude's** numbers for the horizontal water movement given that the kinematic viscosity of the water is 1.31×10^{-6} m^2/s.

    vi) The percentage of particles in the **effluent** having a settling velocity smaller than 0.2 mm/s.

    vii) The **detention time** of the water in the tank.

    viii) What conclusions can you draw from the computations? (U of K, 1984). (Ans. 16 mg/L, 4500 kg/d, 3.8 kg/m^2.d, 6.3 d, 20%, 2.9 hr).

21. What are the main objectives of sedimentation as a unit operation process in water treatment? Indicate the factors that influence the settling velocity of discrete particles in water. Water with a temperature of 10°C contains suspended matter as indicated in the table below:

Concentration of solids, mg/L	Settling velocity, m/hr
90	2.5
60	1.2
30	0.6
20	0.2

1) Draw the cumulative frequency distribution of the settling velocities.

2) The water treatment plant included a sedimentation tank with dimensions of 20, 3 and 2 m for length, breadth and depth respectively. For a flow rate of 60 m^3/hr find:

    1. The efficiency of solids **removal**.

    2. The concentration of **solids** (in ppm) at tank outlet.

    3. Determine the values of **Reynolds** and the **Froudes** numbers for horizontal water movement (at 10°C, $v = 1.31 \times 10^{-6}$ m^2/s). What conclusions could you draw from these values? (U of K, 1987). (Ans. 86%, 28 mg/L).

22. Discuss briefly merits of coagulation/flocculation and sedimentation units in a water works plant. A wastewater treatment plant incorporates four circular sedimentation tanks for treating a

wastewater flow of 1200 m^3/hour. Each of the tanks has a diameter of 21 m. The settling column test conducted for a sample of the wastewater revealed the data tabulated below. Determine the suspended **solids removal** efficiency of the sedimentation units. (OIU, UNESCOC, 1999). (Ans. 31 %).

Sampling depth (m)	Concentration of SS removed (mg/L)				
	Sampling time (hours)				
	0	1	2	3	4
0.2	190	101	166	183	186
0.5	190	57	85	114	124
1	190	53	58	70	86
1.5	190	50	54	60	66
2	190	44	54	56	57
2.5	190	40	50	51	56

23. a] Write briefly about types of settling phenomenon involved in water and wastewater treatment. b] A particle size distribution has been obtained from a sieve analysis of sand particles. For each weight fraction, an average settling velocity has been calculated. The data are as follows:

Settling velocity m/minute					
3.0	1.5	0.6	0.30	0.22	0.15
Weight fraction remaining					
0.55	0.46	0.35	0.21	0.11	0.03

Assuming a type 1 settling, find the overall **removal** for an overflow rate of 4000 m^3/m^2.d.

c) How can you overcome erratic behavior problems in design of settling basins? (SQU, 1992). (Ans. 60 %).

24. Discuss the significance of sedimentation process in water and wastewater treatment plants. A settling-column analysis is performed on a dilute suspension of discrete particles at a temperature of 20 °C. Data collected from samples taken at the 150 cm depth are as follows:

Time [min]	0.7	1.2	2.3	4.6	6.3	8.8
Portion of particles with velocities less than those indicated						
	0.58	0.49	0.34	0.16	0.07	0.03

Find the overall **removal** if the overflow (clarification rate) of the basin is 2.7 cm/s. (SQU, 1991). (Ans. 53 %).

25. What are the main objectives of sedimentation in water treatment? How can one improve the sedimentation of small particles? River water has a suspended solids content of 80 mg/L, consisting of uniform particles with a constant settling velocity of 0.35 mm/s. For the clarification of this water in an amount of 21600 m³/day a rectangular settling tank has been built with a length of 30m, a width of 20 m and a depth of 2.5 m. Calculate for ideal settling conditions the suspended **solids content** of the effluent. On some day the efficiency of this tank equals the reduction in suspended matter content calculated in (c) above. On other days, however, the effluent has a much higher turbidity.

   i. Explain this **erratic behavior** by calculating the values of the relevant parameters.

   ii. Indicate simple means to improve the operation of this tank.

   (For this water assume the kinematic viscosity as equal to 1.31×10^{-6} m²/s) (U of K, 1986). (Ans. 14 mg/L).

Flocculent settling

26. A dilute suspension of flocculating particulate matter was analyzed in a settling column test. The data obtained from the test was as recorded in the following table for three different depths and is expressed as percentage remaining:

Depth (cm)	Sampling time (min)					
	10	20	30	40	60	100
50	69	41	37	30	27	22
150	82	51	39	35	31	25
200	82	60	39	34	32	29

Calculate the overall **removals** in a basin with depth equal to 200 cm with alternative retention times of 25, 40 and 65 minutes. (Ans. 60, 70, 80 %).

27. a) Define briefly the following processes: Sedimentation, plain sedimentation, discrete settling and flocculent settling. Give examples and/or application of each 0f these processes. b) What are the main factors which determine the settling velocity of discrete particles? c) In a laboratory batch settlement test of a flocculating material in a settling column with three sampling points the following results were obtained:

Depth (m)	Suspended solids at set time (mg)				
	0 min	15 min	30 min	45 min	60 min
1	120	91	65	50	42
1.5	120	94	74	65	51
2	120	98	77	67	56

Determine the **removal** efficiency of solids in a horizontal flow tank at 35 min. retention time at a depth of 2m and 3m. (U of K, 1985). (Ans. 72%).

28. Delineate main factors that determine the settling characteristics of flocculating particles in water. Suggest suitable coagulating and flocculating materials to be utilized for rural areas in Sudan. Outline your reasons. In a batch settlement of flocculating material the results tabulated below were recorded:

Depth, m	% suspended solids remaining			
	10 min	25 min	40 min	60 min
1	70	47	35	21
2	75	55	45	32
2.5	80	60	46	35

Compute the percentages of solids that could be **removed** in 30 and 45 minutes at a depth of 2.5 m assuming horizontal flow conditions. (U of K, 1988). (Ans. 60, 73 %).

3.2 Filtration

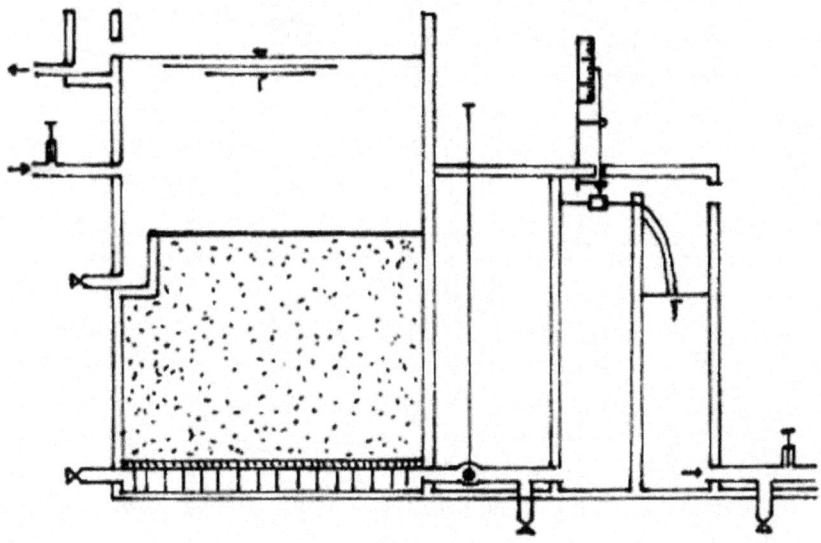

Introduction

Filtration is the separation of liquids from solid particles contained in them by means of a partition with pores of such size that they do not permit the passage of the particles while allowing the free flow of the liquid.

Filtration as a purification process improves water quality by one or more of the following methods:

  a) Removal of suspended and colloidal particles.

  b) Reduction of the population of bacteria and other organisms.

  c) Change of the chemical constituents in the water.

An example of a porous media used is sand with its merits of availability, cheapness, inertness and satisfactory performance.

Example 3.6

1. Making use of the equations developed by Rose and Carman-Kozeny compute the head loss through a 0.8 m sand bed. Assume that the bed consists of uni-sized, spherical sand particles having a diameter of 0.5 mm, the kinematic viscosity is equal to 1.003×10^{-6} m^2/s and the bed porosity is 40 percent while the rate of filtration is 250 l/m^2.min.

2. Writea computer program to find the head loss through a certain sand bed given filter bed shape factor, particles diameter, kinematic viscosity of filtrate, bed porosity, and rate of filtration.
3. Verify your program by solving example 3.6a.

Solution

Rose equation states that:

$$h = 1.067 C_D \frac{1}{p^4} \frac{v^2}{gd\,\varphi} L$$

Where
H_l = Head loss in bed, m.
v= Face velocity, m/s.
d = diameter of the particle, m.
g = Gravitational acceleration, m/s^2.
ϕ = Particle shape factor, dimensionless (= surface area of equivalent
    volume of sphere/actual surface area)
p = bed porosity, dimensionless (= volume of voids/total volume)
l = bed depth, m.
C_D = Newton's drag coefficient.

$$C_D = \frac{24}{Re} + \frac{3}{\sqrt{Re}} + 0.34$$

Re =   Reynolds number, dimensionless. $Re = \dfrac{v_d}{v}$

With the above notations: then L = 0.8m, ϕ = 1, d = 0.5x10^{-3} m, v = 1.003x10^{-6} m^2/s, p = 0.4, v = 250L/m^2.min = 4.17x10^{-3} m^3/m^2.sec, then:

$$Re = \frac{v_d}{v} = \frac{4.17 \times 10^{-3} \times 0.5 \times 10^{-3}}{1.003 \times 10^{-6}} = 2.079$$

$$C_D = \frac{24}{Re} + \frac{3}{\sqrt{Re}} + 0.34 = \frac{24}{2.079} + \frac{3}{\sqrt{2.079}} + 0.34 = 13.96$$

Thus, head loss by using Rose equation:

$$h = 1.067 C_D \frac{1}{p^4} \frac{v^2}{gd\phi} L =$$

$$1.067 \, x \, 13.96 \, x \frac{1}{0.5^4} \frac{\left(4.17 x 10^{-3}\right)^2}{0.5^4 x 9.81 x 0.5 x 10^{-3} x 1} x 0.8$$

$= 1.65m$

Using Carman-Kozeny equation the head loss is given by:

$$h_1 = E \frac{1-p}{p^3} \frac{v^2}{\varphi \, dg} . L$$

Where:

$$E = \frac{150(1-p)}{Re} + 1.75$$

Therefore,

$$E = \frac{150(1-p)}{Re} + 1.75 = \frac{150(1-0.4)}{2.079} + 1.75 = 45.04$$

Hence,

$$h_1 = E \frac{1-p}{p^3} \frac{v^2}{\phi \, dg} . L = 45.4 \frac{1-0.4}{0.4^3} \frac{\left(4.17 x 10^{-3}\right)^2}{1 x 0.5 x 10^{-3} x 9.81} x 0.8$$

$= 1.2m$

Program 3.6 Listing:
Computer program to find head loss through a certain filter bed

```
'******************************
'EXAMPLE 3.6: Head loss
'******************************
Public Class Form1
    Const g = 9.81

    Private Sub Form1_Load(ByVal sender As
        System.Object, ByVal e As
        System.EventArgs) Handles MyBase.Load
        Me.Text = "Example 3.6: Head loss"
        Me.FormBorderStyle =
        Windows.Forms.FormBorderStyle.FixedSingle
        Me.MaximizeBox = False

        Label1.Text = "Particle diameter (d, mm):"
        Label2.Text = "Face velocity (v, m2/s):"
        Label3.Text = "Particle shape factor:"
```

```
        Label4.Text = "Bed depth (L, m):"
        Label5.Text = "Bed porosity:"
        Label6.Text = "Kinematic viscosity (m2/s):"
        Label7.Text = ""
        Button1.Text = "&Calculate"
    End Sub

    Private Sub Button1_Click(ByVal sender As
        System.Object, ByVal e As
        System.EventArgs) Handles Button1.Click
        Dim d, v, phi, L, p, rho, CD, Re As Double
        Dim _E, HL As Double
        d = Val(TextBox1.Text) / 1000    'convert to m
        v = Val(TextBox2.Text)
        phi = Val(TextBox3.Text)
        L = Val(TextBox4.Text)
        p = Val(TextBox5.Text)
        rho = Val(TextBox6.Text)

        Re = v * d / rho
        CD = (24 / Re) + (3 / Math.Sqrt(Re)) + 0.34
        'head loss by Rose equation
        HL = 1.067 * CD * (1 / (p ^ 4)) *
                (v ^ 2 / (g * d * phi)) * L
        Label7.Text = "Head loss by Rose eq. = "
                + FormatNumber(HL, 2) + " m"
        'head loss by Carman Kozeny
        _E = ((150 * (1 - p)) / Re) + 1.75
        HL = (_E * ((1 - p) / (p ^ 3))) *
            ((v ^ 2) / (phi * d * g)) * L
        Label7.Text += vbCrLf + "Head loss by Carman-
            Kozeny eq. = " + FormatNumber(HL, 2) + " m"

    End Sub
End Class
```

Example 3.7

i.   The measured settling velocity of the sand particles of the filter in problem 2.1 was 0.12 m/sec. Determine the depth of expansion of the bed when the filter is being washed at a rate of 450 m^3/m^2.d.

ii. Write a computer program to find the depth of expansion of a bed given the filter backwashing rate, measured settling velocity of particles and filter bed characteristics.

iii. Very your program by solving eple 3.7a.

Solution

Overall bed expansion could be found from the equation:

$$\frac{L_e}{L} = \frac{(1-p)}{1-\left[\dfrac{v_b}{v}\right]^{0.22}}$$

Where
L_e = depth of expanded bed, m
L = bed depth, m
P = bed porosity
v_b = face velocity of the backwash water, m/s
v = particle settling velocity, m/s
With the above notations and given that $v = 0.12$ m/s
$v_b = 450/(60 \times 60 \times 24) = 5.21 \times 10^{-3}$ m/s
Then,

$$\frac{L_e}{L} = \frac{(1-p)}{1-\left[\dfrac{v_b}{v}\right]^{0.22}} = \frac{(1-p)}{1-\left[\dfrac{5.21 \times 10^{-3}}{0.12}\right]^{0.22}}$$

This yields:
$L_e = 0.96$ m

Program 3.7 Listing:
Depth of expansion of a filter bed

```
'**********************************
'EXAMPLE 3.7: Depth of expanded bed
'**********************************
Public Class Form1
    Const g = 9.81

    Private Sub Form1_Load(ByVal sender As
        System.Object, ByVal e As
        System.EventArgs) Handles MyBase.Load
        Me.Text = "Example 3.7"
        Me.FormBorderStyle =
            Windows.Forms.FormBorderStyle.FixedSingle
        Me.MaximizeBox = False

        Label1.Text = "Settling velocity (v, m/s):"
```

```
        Label2.Text = "Bed depth (L, m):"
        Label3.Text = "Bed porosity (p):"
        Label4.Text = "wash rate (vb, m3/m2.d):"
        Label5.Text = ""
        Button1.Text = "&Calculate"
    End Sub

    Private Sub Button1_Click(ByVal sender As
        System.Object, ByVal e As
        System.EventArgs) Handles Button1.Click
        Dim v, vb, L, p, Le As Double

        v = Val(TextBox1.Text)
        L = Val(TextBox2.Text)
        p = Val(TextBox3.Text)
        vb = Val(TextBox4.Text) /
                (60 * 60 * 24) 'convert to m/s

        Le = ((1 - p) / (1 - ((vb / v) ^ 0.22))) * L

        Label5.Text = "Depth of expanded bed = " +
                FormatNumber(Le, 2) + " m"

    End Sub
End Class
```

Example 3.8

1) Determine the clear water head loss in a filter bed that consists of two layers of filter media; a uniform anthracite layer of depth of 0.6 m with an average particle size of 1.6 mm and a specific gravity of 1.5; the other layer is composed of uniform sand 40 cm deep with an average particle size of 0.6 mm and a specific gravity of 2; for a rate of filtration of 150 $L/m^2/min$. The operating temperature was found to be $15°$ C and the porosity is 0.35. Use Rose equation for evaluating the clear water head loss.

2) Write a computer program to determine the clear water head loss in a multi-layer filter bed media given medii depths, average particle size, specific gravity, shape factor, porosity, rate of filtration, operating and operating temperature.

3) Verify your program by solving example 3.8a.

Solution

Rose equation states that:

$$h=1.067C_D\frac{1}{p^4}\frac{v^2}{gd\,\varphi}L$$

$$C_D=\frac{24}{Re}+\frac{3}{\sqrt{Re}}+0.34$$

Given that for anthracite layer: L = 0.6 m, d = 1.6x10^{-3} m

For sand layer: : L = 0.4 m, d = 0.6x10^{-3} m

For a temperature of 15° C the kinematic viscosity equals 1.139x10^{-6} m^2/s, v = 150 L/m^2.min = 2.5x10^{-3} m/sec.

a) <u>Anthracite layer:</u>

$$Re=\frac{v_d}{v}=2.5\times10^{-3}\times1.6\times10^{-3}\,pver\,1.139\times10^{-6}=3.51$$

$$C_D=\frac{24}{Re}+\frac{3}{\sqrt{Re}}+0.34=\frac{24}{3.51}+\frac{3}{\sqrt{3.51}}+0.34=8.78$$

$$h_a=1.067C_D\frac{1}{p^4}\frac{v^2}{gd\phi}L=$$

$$1.067\times8.78\times\frac{1}{0.35^4}\frac{(2.5\times10^{-3})^2}{9.81\times1.6\times10^{-3}\times1}\times0.6$$

$$=0.15m$$

b) <u>Sand layer:</u>
Using the same technique of computation as above, then
Re = 1.32, C$_D$ = 21.13 and h$_s$ = 0.64

Therefore, the clear water head loss in the filter will be equal to the sum of the head loss through anthracite layer and the head loss through the sand layer.

Thus, total head loss = ha + hs = 0.15 + 0.64 = 0.79 m

Program 3.8 Listing:

Clear water head loss in a multi-layer filter bed media

```
'****************************
'EXAMPLE 3.8:
'****************************
Public Class Form1
```

131

```
Const g = 9.81

Private Sub Form1_Load(ByVal sender As
    System.Object, ByVal e As
    System.EventArgs) Handles MyBase.Load
    Me.Text = "Example 3.8"
    Me.FormBorderStyle =
        Windows.Forms.FormBorderStyle.FixedSingle
    Me.MaximizeBox = False
    Label1.Text =
        "Layer 1 Particle diameter (d, mm):"
    Label2.Text = "Layer 1 Bed depth (L, m):"
    Label3.Text =
        "Layer 2 Particle diameter (d, mm):"
    Label4.Text = "Layer 2 Bed depth (L, m):"
    Label5.Text = "Face velocity (v, m/s):"
    Label6.Text = "Particle shape factor:"
    Label7.Text = "Bed porosity:"
    Label8.Text = "Kinematic viscosity (m2/s):"
    Label9.Text = ""
    Button1.Text = "&Calculate"
End Sub

Private Sub Button1_Click(ByVal sender As
    System.Object, ByVal e As
    System.EventArgs) Handles Button1.Click
    Dim d, v, phi, L, p, rho, CD, Re, Ha, Hb
        As Double
    d = Val(TextBox1.Text) / 1000   'convert to m
    L = Val(TextBox2.Text)
    v = Val(TextBox5.Text)
    phi = Val(TextBox6.Text)
    p = Val(TextBox7.Text)
    rho = Val(TextBox8.Text)

    'head loss for first layer by Rose equation
    Re = v * d / rho
    CD = (24 / Re) + (3 / Math.Sqrt(Re)) + 0.34
    Ha = 1.067 * CD * (1 / (p ^ 4)) *
        (v ^ 2 / (g * d * phi)) * L
    Label9.Text = "Head loss (1) by Rose eq. = "
        + FormatNumber(Ha, 2) + " m"
    'head loss for second layer by Rose equation
    d = Val(TextBox3.Text) / 1000   'convert to m
    L = Val(TextBox4.Text)
    Re = v * d / rho
    CD = (24 / Re) + (3 / Math.Sqrt(Re)) + 0.34
    Hb = 1.067 * CD * (1 / (p ^ 4)) *
        (v ^ 2 / (g * d * phi)) * L
    Label9.Text += vbCrLf +
```

```
               "Head loss (2) by Rose eq. = " +
               FormatNumber(Hb, 2) + " m"
           'output total head loss
           Label9.Text += vbCrLf + "Total head loss = "
            + FormatNumber(Ha, 2) + "+"
            + FormatNumber(Hb, 2) + _
             "=" + FormatNumber((Ha + Hb), 2) + " m"
       End Sub
End Class
```

Example 3.9

1. For the final purification of the river water in problem 1.3 the
   sanitary engineer felt it necessary to install a rapid gravity filter.
   The filter bed is 1.20 m thick and consists of uniform sand
   particles having an effective diameter of 0.75 mm. the
   supernatant water depth is 1.50 m while the filtration rate was
   taken to be 5 m³/m²/hour.

   i)　How many filters and which filter bed area will the
   　　sanitary engineer use?

   ii)　Evaluate the initial resistance of the filter bed given that
   　　the measured bed porosity is 35 percent. If at the end of
   　　the filter run the minimum pressure occurs at a depth of
   　　0.3 m beneath the top of the filter bed, compute the
   　　maximum allowable filter resistance. Take temperature
   　　to be 25° C as reported in problem 1.5.

2. Write a computer program to determine number of filters, filter
   bed area, and maximum bed resistance given initial resistance of
   filter bed, media deph, particle effective diameter, supernatant
   water depth, and filtration rate.

3. Verify your program by solving example 3.9a

Solution

For problem 1.3 it is given that $Q = 1500$ m³/hour $= 0.417$ m³/sec.
Data given: thickness of bed, $L = 1.2$ m, $d = 0.75 \times 10^{-3}$ m,
supernatant water depth $= 1.5$ m, $v = 5$ m/hour $= 1.4 \times 10^{-3}$ m/sec

i)　Number of filters could be estimated from the empirical
　　equation:

$$n = 12\sqrt{Q}$$

Where:

n = number of filters

Q = average capacity, m³/sec

Therefore,

$$n = 12\sqrt{Q} = 12\sqrt{0.417} \approx 8$$

Take 10 filters so that when backwashing two filters the remainder will supply the desired amount of water.

The total filtration area = average capacity/filtration rate = 1500/5 = 300 m²

Unit filter area = 300/(10 -2) = 38 m²

ii)

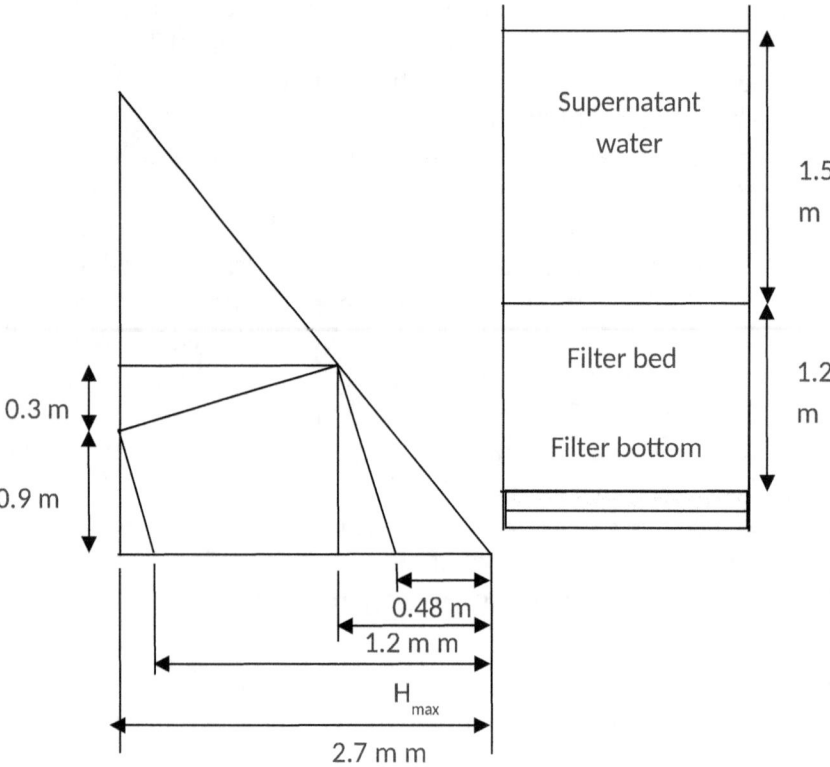

Using Kozeny-Carman equation to determine the bed resistance:

$$H = 180 \frac{v}{g} \frac{(1-p)^2}{p^3} \frac{v}{d^{2^3}} L$$

$$=180\frac{0.89x10^{-6}}{9.81}\frac{(1-0.35)^2}{0.35^3}\frac{1.4x10^{-3}}{(0.75x10^{-3})^2}x1.2$$

$$= 0.48m$$

Maximum bed resistance could be found from the above sketch:

$$H_{max}=2-\frac{0.9}{1.2}(1.2-0.482)=2.16m$$

Exercise 3.3

Head loss

1) A model of sand filter was constructed in the laboratory and it is composed of 15 mm diameter tube filled with sand to a depth of 0.8 m. The sand particles are spherical and uniform and have an effective diameter of 0.6 mm. The porosity has been measured and it amounted to 40 percent. The water temperature was recorded as 20° C. Using both Rose and Carman-Kozeny formulae calculate the **head loss** through the filter if the filtration rate is adjusted at 150 m³/m²/d. (Ans. 0.45, 0.34 m).

Bed expansion

2) The measured settling velocity of the sand particles in the filter in problem (1) was 100 mm/sec. evaluate the filter bed **expansion** if it is known that the filter is washed at a rate of 10 m/sec. (Ans. 1.2 m).

3) a) Discuss the mechanisms which may be important in the filtration of suspensions through sand filters.

   b) A laboratory-scale sand filter consists of a 10 mm diameter tube with a 900 mm deep bed of uniform 0.5 mm diameter spherical sand ($\phi = 1$), porosity 40%. Determine the **head loss** using Rose's formula and the Carman-Kozeny formula when filtering at a rate of 140 m³/m².d. For water at 20°C, $v = 1.01x10^{-3}$ N.s/m².

   c) The measured settling velocity of the sand particles in the filter mentioned above was 100 mm/s. Determine the bed **expansion** when the filter is washed at a rate of 10 mm/s. (U of K, 1983). (Ans. 750, 572 mm, 51%).

4) Sketch a plan and cross section of a rapid sand filter. Describe, in detail, method of operation and the hydraulic factors necessitating cleaning of the filter bed. A filter bed consists of dual filter media, the characteristics of which are as tabulated below:

Characteristics	Anthracite layer	Sand layer
Depth (cm)	60	40
Average particle size (mm)	1.6	0.8
Specific gravity	1.6	2.7
Porosity (%)	40	40
Shape factor	0.70	0.92

i. The clear water **head loss** in the filter bed is desired to be 64 cm. for an operating temperature of 15°C, (coefficient of viscosity is 1.139×10^{-3} Ns/m^2), compute the allowable rate of filtration by using Rose's and Carman-Kozeny's equation.

ii. Comment on the practicability of your answer

iii. Why is the value of the specific gravity for sand higher than that of anthracite in this filter? (Ans. 4.2, 5.8 mm/s).

5) Using Carman-Kozeny equation calculate the initial **head loss** through a filter composed of filter sand particles of depth of an average grain diameter of 0.5 mm and the porosity analysis indicated that it is 0.35. The filter bed depth is 0.5 m and its cross sectional area is 14 m^2. The filter is capable of treating water flow of 850 m^3 each day. The sand particle shape factor could be taken as 0.95 and water temperature is 25° C. If a layer of anthracite of depth 0.6 m is placed on top of this sand bed, determine the total **head loss** through the filter given that the anthracite grain size diameter is 2mm and its porosity is 0.40. (Ans. 0.21 m).

6) For what purpose is the filtration process used? Using the equations (developed by Rose), determine the head loss through a 75 cm sand filter. Assume that the sand bed is composed of spherical uni-sized sand with a diameter of 0.6 mm, the kinematic viscosity is equal to 1.306×10^{-6} m^2/s, the porosity is 40 percent, and the filtration velocity is 240 L/m^2.min.

If a 0.3 m layer of anthracite is placed on top of the sand bed in (b) above, determine the ratio of **head loss** through the anthracite to that of the sand. Assume that the grain size diameter is 2 mm and porosity of the anthracite is 50 percent. (UAE, 1990). (Ans. 1.33, 0.02 m).

7) Rose equation for determining the head loss through a filter states that:

$$h_f = 1.067 \, C_D \frac{v^2 L}{gdfe^4}$$

$$C_D = \frac{24}{Re} + \frac{3}{\sqrt{Re}} 0.34$$

i] Define the terms used in the equation.

ii] A filter of 1.2 m depth treats a flow at a maximum rate of 15 m³/m².hr. The bed has a porosity of 40 percent and it consists of uni-sized spherical sand particles having a diameter of 0.5 rom. For water with a kinematic viscosity of 1.003*10-6 m2js, compute the **head loss** through the filter. (Ans. 2.5 m).

8) What is the purpose of filtration? Rose equation for determining the head loss through a filter states that:

$$h_f = 1.067 \, C_D \frac{v^2 L}{gdfe^4}$$

using this equation, calculate the initial **head loss** through a filter composed of sand particles of an average grain diameter of 0.5 mm and having a porosity of 35%. The filter bed depth is 0.5 m and its cross sectional area is 14 m². The filter is capable of treating a water flow of 850 m³ each day. The sand particle shape factor could be taken as 0.9 and water temperature is 25°C. Briefly describe the operation of a typical rapid sand filter. (SQU, 1993).

9) What is the purpose of filtration? Rose equation for determining the head loss through a filter states that: $h_f = 1.067 \, C_D \, V^2 L/gdn^4\varphi$. $CD = 24/Re + 3/\sqrt{Re} + 0.34$. Using this equation, calculate the initial **head loss** through a filter composed of sand particles of an average diameter of 0.5 mm and having a porosity of 35%. The filter bed depth is 0.5 m and its cross sectional area is 14 m. The filter is capable of treating a water flow of 850 m³ each day. The sands particle shape factor could be taken as 0.9

137

and water temperature is 25°C. Briefly, describe the operation of a typical rapid sand filter. (SQU, 1992). (Ans. 0.3 m).

10) Discuss the mechanisms which may be important in the filtration of suspensions through sand filters. Indicate how the Rose equation:

$$h_f = 1.067\, C_D \frac{v^2}{gd\,\varphi} \cdot \frac{1}{p^4} \cdot L$$

could be used to determine the pressure drop through a rapid gravity filter.

Determine the clear water **head loss** in a filter bed composed of 30 cm of uniform anthracite with an average size of 1.6 mm and 30 cm of uniform sand with an average size of 0.5 mm for a filtration rate of 160 L/m².min. Assume that the operating temperature is 20°C. Use both Rose and Carman-Kozeny equations for computing the head loss. Given: $v = 1.003 \times 10^{-6}$ m²/sec., p = 40% (U of K, 1983). (Ans. 424, 313 mm).

11) Discuss the mechanisms which may be important in the filtration of suspensions through filters. By using the equation developed by Rose, the responsible sanitary engineer determined the head loss through a filter. The filter media consists of a layer of sand 70 cm thick and a 0.4 m layer of anthracite being placed on top of the sand-bed. The bed is assumed to be composed of spherical unisized sand grains with a diameter of 0.6 mm and a porosity of 40%, while the anthracite layer consists of particles of a shape factor of 0.8 and a grain size diameter of 2 mm with a relatively higher porosity of 55 percent. The dynamic viscosity was found to be 0.893 X 10^{-3} Ns/m² and the filtration velocity is 300 l/m².min. Find the error that would have been found if the sanitary engineer has determined the **head loss** by using Carman-Kozeny equation instead. (U of K, 1984). (Ans. 1.1, 0.8 m).

12) Indicate how Rose's equation can be used to determine the clear-water head-loss in a filter bed.

$$h_f = \frac{1.067}{\varphi} \cdot C_D \cdot \frac{1}{e^4} \cdot \frac{L}{d} \cdot \frac{v_f^2}{g}$$

$$C_D = \frac{24}{Re} + \frac{3}{\sqrt{Re}} 0.34$$

Using Rose equation determine the **head-loss** through a 75 cm sand bed. Assume that the bed is composed of spherical unisized sand with a diameter of 0.6 mm. The kinematic viscosity is equal to 1.306×10^{-6} m^2/s, porosity is 0.4 and the filtration velocity is 240 L/m^2.min.

Plot variation of head-loss with filtration velocity for a practical range of variation. (OIU, UNESCOC, 2004).

13) a) Differentiate between slow sand filters and rapid sand filters.

b) Calculate the initial **head loss** through a dual-media filter consisting of a 30 cm layer of uniform anthracite with a grain diameter of 1 mm and a 0.30 m layer of uniform sand with a grain diameter of 0.5 mm at a filtration rate of 2.7 L/m^2. s. The porosity of both media is 42 %, the shape factor of grains for the anthracite is 0.75, and the shape factor of grains for the sand is 0.60. Assume a water viscosity of 1.306×10^{-6} m^2/s. c) List some of the problems associated with detection and identification of viruses. (SQU, 1992). (Ans. 0.8 m).

14) a] Explain how the filtration process improves water quality.

b] Differentiate between "Slow filters" and "Rapid filters".

c] Using the equations [developed by Rose], determine the **head loss** through a 75 cm sand bed. Assume that the sand bed is composed of spherical unisized sand with a diameter of 0.6 mm, the kinematic viscosity is equal to 1.306×10^{-6} m^2/s, the porosity is 40 percent, and the filtration velocity is 240 L/m^2.min.

d] If a 0.3 m layer of anthracite is placed on top of the sand in [c] above, determine the ratio of head loss through the anthracite to that of the sand. Assume that the grain-size diameter is 2 mm and porosity of the anthracite is 50 percent. (SQU, 1991). (Ans. 1.3 m, 2%).

15) Clean water at 20 degree C is passed through a bed of uniform sand at a filtering velocity of 5.0 m/h. The sand grains are 0.35 mm in diameter with a shape factor of 0.85 and a specific gravity of 2.65. The porosity of the bed is 0.4. Calculate the **filter bed depth** to achieve a head loss of 0.6 m through the filter bed. (B.Sc., UoD, 2013)

Filter area and numbers

16) In a village in the Gezira area the responsible sanitary engineer propped using slow sand filters for water purification for the villagers. Determine for an average filter capacity of 1200 m3/hour the:

   a) Total filter surface **area** to be provided by the sanitary engineer.

   b) **Number** and size of individual filter units to be used. (Ans. 3333 m^2, 10).

17) Surface water from a stream contains suspended solids in an amount of 170 mg/l. Consumers in a neighboring village require an amount of water of 1300 m^3/hour. To treat this water the responsible sanitary engineer proposed sedimentation, rapid gravity filtration and chlorination. The filters to be used are composed of uniform sand grains 0.8 mm effective diameter and the filter bed depth amounts to 1.20 m while the supernatant water level is 1.50 m above the filter bed. Filtration rate is kept constant at 4 m/hour:

   1) **How many** filters will the sanitary engineer use?

   2) Which unit filter bed **area** is recommended?

   3) Determine the initial filter bed **resistance** if the porosity is found to be 0.40.

   4) At the end of the filter run the minimum pressure was noted to occur at a depth of 25 cm below the top of the filter bed. What will be the maximum allowable **resistance** offered by the filter in this case?

   5) The villager's water consumption varies with a constant abstraction rate of 2800 m^3/hour during a 9 hour period and 400 m^3/hour constant withdrawals during 1 15 hour period. Compute the minimum volume of **clear water well** needed by the villagers.

   6) When abstracting water at a rate of 2800 m^3/hour, the head supplied by the filtered water pumps was found to be 35 m water column. What **power** is needed from the village generator to drive the pumps? (Ans. 10, 41 m^2, 0.2 m, 1.9 m,13500 m^3, 267 kW).

18) River water contains suspended particles in an amount of 150 g/m^3. The consultant engineer was asked to design a treatment

plant consisting of sedimentation, rapid filtration and chlorination to purify the river water. The amount of water needing purification is 1600 m³/hour. The consultant engineer designed a sedimentation basin that is capable of removing 90% of the suspended load. The effluent of the settling basin was introduced to the filter at a rate of 8.5 m/hour. The filter bed is composed of crushed anthracite with a mass density of 1600 kg/m³ and a porosity of 35 percent and a grain size of 13 mm with a coefficient of permeability of 15 m/hour. The filter bed thickness is 1.3 m, while supernatant water depth was taken as 1.5 m. At the end of the filter run the minimum water pressure was found by the consultant engineer to occur at a depth of 0.35 m beneath the top of the filter bed. The under drainage system of this filter consists of perforated laterals with an inner diameter of 0.00 8 m and 50 openings per unit area.

a) Evaluate the **resistance** of the filter bed at the beginning and at end of the filter run when a negative pressure of 0.45 m water column is permitted.

b) Calculate the filter bed **resistance** during backwashing of the filter if the bed expanded 20 percent. (Ans. 2.6, 0.5 m).

19) A factory in Khartoum City works for 4 days a week at full capacity and 2 days a week at half capacity. The amount of water needed by the factory is 16 million m³ annually. This water is withdrawn from the Nile and is purified by rapid filtration throughout the days of the week at a constant rate of filtration of 5 m³/m²/hour. The filter bed has a thickness of 1.2 m and consists of a porous material of 35 percent porosity and with a grain size of 0.6 mm and a mass density of 3100 kg/m³. It was noticed that during operation the bed porosity decreased. Firstly it was assumed that the clogging of the filter bed is uniform over the top 25 cm and nil over the remaining thickness. The depth of the supernatant water was kept at 1.5 m.

a) Calculate the **number** of filter units used in the factory and their size.

b) Design the **clear well** required to allow continuous filter operation at a constant rate.

c) Evaluate the filter **resistance** when filter run commences and at its end if a negative pressure of 0.4 m water column is allowed. (Ans. 10, 26302 m³, 3 m).

20) Discuss the mechanisms that may be important in the filtration of suspensions through kilter media. Rapid gravity sand filters are to be used to purify surface water in an amount of 19800 litres/min. Each filter is composed of unsized sand grains with an effective diameter of 0.8 mm the filter bed depth amounts to 1.2m while the water percolates at a constant rate of 1.1 mm/s.

    I) Compute the **number** of filters to be used for purifying the water.

    II) Indicate unit filter bed area.

    III) If the head loss through each filter is not to exceed 0.3m (according to Carman-Kozeny equation), compute the bed **porosity**. Take dynamic viscosity (μ) to be equal to 1×10^{-3} Ns/m^2. (Ans. 9, 43 m^2, 40 %).

21) Discuss mechanisms that may be of importance in filtration of suspensions through filters. Differentiate between "slow sand filters" and "rapid sand filters". Explain the working of a rapid gravity filter. A rapid filter plant is to treat 23000 m^3 per day at a rate of 120 m/day. The filter bed is composed of: 30 cm of uniform anthracite medium with an effective size of 0.95 mm and a uniformity coefficient of 1.4; and 30 cm of sand medium with an effective size of 0.6 mm and a uniformity coefficient of 1.2. if the water temperature is 20°C:

    i) Compute required **number** of filters.

    ii) Find the clear water **head loss** in the filter using Rose equation. Assume porosity of 50%.(UAE, 1989). (Ans. 9, 78 mm).

22) Explain how the filtration process improves water quality. A village in the Gezira scheme utilized the nearby canal for drinking purposes. The canal water is of the following characteristics:

Temperature	20°C
pH	6.6
Suspended solids content	150 g/m^3
Calcium (Ca^{++})	72 mg/l
Magnesium (Mg^{++})	48.4 mg/l
Dynamic viscosity at 20°C	1.002×10^{-3}

The consultant engineer was asked to design a treatment plant incorporating sedimentation, rapid filtration & chlorination to treat the canal water. The amount of water needing purification is

1600 m³/hr. The proposed sedimentation unit is capable of removing 70% of the suspended load. The effluent of the settling basin was introduced at the filter at a rate of 8.5 m³/m²/hr. The filter bed is being composed of unisized crushed anthracite with the following characteristics: Mass density = 1600 kg/m³, Porosity = 35%, Average grain size = 1.6mm. The filter bed thickness amounted to 1.2m, while the supernatant water depth measured 1.5m, Compute the:

a) **number** of filters to be used.

b) total filter surface **area** to be provided.

c) unit filter bed **area** needed

d) **head loss** through the filter bed (Use both Rose & Carman-Kozeny equations).

e) minimum volume of **clear water well** needed by the villagers given that the village's water consumption varies with a constant abstraction rate of 3200m³/hr during a 9 hour period & 640 m³/hr constant withdrawal during a 15 hour period. (U of K, 1986). (Ans. 9, 188 m², 24 m², 0.25 m, 0.18 m,14400 m³).

23) The system for treating lake water in an amount of 0.8 m³/sec. consists of intake, raw water pumping station, micro-strainers, slow sand filters, safety chlorination, clear well and filtered water pumping station. At the capacity mentioned above the slow sand filters will operate at a rate of 0.08 mm/sec. Sketch the layout of this system and indicate

a) The **number** of filters to be applied.

b) The unit **area** of the slow filters to be used.

c) The minimum volume of **the clear well** when the filtered water pumping station operates sequentially for 3 hours at a rate of 0.3 m³/sec., 3 hours at a rate of 1 m³/sec., 6 hours at a rate of 1.5 m³/sec., 3 hours at a rate of 1.2 m³/sec, and 9 hours at a rate of 0.3 m³/ sec. (U of K, 1984). (Ans. 14, 769 m², 21600 m³).

24) Comment about purpose of use of sand filters and disinfection in water treatment systems. Using Carman-Kozeny equation calculate the initial head loss through a dual-media filter consisting of a 0.3-m layer of uniform anthracite with a grain diameter of 2.0 mm and a 0.6-m layer of uniform sand with a grain diameter of 1.0 mm at a filtration rate of 1.5 L.s⁻¹.m⁻². The

porosity of both media is 0.40, the shape factor (grain spherity) for the anthracite is 0.75, and the shape factor for the sand is 0.6. Assume a water temperature of 22°C. For a temperature of 22°C, $\mu = 0.9608*10^{-3}$ Ns/m^2, $\rho = 997.77$ kg/m^3(B.Sc. UoD, 2014) (Ans. 38.8 mm)

$$C_D = \frac{24}{Re} + \frac{3}{\sqrt{Re}} 0.34 \qquad E = \frac{150*(1-e)}{Re} + 1.75 \qquad Re = \frac{\rho v_f d}{\mu}$$

Backwashig

25) A filter with dimensions of 4 m by 8 m is used for continuously filtering 8000 m^3 of water daily. The rate of filter backwashing is 25 m^3/m^2/hr for a period of 5 minutes. Calculate the **filtration rate** and the amount of water used for **backwashing** the filter. (Ans. 2.9 L/m2.s, 67 m^3).

26) Indicate the merits and demerits of slow sand filters.

   Why is there such a significant difference in biological activity of a slow sand filter in comparison to a rapid sand filter?

   Which type of filter would you recommend to a small town in Sudan and why?

   Sketch a plan and cross section of an individual filter.

   The system for treating lake water in a daily amount of 8000m^3 consists of intake, raw water pumping station, rapid sand filter safety chlorination, clear well and filtered water pumping station. The rate of filter backwashing is 25 m^3/m^2/hr for a period of 5 minutes. The filter is of dimensions 4m x 8m and depth 0.6m. An analysis of the sand showed that it possesses a porosity of 0.5 with average grain size diameter of 0.6mm and shape factor of 0.92. The supernatant water depth is 1.5m while the water has a viscosity of $1.3x10^{-3}$ Ns/m^2. Sketch the layout of this system and compute:

   i) The **filtration rate**

   ii) The amount of water used for **backwashing** the filter.

   iii) Initial **head loss** through the filter (use both Rose and Carman-Kozeny equations). (U of K, 1985).

Grain diameter

27) Indicate how the Rose equation:

$$h_f = 1.067 \, C_D \frac{v^2}{gd \, \varphi} \frac{1}{e^4} L$$

can be used to determine the pressure drop through a rapid gravity filter. A sand filter bed of depth of 75 cm layer of sand is composed of spherical unsized sand particles with a porosity of 45 percent. For a filtration velocity of 6250 L/m^2/day and a maximum bed head loss (as determined by Rose's equation) to be 12 cm, compute the needed average **grain diameter**. Is this a suitable size? (Assume the kinematic viscosity is equal to $1.3*10^{-6}$ m^2/s). (SUST, 2001). (OIU, UNESCOC, 2000). (Ans. 0.195 mm).

28) What are the mechanisms that may play a significant role in the filtration of suspensions through sand filters?. Illustrate major differences between slow and rapid filters. A filter bed is composed of 75 cm layer of sand with a porosity of 40%. Assume that the bed is composed of spherical unsized sand particles, kinematic viscosity equals to 1.306×10^{-6} m^2/sec, filtration velocity of 240 l/m^2.minute & the maximum head loss (as determined by Rose equation) to be 1.326m. Compute the needed average **grain diameter**. Is this a suitable size? (U of K, 1987). (Ans. 0.6 mm).

29) Elaborate about filtration mechanisms of importance in a sand filter. A sand filter bed consists of unsized grains of porosity, e, of 50 percent. The bed depth amounts to 1 m and it is capable of filtering flow of kinematic viscosity, v, of $1.0105*10^{-6}$ m^2/s at a filtration velocity, v_f, of 0.2 m^3/m^2/hr. The maximum bed head loss, h_f, has been estimated by Rose's equation to be equal to 0.24 m.

  i) Evaluate the average **grain diameter** (mm) needed for the filter bed.

  ii) Comment about suitability of determined grain diameter.

iii)   Calculate the filter bed **head loss** using Carman-Kozeny equation. (B.Sc. UoD, 2013) (Ans. 0.2 mm, 0.17 m).

$$h_f = 1.067\, C_D \frac{v_f^2}{gd\,\varphi} \frac{1}{e^4} L \qquad C_D = \frac{24}{Re} + \frac{3}{\sqrt{Re}} 0.34 \qquad Re = \frac{\rho v_a d}{\mu}$$

$$h_f = E \frac{v_f^2}{gd\,\varphi} \frac{1-e}{e^3} L \qquad E = \frac{150*(1-e)}{Re} + 1.75$$

Rate of filtration

30) a) Sketch a plan and cross section of a rapid sand filter. Describe in detail method of operation and the hydraulic factors necessitating a cleaning to the filter bed.

b) A filter bed consists of dual filter media, the characteristics of which are as indicated below:

Characteristics	Anthracite layer	Sand layer
Depth (cm)	60	40
Average particle size (mm)	1.6	0.8
Specific gravity	1.5	2.6
Porosity (%)	40	40
Shape factor	0.7	0.92

The clear water head loss in the filter bed is desired to be 64cm. For an operating temperature of 15°C, (Coefficient of viscosity is 1.139×10^{-3} Ns/m^2), compute the allowable **rate of filtration** by using:

i) Rose's equation,

ii) Carman-Kozeny's equation.

Comment on the practicability of your answer

Why is the value of the specific gravity for sand higher than that of anthracite in this filter? (U of K, 1988). (Ans. 4.1, 5.8 mm/s).

Slow sand filter

31) a] differentiate between slow sand and rapid sand filters

b] The following design details have been furnished for a **slow sand** filter plant:

Design rate of filtration	763 m^3
Daily demand	0.11 m/hour

Area of filter required $\qquad$ 280 m^2
Size of each filter bed $\qquad$ 14m×10m
Number of filter beds provided   2

As an environmental engineer you are asked to review and comment on this design. (SQU, 1991).

32) a] Differentiate between Slow Sand & Rapid Sand Filters.

b] The following design details have been furnished for a **slow sand filter** plant:

Daily demand $\qquad$ 763m^3
Design rate of filtration $\qquad$ 0.11 m/hour
Area of filter required $\qquad$ 280 m^2
Number of filter beds provided $\qquad$ 2
Size of each filter bed $\qquad$ 14m x 10m

As an environmental engineer you are asked to review & comment on this design. (SQU, 1991).

33) Surface water from a stream contains suspended solids in an amount of 180 mg/L. Consumers in a village in the vicinity demand an amount of water of 1500 m3/hr. To purify this water the responsible sanitary engineer proposed sedimentation, rapid gravity filtration and chlorination. The filters to be used are composed of uniform sand grains with an effective diameter of 0.8 mm. The filter bed depth was selected to be 1.2m while the supernatant water level is 1.5 m above the filter bed. Given that the rate of filtration is 4 m/hr:

1. **How many** filters will be sanitary engineer use?
2. Which unit filter bed **area** is recommended?
3. Determine the initial filter bed **resistance** if the porosity is found to be 0.45.
4. At the end of the filter run the minimum pressure was denoted to occur at a depth of 20 cm below the top of the filter bed. What will be the maximum allowable **resistance** by the filter in this case? (U of K, 1984). (Ans. 10, 47 m^2, 0.14 m, 1.8 m).

34) Discuss the mechanisms which may be important in the filtration of suspensions through sand filters. Surface water from a stream contains suspended solids in an amount of 150 mg/L. The people in the vicinity demand an amount of water of 1300 m^3/hr. To purify this water the district sanitary engineer proposed

screening, sedimentation, rapid gravity filtration and chlorination. The filters to be used are composed of uni-sized sand grains with an effective diameter of 0.8 mm. The filter bed depth amounts to 1.2 m while the supernatant water level is 1.5 m above the filter bed. Filtration is kept constant at a rate of 1.1 mm/s.

1. **How many** filters will the sanitary engineer use?
2. Which unit filter bed area is recommended?
3. Determine the values of the initial bed **resistance** if the porosity is found to be 35%. (Assume viscosity to be 0.893×10^{-6} m^2/s)
4. At the end of the filter run the minimum pressure was denoted to occur at a depth of 0.2m below the top of the filter bed. What will be the maximum **resistance** offered by the filter in this case?
5. The villagers' water consumption varies with a constant abstraction rate of 2500 m^3/hr during a 9 hour period and 580 m^3/hr constant withdrawals; during a 15 hour period. Compute the minimum volume of **clear water well** needed by the villagers. (U of K, 1986). (Ans. 10, 41 m^2, 0.3 m, 2 m, 10800 m^3).

35) To reduce building costs, rapid filters have sometimes been constructed with a shallow depth of water on top of the sand bed. Discuss this statement and point out the disadvantages of the construction. (U of K, 1987).

An upflow filter is provided with a bed of uniform spherical sand grains of 1.2 mm in diameter in a thickness of 1.5 m, above which filtered water is present in a depth of 0.8 m. the filter is operated at a constant rate of 2 mm/s, for which resistance of the clean bed amounts to 0.3 m.

1. What is the maximum allowable filter **resistance**?
2. Sketch the pressure distribution at the beginning and at the end of the filter run.

1) Derive expressions for rate of removal of solids in a filter and rate of deposition of solids assuming constant coefficient of filtration. Comment about your assumptions.

2) The filter mentioned in (b) above is used for filtering water of concentration of impurities of 15 mg/L. if the

constant characteristics of the bed is 6 per m, compute solids concentration at half depth and rate of **deposition of solids per unit time.** ($v = 1.3 \times 10^{-6}$ m^2/s) (U of K, 1987). (Ans. 1.5m, 0.17 mg/L, 0.002).

3.3 Aeration

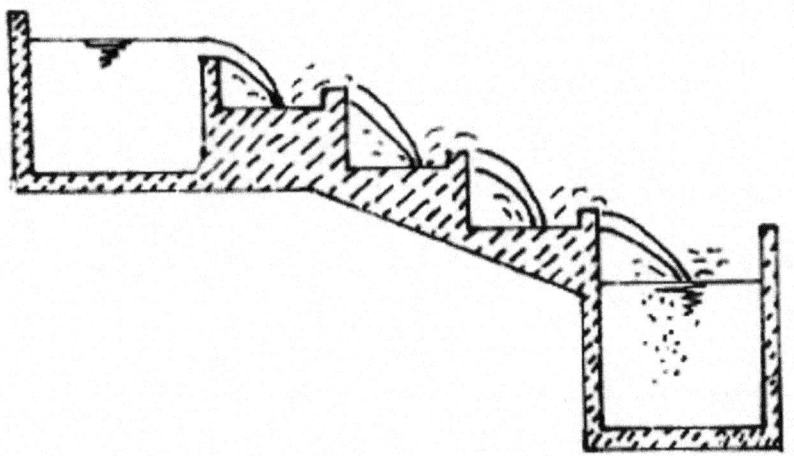

Aeration is the process in which air is being added to water or sewage in order to raise its dissolved oxygen level or to remove odor – and taste – producing substances or to motivate and propagate reactions.. The process should be continuous for interface renewal and this should be effectuated with the minimum of initial and operational expenses.

Exercise 3.10

1. Calculate the saturation concentration of oxygen in pure water at 20° C exposed to air at a pressure of 101.3 kPa. Assume that dry air contains 20.948 percent oxygen by volume. Determine also Henry's constant for these conditions and Bunsen absorption coefficient.
2. Write a computer program to calculate the saturation concentration of oxygen in pure water, Henry's constant and Bunsen absorption coefficient given temperature of fluid, air pressure, percent dry air oxygen composition by volume
3. Verify your program by solving example 3.10a.

Solution

The saturation concentration is given by:

$$C_S = k_D \frac{MW}{RT} x (p - p_w)$$

Where:

C_s = Saturation concentration, g/m^3.
K_D = Distribution coefficient.
MW = molecular weight of the gas, g.
R   = Universal gas constant (= 8.3143 J/K.mole).
T   = Absolute temperature, K.
x   = volumetric composition.
P   = Partial pressure of the respective gas in the gas phase, Pa.
P_w = water vapor pressure, Pa.

Therefore, T = 20 + 273.16 = 293.16 K; for this temperature from tables for oxygen the distribution coefficient k_D = 0.0337 and p_w = 2.33 kPa, MW of oxygen = 32.

Thus,

$$C_S = 0.0337 \frac{32}{8.3143 \times 293.16} x 0.20948 x (101.3 - 2.33) x 10^3$$
$$= 9.2 g/m^3$$

Henry's constant could be found from the formula:

$$\left(k_H = \frac{k_D MW}{RT} \right)$$

Where
k_H = Henry's constant, g/m^3*Pa, g/J
Thus,

$$k_H = \frac{k_D MW}{RT} = 0.0337 \, x \, \frac{32}{8.3143 \, x \, 293.16} = 4.4 \, x \, 10^{-4} \, g/J$$

Bunsen absorption coefficient is given by:

$$\left(k_b = \frac{k_D T_o}{T} \right)$$

Where:
K_b = Bunsen absorption coefficient.

Thus,

$$k_b = \frac{k_D T_o}{T} = 0.0337 \, x \, \frac{273.16}{293.16} = 0.0314$$

Program 3.10 Listing:

Saturation concentration of oxygen in pure water, Henry's constant and Bunsen absorption coefficient

```
'*********************************************
'EXAMPLE 3.10: Saturation concentration
'*********************************************
Public Class Form1
    Dim kDTable(3) As Double
    Dim pwTable(6) As Double
    Const R = 8.3143

    Private Sub Form1_Load(ByVal sender As
        System.Object, ByVal e As
        System.EventArgs) Handles MyBase.Load
        Me.Text = "Example 3.10"
        Me.FormBorderStyle =
            Windows.Forms.FormBorderStyle.FixedSingle
        Me.MaximizeBox = False

        'set up distribution coefficient (kD) table
        kDTable(0) = 0.0493    'at 0 degrees C
        kDTable(1) = 0.0398    'at 10 degrees C
        kDTable(2) = 0.0337    'at 20 degrees C
        kDTable(3) = 0.0296    'at 30 degrees C
        'set up vapor pressure of water (pw) table
        pwTable(0) = 0.611    'at 0 degrees C
        pwTable(1) = 0.872    'at 5 degrees C
        pwTable(2) = 1.23     'at 10 degrees C
        pwTable(3) = 1.71     'at 15 degrees C
        pwTable(4) = 2.33     'at 20 degrees C
        pwTable(5) = 3.17     'at 25 degrees C
        pwTable(6) = 4.24     'at 30 degrees C

        'set up user interface
        Label1.Text = "Mol. Wt of gas (MW):"
        Label2.Text = "Temp. (T) in C:"
        Label3.Text = "Volumetric composition (x):"
        Label4.Text = "Partial pressure of gas (Pa):"
        Label5.Text = ""
        Button1.Text = "&Calculate"
    End Sub
```

```vb
Private Sub Button1_Click(ByVal sender As
    System.Object, ByVal e As
    System.EventArgs) Handles Button1.Click
    Dim MW, T, x, Pa As Double
    Dim kD, pw, Cs, kH, kB As Double

    MW = Val(TextBox1.Text)
    T = Val(TextBox2.Text)
    x = Val(TextBox3.Text)
    Pa = Val(TextBox4.Text)
    'find kD in the Table
    If T = 0 Then kD = kDTable(0) _
    Else kD = kDTable(T / 10)
    'find pw in the Table
    If T = 0 Then pw = pwTable(0) _
    Else pw = pwTable(T / 5)
    T += 273.16    'convert to Kelvin
    If x > 1 Then x /= 100

    'Calculate saturation concentration
    Cs = ((kD * MW * x) / (R * T)) * (Pa - pw)
        * (10 ^ 3)
    'Calculate Henry's constant
    kH = (kD * MW) / (R * T)
    'Calculate Bunsen absorption coefficient
    kB = (kD * 273.16) / T
    'Output the results
    Label5.Text =
        "Saturation concentration (Cs) = "
        + FormatNumber(Cs, 4) + " g/m3"
    Label5.Text += vbCrLf +
        "Henry's constant (kH) = " _
        + FormatNumber(kH, 8) + " g/J"
    Label5.Text += vbCrLf +
        "Bunsen coefficient (kB) = "
        + FormatNumber(kB, 4)
End Sub
End Class
```

Exercise 3.11 (see Program 3.10 Listing)

1. Assuming that air at atmospheric pressure contains 21 percent oxygen by volume. Determine the oxygen saturation concentration in water at a depth of 10 m given that water temperature is $30^{\circ}C$.
2. Write a computer program to determine the oxygen saturation concentration in water given its depth, and water temperature.
3. Verify your program by solving example 3.11a.

Solution

Water temperature = $30°$ C = 303.16 K

For this temperature, $k_D = 0.0296$ and vapor pressure, $p_w = 4.24$ kPa

Therefore, gas pressure = $x(P - p_w) = 0.21(10^3 x9.81x8 + 101.3x10^3 - 4.24x10^3) = 36.86x10^3$ Pa

Using the above equation for saturation concentration:

$$C_s = 0.0296\frac{32}{8.3143 \times 303.16} \times 36.86 \times 10^3 = 13.9\, g/m^3$$

Exercise 3.12

1) In a water installation, ground water with oxygen content of 10 percent saturation and at a temperature of $25°$ C is aerated by using a cascade aerator, consisting of 3 identical steps, which is capable of raising its content from zero to 30 percent saturation. Determine the oxygen concentration of the effluent.
2) Write a computer program to determine the oxygen concentration of the effluent of a cascade aerator given influent water oxygen content, temperature, number of steps, and efficiency of cascade.
3) Verify your program by solving example 3.12a.

Solution

From the penetration theory, the efficiency coefficient of the cascade is given by:

$$K = \frac{C_e - C_i}{C_s - C_i}$$

Where:

K = efficiency coefficient.

C_e = Effluent concentration, g/m^3.

C_i = Influent concentration, g/m^3.

C_s = Saturation concentration, g/m^3.

Thus, given that for each step: $C_e = 0.3C_s$, $C_o = 0$, then:

$$K = \frac{C_e - C_i}{C_s - C_i} = \frac{0.3C_s - 0}{C_s - 0} = 0.3$$

The above equation could be written as:

$$C_e = Ci + K (Cs - Ci)$$

Effluent from the first step

$$C_{e1} = 0.1Cs + 0.3 (Cs - 0.1Cs) = 0.37 C_s$$

Effluent from the second step

$$C_{e2} = 0.37Cs + 0.3 (Cs - 0.37Cs) = 0.559 C_s$$

Effluent from the third step

$$C_{e3} = 0.559Cs + 0.3 (Cs - 0.559Cs) = 0.691 C_s$$

But at 25° C, from tables, Cs = 8.4 g/m^3

Therefore, the effluent from the third step of the cascade will have a concentration of 0.691x8.4 = 5.8 g/m^3.

Program 3.12 Listing:
Oxygen concentration of effluent from a cascade aerator

```
'******************
'EXAMPLE 3.12
'******************
Public Class Form1
    Dim Cs() As Double

    Private Sub Form1_Load(ByVal sender As
        System.Object, ByVal e As
        System.EventArgs) Handles MyBase.Load
        Me.Text = "Example 3.12"
        Me.FormBorderStyle =
            Windows.Forms.FormBorderStyle.FixedSingle
        Me.MaximizeBox = False

        Label1.Text = "Oxygen content (%)"
        Label2.Text = "Temperature (C)"
        Label3.Text = "No. of steps"
        Label4.Text = "Influent conc. (g/m3)"
        Label5.Text = "Effluent conc. (g/m3)"
        Label6.Text = ""
        Button1.Text = "&Calculate"

        'setup oxygen saturation (Cs) table
        ReDim Cs(30)
        Cs(0) = 14.6
        Cs(1) = 14.2
        Cs(2) = 13.8
        Cs(3) = 13.5
```

```
        Cs(4) = 13.1
        Cs(5) = 12.8
        Cs(6) = 12.5
        Cs(7) = 12.2
        Cs(8) = 11.9
        Cs(9) = 11.6
        Cs(10) = 11.3
        Cs(11) = 11.1
        Cs(12) = 10.8
        Cs(13) = 10.6
        Cs(14) = 10.4
        Cs(15) = 10.2
        Cs(16) = 10.0
        Cs(17) = 9.7
        Cs(18) = 9.5
        Cs(19) = 9.4
        Cs(20) = 9.2
        Cs(21) = 9.0
        Cs(22) = 8.8
        Cs(23) = 8.7
        Cs(24) = 8.5
        Cs(25) = 8.4
        Cs(26) = 8.2
        Cs(27) = 8.1
        Cs(28) = 7.9
        Cs(29) = 7.8
        Cs(30) = 7.6
End Sub

Private Sub Button1_Click(ByVal sender As
    System.Object, ByVal e As
    System.EventArgs) Handles Button1.Click
    Dim steps, O2, T, k, Ce, Ci As Double
    O2 = Val(TextBox1.Text)
    T = Val(TextBox2.Text)
    steps = Val(TextBox3.Text)
    Ci = Val(TextBox4.Text)
    Ce = Val(TextBox5.Text)

    If T > 30 Or T < 0 Then
        MsgBox("Please enter a temp. value
            between 0 and 30 C.", vbOKOnly,
            "Enter value temp.")
        Exit Sub
    End If

    If Ce > 1 Then Ce /= 100
    If Ci > 1 Then Ci /= 100
    k = (Ce - Ci) / (1 - Ci)
```

```
        If O2 > 1 Then O2 /= 100
        Ci = O2
        For i = 1 To steps
            Ce = Ci + (k * (1 - Ci))
            Ci = Ce
        Next
        Ce = Ce * Cs(T)
        Label6.Text = "Cascade of " + steps.ToString
                 + " steps:" + vbCrLf
        Label6.Text +=
            "Effluent will have concentration of "
            + FormatNumber(Ce, 4) + " g/m3"
    End Sub
End Class
```

Exercise 3.13

1) A cascade aerator is capable of lowering the carbon dioxide of groundwater from 20 to 12 g/m3, and a spray aerator is able to lower the carbon dioxide content of the same water to 8 g/m3. The air contains 0.032 percent carbon dioxide with solubility in water of 2500 g/m3 per atmosphere. Find the level of carbon dioxide that would be attained when both aerators are used in series. Which arrangement of aerators is to be favored?

2) Write a computer program to predict arrangement of aerators is to be favored given types of aerators used, initial concentration and related gas characteristics.

3) Verify your program by solving example 3.13a.

Solution

From the given data: $Cs = (0.032/100) \times 2500 = 0.8$ g/m^3

The efficiency coefficient for both aerators could be determined by using the above equation. Therefore, if K_c and K_s represent the efficiency coefficients of the cascade aerator and spray aerator respectively, then:

$$K_c = \frac{C_e - C_i}{C_s - C_i} = \frac{12 - 20}{0.8 - 20} = 0.42$$

And

$$K_S = \frac{C_e - C_i}{C_S - C_i} = \frac{8 - 20}{0.8 - 20} = 0.62$$

If the cascade aerator is to be used first, then the effluent concentration:

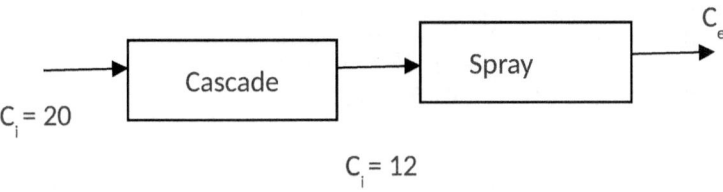

$C_i = 20$

$C_i = 12$

$C_e = Ci + K (Cs - Ci) = 12 + 0.62(0.8 - 12) = 5.1$ g/m3

If the spray aerator is to be used first, then the effluent concentration:
$C_e = Ci + K (Cs - Ci) = 8 + 0.42(0.8 - 8) = 5$ g/m^3
The effluent from both aerators seems nearly the same, therefore any sequence of aerators may be used.

Program 3.13 Listing:
Prediction of arrangement of aerators to be favored

```
'*******************
'EXAMPLE 3.13
'*******************
Public Class Form1

    Private Sub Form1_Load(ByVal sender As
        System.Object, ByVal e As
        System.EventArgs) Handles MyBase.Load
        Me.Text = "Example 3.13"
        Me.FormBorderStyle =
            Windows.Forms.FormBorderStyle.FixedSingle
        Me.MaximizeBox = False

        Label1.Text = "Gas content (%)"
        Label2.Text = "Gas solubility in water (g/m3)"
        Label3.Text = "Influent conc. (g/m3)"
        Label4.Text =
            "Effluent conc. of 1st aerator (g/m3)"
        Label5.Text =
            "Effluent conc. of 2nd aerator (g/m3)"
        Label6.Text = ""
        Button1.Text = "&Calculate"
```

```
    End Sub

    Private Sub Button1_Click(ByVal sender As
        System.Object, ByVal e As
        System.EventArgs) Handles Button1.Click
        Dim Gas, sol, k1, k2, Ce1, Ce2, Ci, Cs
                As Double
        Gas = Val(TextBox1.Text)
        sol = Val(TextBox2.Text)
        Ci = Val(TextBox3.Text)
        Ce1 = Val(TextBox4.Text)
        Ce2 = Val(TextBox5.Text)

        If Gas > 1 Then Gas /= 100
        Cs = (Gas / 100) * sol

        k1 = (Ce1 - Ci) / (Cs - Ci)
        k2 = (Ce2 - Ci) / (Cs - Ci)
        Dim tmp, tmp2 As Double
        'calculate Ce if aerator (1) is used first
        tmp = Ce1 + (k2 * (Cs - Ce1))
        'calculate Ce if aerator (2) is used first
        tmp2 = Ce2 + (k1 * (Cs - Ce2))

        Label6.Text = "Cs = " + FormatNumber(Cs, 2)
                + " g/m3" + vbCrLf
        Label6.Text += "k (first aerator) = "
                + FormatNumber(k1, 2) + vbCrLf
        Label6.Text += "k (second aerator) = "
                + FormatNumber(k2, 2) + vbCrLf
        Label6.Text += "Aerator (1) used first:
            Effluent conc. = " + _
            FormatNumber(tmp, 4) + " g/m3" + vbCrLf
        Label6.Text += "Aerator (2) used first:
            Effluent conc. = " + _
            FormatNumber(tmp2, 4) + " g/m3" + vbCrLf
    End Sub
End Class
```

Exercise 3.14

1) A cascade aerator is to be used for the aeration of 500 m^3
    groundwater per hour with an oxygen content of 2.3 g/m^3. The
    aerated water has an oxygen concentration of 9.3 g/m^3 at 10° C.

    a) Calculate the expected oxygen concentration at the over
        flow if ground water is anaerobic.

b) Find required electric power needed when the head loss in the aerator is 1.25 m. The efficiency of the pump is 70% and the efficiency of the motor is 90%.
c) Calculate the oxygenation efficiency (OE) in kg O_2/kWh of the aerator at zero oxygen concentration, taking into consideration the efficiency of pump and motor.

2) Write a computer program to calculate expected oxygen concentration at the over flow, electric power needed, and oxygenation efficiency given flow rate, influent oxygen content, temperature, head loss in the aerator, and efficiency of pump and motor.

3) Verify your program by solving example 3.14a.

Solution

For a temperature of $10°$ C the saturation concentration is 11.3 g/m^3, efficiency coefficient of the aerator:

$$K = \frac{C_e - C_i}{C_s - C_i} = \frac{9.3 - 2.3}{11.3 - 2.3} = 0.78$$

1) $C_e = C_i + K (C_s - C_i) = 0 + 0.78(11.3 - 0) = 8.8$ g/m^3
2) Gross power:

$$Grosspower = \frac{\rho g Q h}{\eta_p \eta_m}$$

Where:
ρ = density, kg/m^3
g = gravitational acceleration, m/s^2
Q = discharge, m^3/s
H = head loss, m
η_p = efficiency of pump
η_m = efficiency of motor
Thus,

$$Grosspower = \frac{\rho g Q h}{\eta_p \eta_m} = \frac{1000 \times 9.81 \times \dfrac{500}{60 \times 60} \times 1.25}{0.7 \times 0.9} = 2.7 \, kW$$

159

3) Oxygenation efficiency (OE) is given by:

$$OE = \frac{QKC_S}{\dfrac{\rho gQh}{\eta_p \eta_m}} \quad \text{notations as defined above}$$

$$OE = \frac{QKC_S}{\dfrac{\rho gQh}{\eta_p \eta_m}} = \frac{500 \times 0.78 \times 11.3 \times 10^{-3}\, kgO_2/hr}{2.7\, kWh/h}$$

= 1.63 kgO$_2$/kWh

Program 3.14 Listing:
Expected oxygen concentration at the over flow, electric power needed, and oxygenation efficiency

```
'******************
'EXAMPLE 3.14
'******************
Public Class Form1
    Dim Cs() As Double
    Const g = 9.81     'gravitational acceleration

    Private Sub Form1_Load(ByVal sender As
        System.Object, ByVal e As
        System.EventArgs) Handles MyBase.Load
        Me.Text = "Example 3.14"
        Me.FormBorderStyle =
            Windows.Forms.FormBorderStyle.FixedSingle
        Me.MaximizeBox = False

        Label9.Text = "Select an option:"
        Button1.Text = "&Calculate"
        ComboBox1.Items.Add("Calculate Efficiency
                coefficient k")
        ComboBox1.Items.Add("Calculate Gross power GP")
        ComboBox1.Items.Add("Calculate Oxygen
                efficiency OE")
        ComboBox1.SelectedIndex = 0

        'setup oxygen saturation (Cs) table
        ReDim Cs(30)
        Cs(0) = 14.6
        Cs(1) = 14.2
        Cs(2) = 13.8
        Cs(3) = 13.5
```

```
        Cs(4) = 13.1
        Cs(5) = 12.8
        Cs(6) = 12.5
        Cs(7) = 12.2
        Cs(8) = 11.9
        Cs(9) = 11.6
        Cs(10) = 11.3
        Cs(11) = 11.1
        Cs(12) = 10.8
        Cs(13) = 10.6
        Cs(14) = 10.4
        Cs(15) = 10.2
        Cs(16) = 10.0
        Cs(17) = 9.7
        Cs(18) = 9.5
        Cs(19) = 9.4
        Cs(20) = 9.2
        Cs(21) = 9.0
        Cs(22) = 8.8
        Cs(23) = 8.7
        Cs(24) = 8.5
        Cs(25) = 8.4
        Cs(26) = 8.2
        Cs(27) = 8.1
        Cs(28) = 7.9
        Cs(29) = 7.8
        Cs(30) = 7.6
End Sub

Private Sub Button1_Click(ByVal sender As
    System.Object, ByVal e As
    System.EventArgs) Handles Button1.Click
    Dim T, k, Ce, Ci As Double
    Dim GP, Q, H, rho, effPump, effMotor As Double
    Dim OE As Double

    Select Case ComboBox1.SelectedIndex
        Case 0  'selected 'calculate k'
            Ci = Val(TextBox1.Text)
            Ce = Val(TextBox2.Text)
            T = Val(TextBox3.Text)
            If T > 30 Or T < 0 Then
                MsgBox("Please enter a temp.
                    value between 0 and 30 C.",
                    vbOKOnly, "Enter value temp.")
                Exit Sub
            End If
            k = (Ce - Ci) / (Cs(T) - Ci)
            Label8.Text = "k = " +
                FormatNumber(k, 2) + vbCrLf
```

161

```vbnet
                    Label8.Text += "If water is anaerobic,
                        Ce = " + _
                        FormatNumber((k * Cs(T)), 2) +
                        " g/m3" + vbCrLf
            Case 1  'selected çalculate GP
                rho = Val(TextBox1.Text)
                Q = Val(TextBox2.Text)
                H = Val(TextBox3.Text)
                effPump = Val(TextBox4.Text)
                effMotor = Val(TextBox5.Text)
                Q /= (60 * 60)  'convert to m3/s
                If effPump > 1 Then effPump /= 100
                If effMotor > 1 Then effMotor /= 100
                GP = (rho * g * Q * H) /
                        (effPump * effMotor)
                Label8.Text = "GP = " +
                    FormatNumber(GP, 2) + "kW"
            Case 2  'selected çalculate OE
                rho = Val(TextBox1.Text)
                Q = Val(TextBox2.Text)
                H = Val(TextBox3.Text)
                effPump = Val(TextBox4.Text)
                effMotor = Val(TextBox5.Text)
                T = Val(TextBox6.Text)
                k = Val(TextBox7.Text)
                If T > 30 Or T < 0 Then
                    MsgBox("Please enter a temp.
                        value between 0 and 30 C.",
                        vbOKOnly, "Enter value temp.")
                    Exit Sub
                End If
                If effPump > 1 Then effPump /= 100
                If effMotor > 1 Then effMotor /= 100
                'Q /= (60 * 60)  'convert to m3/s
                GP = (rho * g * (Q / (60 * 60)) * H)
                        / (effPump * effMotor)
                OE = (Q * k * Cs(T)) / GP
                Label8.Text = "OE = " +
                    FormatNumber(OE, 2) + "kgO2/kW"
        End Select
    End Sub

Private Sub ComboBox1_SelectedIndexChanged(ByVal
    sender As System.Object, ByVal
    e As System.EventArgs)
    Handles ComboBox1.SelectedIndexChanged
        'change user interface according to
        'selected option
        Select Case ComboBox1.SelectedIndex
            Case 0
```

```
            Label1.Text =
                "Influent conc., Ci (g/m3)"
            Label2.Text =
            "Effluent conc., Ce (g/m3)"
            Label3.Text = "Temperature, T (C)"
            Label4.Text = ""
            Label5.Text = ""
            Label6.Text = ""
            Label7.Text = ""
            TextBox4.Visible = False
            TextBox5.Visible = False
            TextBox6.Visible = False
            TextBox7.Visible = False
        Case 1
            Label1.Text = "Density (kg/m3)"
            Label2.Text = "Discharge, Q (m3/hr)"
            Label3.Text = "Head loss, H (m)"
            Label4.Text = "Efficiency of pump (%)"
            Label5.Text =
                "Efficiency of motor (%)"
            Label6.Text = ""
            Label7.Text = ""
            TextBox4.Visible = True
            TextBox5.Visible = True
            TextBox6.Visible = False
            TextBox7.Visible = False
        Case 2
            Label1.Text = "Density (kg/m3)"
            Label2.Text = "Discharge, Q (m3/hr)"
            Label3.Text = "Head loss, H (m)"
            Label4.Text = "Efficiency of pump (%)"
            Label5.Text =
                "Efficiency of motor (%)"
            Label6.Text = "Temperature (C)"
            Label7.Text =
                "Efficiency coefficient (k)"
            TextBox4.Visible = True
            TextBox5.Visible = True
            TextBox6.Visible = True
            TextBox7.Visible = True
        End Select
        Label8.Text = ""
    End Sub
End Class
```

Exercise 3.15

1) In a water treatment plant the responsible sanitary engineer designed a cone aerator for aeration of the water. According to the manyufacturer's specification the cone has an oxygentation capacity of 40 g oxygen per second. Determine the level of oxygen that could be transefered by the cone under working conditions at 24 ^{o}C given that the level of dissolved oxygen to be established is 1.6 mg/l.

2) Write a computer program to calculate the level of oxygen that could be transefered by a cone aerator for aeration of water under working conditions given temperature, level of dissolved oxygen to be established, and cone oxygentation capacity.

3) Verify your program by solving example 3.15a.

Solution

Oxygentation capacity defined at rate of oxygen transfer at 10 ^{o}C and 101.3 kPa and an oxygen concentration of zero is:

$$OC = k' \, C' \ (g/m^2.s^3)$$

Where:

C's = Oxygen saturation concentration in pure water at $10\ ^oC$ and

101.3 kPa

The oxygentation capacity OC (g/s) of an aeration system is:

$$OC = k'^2\ V C'^s$$

where:

V = volume of water, m^3
k_2 = overall gas transfer coefficient.
The influence of temperature is accounted for by:

$$(k_2)_{10}=(k_2)_T\sqrt{\frac{D_{10}}{D_T}}$$

Where correction factor has been deduced experimentally {17} as:

$$\sqrt{\frac{D_{10}}{D_T}}=1.0188[10^oC-T]$$

Given that OC = 40 gO_2/s and C = 1.6 g/m^3
Then,

$$OC^{24} = (k'^2)^{24}\ (C'^s - C).V$$

But

$$OC^{10} = (k'^2)^{10}\ (C'^s)^{10}\ .V = 30$$

Thus,

$$(k'^2)^{10}\ .V = 30/11.3 = 2.655$$

$$(k_2)_{10}=(k_2)_{24}\sqrt{\frac{D_{10}}{D_{24}}}$$

But,

$$\sqrt{\frac{D_{10}}{D_{24}}} = 1.0188^{10-24} = 0.7702$$

Then,

$$V.(k)^{2\ 24} = 2.655/0.7702 = 3.447$$

Thus,

$$OC^{24} = 3.447(8.5 - 1.6) = 23.8\ gO^2 /s$$

Exercise 3.4

Pressure

1) Determine the **depth** of water at which the oxygen partial pressure is 40kPa. Find the oxygen concentration at this depth given that the water is 80 percent saturated with oxygen at a temperature of 10° C. (Ans. 9.1 m, 17.3 mg/L).

Efficiency, power and oxygenation efficiency

2) The oxygen content of a groundwater is 1.8 g/m^3 and its temperature is recorded at 14° C. The responsible sanitary engineer propped aerating this water to be used as a supply to a nearby town. A cascade aerator has been selected and it consists of four steps each 0.35 m high. The oxygen concentration of the effluent of the system was found to be 8 mg/l. After some time of operation the cascade was badly damaged whence it was deemed necessary to rebuild it. The engineer suggested designing a cascade consisting of two steps and having a height equal to half that of the old.

a) What is the **efficiency** coefficient of the existing cascade?
b) What is the expected oxygen concentration of the **effluent** of the new cascade?

c) Compute **power** needed to aerate 8 m^3 per second of water given that the overall efficiency of pump and motor amounts to 85 percent.

d) Estimate energy **expenses** of this aeration knowing that the all-inclusive price of one kilowatt-hour equals 8 pts. (Ans. 0.3, 7.4 mg/L, 1297 kJ, L10).

3) A cascade aerator consisting of three identical steps each has a height of 0.3 m high increases the oxygen concentration of groundwater from 2 to 6 gram oxygen per m^3 The saturation concentration of the water is 9.7 mg/l. Due to severe damage to some of its parts, the cascade is to be reconstructed. To promote gas exchange phenomenon the rebuilt cascade consists of a single step 0.9 m high.

a) Determine the **efficiency** coefficient of the existing cascade?

b) Determine the **oxygen concentration** of the effluent of the rebuilt cascade?

   Find the **oxygenation efficiency** of both cascades. (Ans. 0.52, 6.5 mg/L, 0.57 mgO$_2$/J,

   0.07 mgO$_2$/J).

4) Groundwater with an oxygen content of 1.4 g/m^3 is to be used for consumption. A cascade aerator is to be used for the aeration process and it consists of two similar steps. The aerated water was noted to have an oxygen content of 7.5 g/m^3 at 18o C. The sanitary engineer responsible proposed improving the aeration

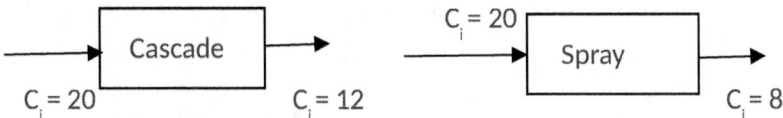

system by installing a third similar step. Compute the expected **oxygen content** at the outflow of the three step aerator. Quantify also the **oxygen content** at the outflow of the three step aerator if the groundwater is anaerobic. (Ans. 8.5, 8.3 mg/L).

Number of steps

5) For the aeration of groundwater in the Gezira area in an amount of 6.8 m^3 per second, a cascade aerator is to be used. The aerator is composed of a number of steps, each 0.4 m high. Each step is capable of increasing the oxygen content from 15 to 35 percent

of the saturation concentration. Determine the **number** of steps stipulated to raise the oxygen content from nil to 75 percent of the saturation value. (Ans. 6).

6) a) For what purpose is the aeration process used? b) For the aeration of groundwater in an amount of 396 m3/min., a cascade aerator will be used, consisting of a number of steps, each 0.3 m high. According to laboratory tests one step is able to raise the oxygen content from 10 to 30 percent of the saturation value. **Howmany steps** are required to raise the oxygen content from zero to 70 percent saturation? What are the energy costs of this aeration when the overall efficiency of pump and motor amount to 0.7 and the all inclusive price of one kwh equals to 10 pts. ? (Water temperature is 20°C.) (U of K, 1983). (Ans. 5, L333/d).

7) The oxygentation capacity OC of an aeration system is 20 g oxygen per hour. The aeration tank, which is completely mixed, is provided with aurafce aerator that consumes 400 kWh per day at steady conditions. The aeration tank bvomlume is 250 m3. This aeration system is being used for aeration of pure water at 10o C with an oxygen concentration of 3.5 mg/l.

    a. Calculate the maximum **flow** of water that can be aerated by this aeration system in order to acquire an effluent conmentration of 7.5 mgO$_2$/l/

    b. Determine the **detention** time of the water in the aeration tank under the prevailing conditions of the first part of the question.

    c. Compute the **oxygenation efficiency,** OE under standard conditions. (Ans. 1683 m^3/hr, 0.15 hr, 0.33 mg/J).

8) An aeration tank having a volume of 140 m^3 is equipped with a cone aerator. The system is checked with pure water at 10° C. Water flow is recorded as zero. When complete mixing has been assumed, the test results were as tabulated below:

Time (min)	Oxygen concentration (g/m^3)
0	3
8	8

a) Compute the anticipated **oxygen concentration** at time of 14 minutes.

b) Determine the **oxygenation capacity** of the system. (Ans. 9.7 mg/L, 3 gO$_2$/s).

9) "Use of aeration for well waters was questioned because of possible contamination by air borne impurities including pathogens". Discuss this statement. At what **depth** is the saturation concentration of oxygen 16.5 mg/L, given that the water is at a temperature of 20°C and that air at atmospheric pressure contains 21 percent of oxygen by volume? A cascade aerator of efficiency 50% is to be used for the aeration of 8 m^3 of ground water is at a temperature of 10°C, calculate:

    i)   The expected **oxygen concentration** at the overflow, if the groundwater is anaerobic.

    ii)   The **number** of steps required to increase the oxygen concentration of the water mentioned in (1) above to 75 percent of the saturation value. (Ans. 8m, 5.7 mg/L, 2).

10) a) In a village in the western part of Sudan deep ground water has been utilized for drinking purposes. Laboratory experiments indicated that this water contains Ammonia and a very large number of E .Coli. Indicate possible socio-economical ways for treating this water. What are the merits of the chosen processes?

b) For the aeration of deep ground water in El Fashir area in an amount of 10 m^3/*second*, a cascade aerator is to be used. The aerator is composed of a number of steps, each 0.4 m high. Each step is capable of increasing the oxygen content from 10 to 40 percent of the saturation concentration. Determine:

    i) The **number** of steps stipulated to raise the oxygen content from anaerobic conditions to 85 percent of the saturation value.

    ii) The required electric **power** when the efficiency of the pump is 70 percent and that of the motor is 90 percent. (U of K, 1984). (Ans. 5, 3114 kW).

11) Ground water from a catchment area contains 3.5 mg/l ironll (Fe++) and one mg/l NH4+. The water is treated by spray-aeration, rapid filtration and cascade aeration. The Iron-II and the ammonia are completely oxidized. Experiments have shown that the spray aerator can decrease the O$_2$-deficiency from 90% to 5%, whereas the cascade aeration can increase the O$_2$-content from zero to 50%. What will be the expected **oxygen content** of

the treated water, if the saturation concentration of oxygen at the prevailing temperature is 9 mg/l? (Given: 0.17 mg oxygen are needed to oxidize 1 mg Fe^{++}, while 3.6 mg oxygen are needed to oxidize 1 mg NH_4^+).(U of K, 1984). (Ans. 6.6 mg/J).

12) a) Why do you need to aerate groundwater that contains ammonia, iron & manganese before filtration? b) Groundwater of 8°C containing 2mg O_2/l and 40g CO_2/m^3 is aerated by means of spray aeration. The effluent oxygen concentration amounts to 8g O_2/m^3. Compute:

    i) The **effluent** CO_2 **concentration**, assuming spray aeration to be equally efficient for oxygenation & removal of CO_2.

    ii) The effluent **oxygen** & CO_2 concentration when the above effluent is passed over a second spraying stage of equal efficiency.

    In the case under consideration, the air contains 0.04% carbon dioxide with solubility in water of 2500mg/l per atmosphere. While the oxygen saturation concentration at the prevailing conditions is 11.9g oxygen per m^3. (U of K, 1985). (Ans. 16.4, 10.4, 7.1 mg/L).

13) Groundwater is used as a supply to a town in Darfour region. The laboratory analysis revealed that its oxygen content amounted to 1.8 mg/l while the recorded temperature is 14°C. The selected aeration system to improve the oxygen content is a 4-stem cascade, which was able to raise the oxygen concentration to 8mg/l & each step of the cascade is 0.35m high. This cascade was awfully damaged & it was deemed necessary to rebuild a new one. The responsible sanitary engineer proposed designing a 2-stem cascade with a height equal to that of the old one. Compute:

    a) **efficiency** coefficient of the existing cascade.

    b) the expected oxygen saturation of the **effluent** of the new cascade.

    c) resulting % change in **power** to aerate 8m^3 per second of water after building the new cascade given that the efficiency of pump is 90% while that of the motor is 92%.

    d) The energy **expenses** of the new aeration unit knowing that the all-inclusive price of 1 kWh equals 15 pts. (U of K, 1986). (Ans. 0.41, 7.4 mg/L, 133 kW, L20).

14) a) For what purpose is the aeration process used? Indicate examples and/or application of this process in the field of water purification. b) Laboratory analysis of a sample of ground water showed that it contains 4 mg/l Iron II and 1.2 pp Ammonia. The responsible sanitary engineer advocated treatment that incorporates spray aeration, rapid filtration & cascade aeration. The iron and ammonia are completely oxidized by the chosen system. Further experiments have indicated that the spray aerator can decrease the oxygen-deficiency from 90% to 5% of saturation value; whereas, the cascade aeration can increase the oxygen content from zero to 50% of saturation concentration.

i) What will be the expected **oxygen content** of the treated water, if the saturation concentration of oxygen at the water temperature is 9.2 g/m^3? (Given: O.17mg oxygen are needed to oxidize 1mg Iron II; 3.6mg oxygen are required to oxidize 1mg ammonia.)

ii) Give justification for the sanitary engineer's suggestion. (U of K, 1987).

(Atomic weights: Ca=40, C=12, H=1, Fe=56, Mg=24.3, N=14, O=16, C1=35.5) (Ans. 6.4 mg/L).

15) a) Briefly describe the factors that influence the saturation concentration of a gas in water.

b) In an experiment of oxygenation, the concentration of oxygen was determined in intervals of 2 minutes as stated in the following table: (U of K, 1988).

Time, t (minute)	Oxygen concentration, c (mg/l)
0	3.8
2	5.2
4	6.3
6	7.2
8	7.9
10	8.4
12	8.8
14	9.2

Assuming that the oxygen saturation concentration is taken as 10.5 mg/l, find:

i) The overall gas transfer **coefficient** (k_2)

ii) Whether the above mentioned assumption is valid? (Ans. 0.002 /s).

16) A cascade aerator is capable of lowering the carbon dioxide content of groundwater from 20 to 12 mg/l and a spray aerator is able to lower the carbon dioxide content of the same water to 8 g/m^3. When both aerators are used in series, with cascade first, the effluent gas concentration was found to be 5 mg/l. If percentage of carbon dioxide in air is 0.032, find the **solubility** of the gas in water per atmospheric pressure. (U of K, 1988). (Ans.2500 mg/L).

17) a) For what purpose is the aeration process used? b) A cascade aerator is used for the aeration of 500 m^3 ground water per hour with an oxygen content of 2.3 g/m^3. The aerated water has an oxygen concentration of 9.3 g/m^3 at 10°C.

   a) Calculate the expected oxygen **concentration** at the over flow if the oxygen concentration of the groundwater is zero.
   b) Calculate the required electric **power** when the loss of head in the aerator is 1.25m. The efficiency of the pump is 70 percent. The efficiency of the motor is 90 percent.
   c) Calculate the **oxygenation efficiency** (OE) in kgO2/kwh of the aerator at zero oxygen concentration, taking into account the efficiency of pump and motor. (U of K, 1983). (Ans. 8.8 mg/L, 2.7 kW, 1.6 kgO$_2$/kWh).

18) An aeration system is being used for aeration of pure water at 10oC with an oxygen concentration of 3.5 mg/L. The aeration tank has a volume of 250 m^3 and is completely mixed. This tank is provided with a surface aerator that consumes 400 kWh per day at steady conditions. Given that the oxygenation capacity OC of the aeration system is 20Kg O$_2$/hr, compute the following:

   i)   The maximum **flow** of water that can be aerated by this aeration system in order to acquire an effluent concentration of 7.5 mg O$_2$/L.
   ii)   The **detention time** of the water in the aeration tank under the prevailing conditions of the first part of the question.
   iii)   The **oxygentation efficiency** OE under the standard conditions. (U of K, 1984). (Ans. 1681 m3/hr, 0.15 hr, 0.33 kgO$_2$/Jmg/L).

19) Discuss the factors that influence the saturation concentration of a gas in water.

Determine the **saturation concentration** of oxygen in pure water at $10^{\circ}C$ exposed to air sat a pressure of 101.3 kPa. Assume that dry air contains 20.948 percent oxygen by volume. Determine also **Henry's** constant for these conditions.

Groundwater, at a temperature of $10^{\circ}C$, C_s = 11.3 mg/L, is aerated by means of bubble aeration to increase the oxygen content from 1.5 to 9.5 mg O_2/L. In the aeration tank air is diffused at a rate of 4 m^3 air per m^3 of water. The retention time of the water in the tank is 10 minutes. The initial carbon dioxide content amounts to 50 mg CO_2/L. Assuming complete mixing conditions in the aeration tank, determine:

   i. The aeration **coefficient**, k_2

   ii. The **oxygenation capacity** of the system. (U of K, 1987). (Ans. $7.4*10^{-3}$ /s, 0.08 gO^2/m^3.s).

20) Discuss the importance of aeration as a unit process in water works. An oxygenation experiment revealed the following results:

Time, t, seconds	Concentration of oxygen, C, ppm
0	3.8
120	5.2
240	6.3
360	7.2
480	7.9
600	8.4
720	8.8
840	9.2

Oxygen saturation concentration is assumed to be 10.5. For the given data:

1) Compute the overall gas transfer **coefficient**.

2) Derive the **relationship** between oxygen concentration and rate of absorption by transfer.

3) Show whether the mentioned assumption is justified. (U of K, 1988). (Ans. 0.002 /s).

3.4 Water Desalination
Example 3.16

a) Indicate merits and demerits of Reverse Osmosis process.
b) The molar concentration of the major ions in a brackish groundwater supply are as follows:

Na^+	0.02	Cl^-	Missing
Mg^{++}	0.005	HCO_3^-	0.001
Ca^{++}	0.01	NO_3^-	0.002
K^+	0.001	SO_4^{--}	0.012

Reverse osmosis process was used to desalinate the water. The osmotic pressure difference across a semi permeable membrane that had brackish water on one side and mineral-free water on the other is 26.9 psi (1.83 atm).

   a)   Assuming a temperature of 25°C, compute the missing chloride ion concentration for the conditions outlined.

   b)   If in (i) above, a yield of 80% fresh water were desired, what minimum pressure would be required to balance the osmotic pressure difference that will develop? (UAE, 1990).

c) Write a computer program to compute the missing ion concentration and minimum pressure that would be required to balance the developed osmotic pressure difference for a water smple given molar concentration of the major ions, osmotic pressure difference across the semi permeable membrane and percent yiedld desired.

d) Verify your program by solving example 3.16a.

Solution:

1)

Advantages	Disadvantages
- treatment of water from 700 ppm upwards	- need of pretreatment
- reduction of TDS (99%)	- device plugging
- rejection of bacteria, viruses	- colloidal fouling
- removal of dissolved organics	- metal oxide precipitation

| - removal of colloidal & biological matter | - biological fouling |
| - removal of turbidity | - expensive |

2) $P_{osm} = 1.83$ atm = CRT

i) R = 0.082 L.atm/k/mole,

T = 273.16 + 25 = 298.16 K

$\therefore$C = 1.83÷(0.082×298.16) = 0.075 M

= 0.02 + 0.005 + 0.01 + 0.001 + Cl$^-$ + 0.001 + 0.002 + 0.012

$\therefore$0.075 = Cl$^-$ + 0.051

$\therefore$ Cl$^-$ = 0.024 M

= 0.024×35.5 = 0.85 mg/L Cl$^-$

ii) Yield = 80%

The salts originally present in 5 volumes of brackish water would be concentrated in one volume of brackish water left behind in the membrane after four volumes of fresh water have passed throw membrane. $\therefore$ The particle concentration in remaining brackish water would be five times of that original brackish water, or:

C = 5×0.075 = 0.375 M

$\therefore$$P_{osm}$ = 0.375×0.082×298.16 = 9.2 atm

$\therefore$pressure required to push fresh water throw membrane would be in excess of P_{osm}.

Program 3.16 Listing:
Missing ion concentration and minimum pressure that would be required to balance developed osmotic pressure difference

```
'********************
'EXAMPLE 3.16
'********************
Public Class Form1
    Dim MW(17) As Double
    Dim Symbol(17) As String
    Const R = 0.082
```

```vb
Private Sub Form1_Load(ByVal sender As
    System.Object, ByVal e As
    System.EventArgs) Handles MyBase.Load
    loadSymbols()
    loadMWs()

    Me.Text = "Example 3.16"
    Me.FormBorderStyle =
        Windows.Forms.FormBorderStyle.FixedSingle
    Me.MaximizeBox = False

    Label1.Text =
        "Osmotic pressure diff., Posm (atm)"
    Label2.Text = "Temp. (C)"
    Label3.Text = "Enter component ions conc.,
            leave the missing one empty:"
    Label4.Text = "Na+"
    Label5.Text = "Mg++"
    Label6.Text = "Ca++"
    Label7.Text = "K+"
    Label8.Text = "Cl-"
    Label9.Text = "HCO3-"
    Label10.Text = "NO3-"
    Label11.Text = "SO4-"
    Label12.Text = ""
    Button1.Text = "&Calculate"

End Sub

Sub loadSymbols()
    'This subroutine loads an array of symbols.
    'It is useful as an index to the array of
    'molecular weights.
    'although it is not used in this particular
    'program, it is useful if one wants to
    'expand the program.
    Symbol(0) = "Al"
    Symbol(1) = "Ba"
    Symbol(2) = "Ca"
    Symbol(3) = "C"
    Symbol(4) = "Cl"
    Symbol(5) = "Cu"
    Symbol(6) = "H"
    Symbol(7) = "I"
    Symbol(8) = "Fe"
    Symbol(9) = "Mg"
    Symbol(10) = "N"
    Symbol(11) = "O"
    Symbol(12) = "P"
    Symbol(13) = "K"
```

```
        Symbol(14) = "Ag"
        Symbol(15) = "Na"
        Symbol(16) = "S"
        Symbol(17) = "Zn"
End Sub

Sub loadMWs()
     MW(0) = 27
     MW(1) = 137
     MW(2) = 40
     MW(3) = 12
     MW(4) = 35.5
     MW(5) = 63.5
     MW(6) = 1
     MW(7) = 126.9
     MW(8) = 56
     MW(9) = 24.3
     MW(10) = 14
     MW(11) = 16
     MW(12) = 31
     MW(13) = 39
     MW(14) = 108
     MW(15) = 23
     MW(16) = 32
     MW(17) = 65
End Sub

Private Sub Button1_Click(ByVal sender As
     System.Object, ByVal e As
     System.EventArgs) Handles Button1.Click
     Dim Posm, T, C As Double
     Dim total, missingConc As Double
     Posm = Val(TextBox1.Text)
     T = Val(TextBox2.Text)
     T += 273.16    'convert to kelvin

     C = Posm / (R * T)
     total = Val(TextBox3.Text)
     total += Val(TextBox4.Text)
     total += Val(TextBox5.Text)
     total += Val(TextBox6.Text)
     total += Val(TextBox7.Text)
     total += Val(TextBox8.Text)
     total += Val(TextBox9.Text)
     total += Val(TextBox10.Text)
     missingConc = C - total
     If Val(TextBox3.Text) = 0 Then
         missingConc *= MW(15)
                'Na is the missing ion
         Label12.Text = "Na conc. = " +
```

177

```
                    FormatNumber(missingConc, 2) + " mg/L"
            ElseIf Val(TextBox4.Text) = 0 Then
                missingConc *= MW(9)
                        'Mg is the missing ion
                Label12.Text = "Mg conc. = " +
                    FormatNumber(missingConc, 2) + " mg/L"
            ElseIf Val(TextBox5.Text) = 0 Then
                missingConc *= MW(2)
                        'Ca is the missing ion
                Label12.Text = "Ca conc. = " +
                    FormatNumber(missingConc, 2) + " mg/L"
            ElseIf Val(TextBox6.Text) = 0 Then
                missingConc *= MW(13)
                        'K is the missing ion
                Label12.Text = "K conc. = " +
                    FormatNumber(missingConc, 2) + " mg/L"
            ElseIf Val(TextBox7.Text) = 0 Then
                missingConc *= MW(4)
                        'Cl is the missing ion
                Label12.Text = "Cl conc. = " +
                    FormatNumber(missingConc, 2) + " mg/L"
            Else
                Label12.Text = "Missing conc. = " +
                    FormatNumber(missingConc, 2) + " M"
            End If
        End Sub
    End Class
```

Example 3.17

a) Write briefly about methods used for desalting sea brackish water.

b) What approximate osmotic pressure would be created across a semi permeable membrane if water containing 0.01 M Na_2SO_4, 0.01 M $MgCl_2$, and 0.03 M $CaCl_2$ were placed on one side of the membrane and distilled water were on the other side? (UAE, 1989).

c) Write a computer program to compute osmotic pressure that would be created across a semi permeable membrane given molar concentration of salts water contains when placed on one side of a membrane and distilled water placed on the other side.

d) Verify your program by solving example 3.17a.

Solution:

a) Distillation, solar evaporation, freezing, RO, electrodialysis, ion exchange.

b) $P_{osm} = CRT$

C = molar concentration of particles.

R = universal gas const. = 0.082 Latm/mole.k

T = temp. = 25°C assumed

= 25+273.16 = 298.16 k

Ions formed in water are:

$2Na^+$, SO_4^{--}, Mg^{++}, $2Cl^-$, Ca^{++}, $2Cl^-$

of concentrations:

0.01×2, 0.01, 0.02, 0.02×2, 0.03, 0.03×2

Respectively.

∴C = 0.01×2 + 0.01 + 0.02 + 0.02×2 + 0.03 + 0.03×2 = 0.18 M

P_{osm} = 0.18×298.16×0.082 = 4.4 atm

Theoretical exercises 3.4a

1) Answer any THREE of the following:

a) "The only effective way to keep sanitary wastewater and storm flows apart is by means of **separate** systems". Discuss this statement and outline major merits and demerits of a separate system.

b) "Each type of sewer pipe, its advantages and limitations should be evaluated carefully in the selection of **pipe material** for given applications". Discuss this statement and outline pipes that may be used beneficially in the UAE.

c) "**Solar distillation** could ideally suit distillation plants". Discuss this statement indicating advantages and limitations of this method.

d) "The fundamental thesis governing the disposal of effluents and the regulation of pollution is to make the **treatment plants** part of the work and to let nature complete it". Discuss this statement.

2) Briefly outline the most important "Desalination **methods**".

3) Raoult's law assumes: P_{osm} = C*R*T. Indicate **assumptions** involved in the law, and show how it is used for estimating osmotic pressure

4) Briefly outline "Desalination **methods**". Which method would you recommend for a town in the Middle East.

Practical exercises 3.4b

1) A semi-permeable membrane separates a mineral-free water from a water that contains the following impurities:

Impurity	Concentration, mg/L
Calcium Chloride, CaCl2	2
Magnesium Chloride, MgCl2	2
Sodium Sulfate, Na2SO4	3

Determine the **osmotic pressure** that would be required across

this semi-permeable membrane at a temperature of 20^oC. (SUST, 2001). (Ans. 10 atm).

2) A semi-permeable membrane separates a mineral-free water from a water that contains the following impurities:

Impurity	Concentration, mg/L
Calcium Chloride, CaCl2	1
Magnesium Chloride, MgCl2	1
Sodium Sulfate, Na2SO4	4

Determine the **osmotic pressure** that would be required across

this semi-permeable membrane at a temperature of 20^oC. (OIU, UNESCOC, 2000). (Ans. 7.8 atm).

3.5 Water Softening

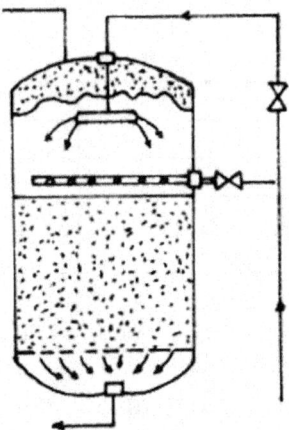

Water softening connotes the partial or total removal of water hardness by introduction of chemicals to the water concerned, or by ion exchangers.

Hardness is a tribulation to both homes and industrial sectors. It prevents soap lathering, creates undesirable scums and deposits scales in hot water installations. It is caused normally by calcium and magnesium salts, and also by salts of iron or aluminum, which react with soap to formulate typical scum observed around baths and wash basins.

Hardness can be divided into:
- Temporary hardness: due to carbonates or bicarbonates of calcium or magnesium.
- Permanent hardness: due to chlorides or sulfates of calcium or magnesium.

Softening of drinking water is not an essential requirement as far as hygiene situations are concerned. Not only that, but soft water is even suspected of promoting cardiovascular disease.

Example 3.18

- Laboratory tests on a sample of water gave the following tabulated results:

Determinations	Concentration, g/m^3
Ca^{++}	140
Mg^{++}	24.3
HCO_3^-	305
CO_2	12

Calculate the water total hardness, carbonate and noncarbonated hardness.
- Write a computer program to compute water total hardness, carbonate and noncarbonated hardness given concentrations of cations and anions.
- Verify your program by solving example 3.18a.

Solution

The concentrations in eq./m^3 can be found by using the formula eq/m^3 = (g/m^3)/equivalent weight, as seen from the table below:

Determinations	Concentration, g/m^3	Equivalent weight	Concentration, eq./m^3
Ca^{++}	140	20	7
Mg^{++}	24.3	12.15	2
HCO_3^-	305	61	5
CO_2	12		

Total hardness = carbonate hardness + non-carbonate hardness = 7 + 2 = 9 eq/m^3
Carbonate hardness = 5 eq/m^3
Non-carbonate hardness = 9 - 5 = 4 eq/m^3

Program 3.18 Listing:
Water total hardness, carbonate and noncarbonated hardness

```
'*****************************
'EXAMPLE 3.18: Water Hardness
'*****************************
Public Class Form1
    'define molecular weights for Ca, Mg, HCO3
    Const MWCa = 20
    Const MWMg = 12.15
    Const MWHCO3 = 61

    Private Sub Form1_Load(ByVal sender As
        System.Object, ByVal e As
        System.EventArgs) Handles MyBase.Load
        Me.Text = "EXAMPLE 3.18: Water Harness"
        Me.FormBorderStyle =
            Windows.Forms.FormBorderStyle.FixedSingle
        Me.MaximizeBox = False

        Label1.Text = "Select option:"
        Label2.Text = "Ca++"
        Label3.Text = "Mg++"
        Label4.Text = "HCO3-"
        Label5.Text = "CO2"
        Label6.Text = ""
        Button1.Text = "&Calculate hardness"
        ComboBox1.Items.Clear()
        ComboBox1.Items.Add("Conc. as g/m3")
        ComboBox1.Items.Add("Conc. as mg/L")
        ComboBox1.SelectedIndex = 0
    End Sub

    Private Sub Button1_Click(ByVal sender As
        System.Object, ByVal e As
        System.EventArgs) Handles Button1.Click
        Dim totalHard, carbHard, nonCarbHard As Double
        Dim concCa, concMg, concHCO3, concCO2 As Double

        concCa = Val(TextBox1.Text)
        concMg = Val(TextBox2.Text)
        concHCO3 = Val(TextBox3.Text)
        concCO2 = Val(TextBox4.Text)

        Select Case ComboBox1.SelectedIndex
            Case 0  'conc. as g/m3
                'Total hardness = carbonate hard +
                'non-carb hard
                totalHard = (concCa / MWCa) +
```

```vb
                    (concMg / MWMg)
            carbHard = concHCO3 / MWHCO3
            nonCarbHard = totalHard - carbHard
            Label6.Text = "Total hardness = " +
                totalHard.ToString + " eq/m3"
            Label6.Text += vbCrLf
            Label6.Text += "Carb. hardness = " +
                carbHard.ToString + " eq/m3"
            Label6.Text += vbCrLf
            Label6.Text += "Non-Carb. hardness = "
                + nonCarbHard.ToString + " eq/m3"
        Case 1  'conc. as mg/L
            'Total hardness = carbonate hard +
            'non-carb hard
            totalHard = ((concCa / MWCa) * 50) +
                ((concMg / MWMg) * 50)
            carbHard = (concHCO3 / MWHCO3) * 50
            nonCarbHard = totalHard - carbHard
            Label6.Text = "Total hardness = " +
                totalHard.ToString + " mg/L CaCO3"
            Label6.Text += vbCrLf
            Label6.Text += "Carb. hardness = " +
                carbHard.ToString + " mg/L CaCO3"
            Label6.Text += vbCrLf
            Label6.Text += "Non-Carb. Hardness = "
                + nonCarbHard.ToString
                + " mg/L CaCO3"
    End Select
End Sub

Private Sub ComboBox1_SelectedIndexChanged(ByVal
        sender As System.Object, ByVal
        e As System.EventArgs)
    'clear the fields whenever selection is changed
    TextBox1.Text = ""
    TextBox2.Text = ""
    TextBox3.Text = ""
    TextBox4.Text = ""
End Sub
End Class
```

184

Example 3.19 (see Program 3.18 Listing)

Given the tabulated analytical results for a water sample:

Determination	Concentration, mg/L
Ca^{++}	80
Mg^{++}	12.15
HCO_3^-	305
CO_2	12

Compute the water total hardness, carbonate and noncarbonated hardness expressed as mg/L $CaCO_3$ of this water sample.

Solution

Concentrations expressed as mg/L $CaCO_3$ can be computed from the equations:

mg/L $CaCO_3$ =mg/L x (50/equivalent weight) = meq. X 50

Determination	Concentration, mg/L	Equivalent weight	Concentration, meq./L	Concentration, mg./L as $CaCO_3$
Ca^{++}	80	20	4	200
Mg^{++}	12.15	12.15	1	50
HCO_3^-	305	61	5	250
CO_2	12	222	0.55	28

Total hardness = 200 + 50 = 250 mg/L $CaCO_3$

Carbonate hardness = 250 mg/L $CaCO_3$

Non-carbonate hardness = 250 - 250 = 0

Example 3.20

Raw water with analysis as tabulated below is to be softened by the lime-soda process. Calculate the quantity of lime required for softening knowing that 90 percent of the lime commercially available is pure.

Analysis	Concentration, g/m^3
Calcium	140
Magnesium	24.3
Bicarbonate	305
Carbon dioxide	12

Also compute total hardness and noncarbonated hardness.

Solution
Concentration could be expressed in eq/m^3 as.

Determinations	Concentration, g/m^3	Equivalent weight	Concentration, eq./m^3
Ca^{++}	140	20	7
Mg^{++}	24.3	12.15	2
HCO$_3^-$	305	61	5
CO$_2$	12		0.55

Total hardness = 7 + 2 = 9 eq/m^3 = 450 mg/L CaCO$_3$
Carbonate hardness = 59 eq/m^3 = 250 mg/L CaCO$_3$
Non-carbonate hardness = 9 - 5 = 4 eq/m^3 = 200 mg/L CaCO$_3$
Basic reactions for the lime-soda process using hydrated lime are:
a) $CO_2 + Ca(HCO_3)_2 \rightarrow CaCO_3\downarrow + H_2O$
b) $Ca(HCO_3)_2 + Ca(OH)_2 \rightarrow 2CaCO_3\downarrow + 2H_2O$
A bar diagram of the raw water is as shown below:

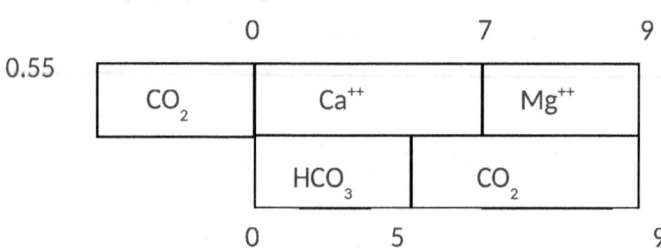

Amount of hydrated lime needed could be calculated as:
Lime = (CO2 + Ca(HCO$_3$)$_2$) g/m^3, in this case.
Therefore, Lime = 0.55 + 5 = 5.55 eq./m^3
Excess lime should be used to compensate solubility of calcium carbonate etc.
Therefore, use 215 g/m3 of lime, but since only 90 percent of lime is pure, then amount of lime from market = 215x100/90 = 239 g lime/m^3

Exercise 3.5

1) Discuss the significance of water **hardness**. A water sample was taken for analysis. The laboratory investigations revealed that it contains 100 mg/L calcium as Ca^{++}, 240 mg/L of bicarbonates as HCO_3^- and 52 mg/L of sulfates as SO_4^{++}. No other ions were detected in the water. What is the concentration of **Ca^{++}** expressed as mg/L calcium carbonate? How much **lime** $Ca(OH)_2$ and **soda ash** Na_2CO_3 need to be added in iron der to react with all the hardness? (Ans. 220, 59 mg/L).

2) Make up a **bar diagram** in terms of calcium carbonate for a water sample with the following tabulated analysis:

Analysis	Concentration, g/m^3
Calcium	101
Magnesium	4.75
Sodium	14
Bicarbonate	220
Sulfate	88.4
Chloride	21.3

Compute the **total hardness, the carbonate** and **noncarbonated** hardness of the water. Evaluate the amount of **lime** $Ca(OH)_2$ required to soften this water if the lime commercially available is 85 percent pure. (Ans. 272, 181, 92 mg/L $CaCO_3$, 218 mg/L).

3) Analysis of a hard water sample revealed that the total hardness is 280 mg/l as CaO, Magnesium hardness is 12.5°F, alkalinity of 300 mg/l, and acidity of 15 mg/l as $CaCo_3$. Compute the amount of **lime** needed for **softening**. (U of K, 1988). (Ans. 326 mg/L).

4) Surface water has the following analysis:
   calcium 82 mg/L as Ca^{++}, Magnesium 61 mg/L as Mg^{++}, Sodium 9.2 mg/L as Na^+, bicarbonate 366 mg/L as HCO_3^- and sulfate 120 mg/L as SO_4^{++}, chloride 35.5 mg/L as Cl^-.

   1. Calculate the number of **milliequivalents** per litre of each substance.

   2. Compute the **total, carbonate** and **non-carbonate** hardness of the water expressed as mg/L calcium carbonate.

3. Draw a **bar diagram** of the surface water.

4. How much **lime** is required to cater for all the hardness of the water? (Ans. 455, 300, 155 mg/L $CaCO_3$, 368 mg/L).

5) Surface water was analyzed in the laboratory and the analysis performed indicated that the water has a total hardness of 400 mg/L expressed as calcium carbonate; magnesium hardness was found to be 165 mg/L as calcium carbonate, and the amount of carbon dioxide present equals 10 mg/L expressed as CO_2. Determine the right quantities of **soda ash and hydrated lime** needed to soften the water if it is known that the water total alkalinity amounts to 250 mg/L expressed as calcium carbonate. (Ans. 210, 159 mg/L).

6) Plot a **bar diagram** of a water sample with the following characteristics:

Analysis	Concentration, g/m^3
Ca^{++}	84
Mg^{++}	62.2
Na^+	335.5
SO_4^{--}	148.8
HCO_3^-	305
Cl^-	42.6
CO_2	15

Also determine the amount of total, **carbonate** and **noncarbonated hardness**. How much **commercially** available **lime** will be needed to react with all the hardness if it is 90% pure? (Ans. 465, 275, 190 mg/L $CaCO_3$, 347 mg/L).

7) Analysis of a hard water sample revealed that the total hardness is 250 mg/L as calcium carbonate, calcium hardness 130 mg/L as calcium carbonate, alkalinity 200 mg/L as calcium carbonate and the acidity is 0.3 mequivalents/L. Compute the amount of **lime and soda ash** needed for softening this water.

8) a) Indicate different methods that may be used for water **softening**. b) Would you recommend water softening for a village in the United Arab Emirates? Give reasons.

c) Analysis of a sample of water showed the following results:

Item	Concentration
Ca^{++}	200 mg $CaCO_3$/L
Mg^{++}	18.225 mg Mg^{++}/L
HCO_3^-	219.6 mg HCO_3^- /L
Free CO_2 and carbonic acid	66 mg CO_2/L

Find the **total, carbonate** and **non-carbonate** hardness for this water sample.

- Plot the **bar diagram** for the water
- Determine the required **chemical dosages to soften** this water as much as possible without removing magnesium
- What will be the final **hardness** of the water. (Ans. 275, 180, 95 mg/L $CaCO_3$, 244, 21 mg/L).

9) a) What are the advantages and disadvantages of water **softening**?

b) The laboratory analysed of a sample of water medicated the following results:-

Analysis	Concentration
Suspended solids	100 mg/L
Total hardness	252 mg coo/L
Magnesium hardness	48.6 mg Mg/L
CO_2	10 mg CO_2/L
Total alkalinity	300 mg $CaCO_3$/L

I) Construct the **bar diagram** of this water.

II) Compute the necessary quantities of **hydrated lime and soda ash** needed to soften this water, knowing that the **commercially** available lime is 90 percent pure.

[Ca=40, Mg=24.3, H=1, C=12,O=16, Na=23, Cl=35.5, S=32, Fe=56]. (Ans. 370, 177 mg/L).

10) Indicate the merits & demerits of water **hardness**. Surface water was analyzed in the laboratory & the analysis performed indicated the following:

Mg^{++} & Ca^{++} hardness	448 mg/l CaO
Mg^{++}	165 mg/l $CaCO_3$
CO_2	10 mg/l CO_2
Total alkalinity	450 mg/l $CaCO_3$
pH	6.8

a) Compute the **calcium hardness** in $^\circ$D.

b) Plot the **bar diagram** of this water indicating the cation-anion balance.

c) Compute the right quantities of **commercially available soda ash & hydrated lime** needed to soften this water knowing that the chemicals are 90% pure.

d) Indicate whether this water is **aggressive** to asbestos-cement pipes. (U of K, 1986). (Ans. 36 $^\circ$D, 12.3).

11) a. What are the advantages and disadvantages of water **softening**?

b. Indicate two chemical methods to be used for water softening with illustrations of reactions involved.

c. Would you recommend water softening for a village in rural Sudan? Give reasons.

d. A village in the Rahad scheme utilizes water for domestic consumption from a nearby canal. The laboratory analysis of a sample of the water indicated the following results:

Analysis	Concentration
Suspended solids content	100 mg/l
Total hardness	252 mg CaO/l
Magnesium hardness	48.6 mg Mg/l
CO_2	10 mg CO_2/l
Total alkalinity	300 mg $CaCO_3$/l
Iron	48.8 mg Fe^{++}/l

i) Construct the **bar diagram** of this water.

ii) Compute the necessary quantities of **Soda ash and hydrated lime** needed to soften this water, knowing that the **commercially** available lime is 90 percent pure. (U of K, 1986).

12) a) Indicate the merits and demerits of water **hardness**. Give examples for process application.

b) Surface water was analyzed in the laboratory and the analysis performed revealed the following results:

Total alkalinity	450 mg/l as $CaCO_3$
Acidity	10 mg/l as CO_2
Mg^{++} & Ca^{++} hardness	448 mg/l as CaO
Mg^{++}	165 mg/l as $CaCO_3$
pH	6.7

i) Compute the **calcium** hardness in ppm & $^\circ$D.

ii) Plot the **bar diagram** for this water.

iii) Indicate how you would remove the hardness & compute concentrations of **chemicals** to be used.

190

iv) Is this water **aggressive** to asbestos-cement pipes? (U of K, 1987). (Ans. 350 mg/L, 12.2).

13) a) Discuss the significance of water **hardness**.

b) An analysis of hard water revealed the following data:

The total hardness	250 mg/l as $CaCO_3$
Calcium hardness	130 mg/l as $CaCO_3$
Alkalinity	200 mg/l as $CaCO_3$
Acidity	0.3 mequivalent/L

i) Construct the **bar diagram** of this water.

ii) Compute the amount of **lime and soda ash** (in mg/l) needed for softening this water.

(Ca = 40, O = 16, H = 1, Mg = 24.3, Na = 23, Cl = 35.5, S = 32, C = 12)

c) Indicate whether this water is **aggressive** to asbestos-cement pipes (Use an appropriate corrosion potential and assume the hydrogen ion concentration to be 2.5×10^{-7} mole/l). (U of K, 1985). (Ans. 196, 53 mg/L 11).

3.6 Disinfection

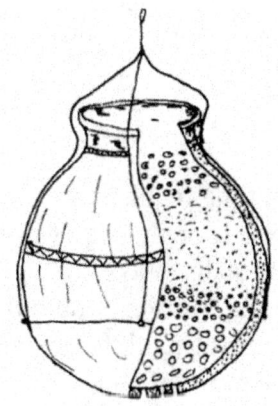

Disinfection is the destruction of pathogenic microorganisms that cause diseases. It differs from sterilization which is the destruction of all microscopic life.

Disinfection could be applied by physical methods such as heat treatment, pasteurization, ultra-violet radiation and metal ions such as silver and copper. Disinfection may also be applied by chemical methods by introducing oxidants to disinfect the water such as chlorine, ozone, chlorine dioxide, bleaching powder, iodine, etc.

An effective chemical disinfectant should inherit certain characteristics such as being:
- Effective and rapid in killing pathogens.
- Easy to dissolve in water in concentration needed for disinfection.
- Able to provide a residual.
- Inactivate as far as odor, color or taste of the water is concerned.
- Easy detection and measurement in water.
- Non-toxic to human and animal life.
- Easy to detect, quantify, handle, transport, use and control.
- Inexpensive.

The disinfection process depends on several factors such as the type and concentration of organisms and disinfectant utilized, existence of oxidant consuming substances, temperature, pH, dose and time. Each species of microorganisms has a different sensitivity to each disinfectant.

Example 3.21

1   Compute the contact time required to achieve a 99.99 percent kill for disinfecting a microorganism system, given that the rate constant (to base 10) equals 0.06 sec^{-1}.
2   Write a computer program to compute the contact time required to achieve a a certain percent kill for disinfecting a microorganism system, given that the rate constant.
3   Verify your program by solving example 3.21a.

Solution

Chick's law assumes that the distribution of microorganisms is controlled by process of diffusion, and it states that:

$$-\frac{dN}{dt} = k'N$$

where:

N =   Number of viable microorganisms of one type .

t =   Contact time.

K' =   Rate constant.

Integration for $N = N_o$ at $t = o$ yield,

$$\log \frac{N}{N_o} = -k't$$

Or $N/N_o = e^{-k't}$

$N_o =$ Number of viable microorganisms of one type at time $t = 0$.

Given that $\dfrac{N}{N_o} = \dfrac{100 - 99.99}{100} = 10^{-4}$

And k' to base 10 equals 0.06 sec-1 then Contact time, $t = -\dfrac{1}{k'}\log \dfrac{N}{N_o}$

$$contact\ time = t = -\frac{1}{k'}\log \frac{N}{N_o} = -\frac{1}{0.06}\log 10^{-4} = 67\ seconds$$

Program 3.21 Listing:

Contact time to achieve a certain percent kill for disinfecting a certain microorganism system

```
'*********************************
'Example 3.21: Chick's Law
'*********************************
Public Class Form1

    Private Sub Form1_Load(ByVal sender As
        System.Object, ByVal e As
        System.EventArgs) Handles MyBase.Load
        Me.Text = "Example 3.21: Chick's Law"
        Me.FormBorderStyle =
            Windows.Forms.FormBorderStyle.FixedSingle
        Me.MaximizeBox = False

        Label4.Text =
            "This example will use Chick's law to"
        Label4.Text += vbCrLf + "calculate time needed
            to achieve percentage killing"
        Label1.Text =
            "Percentage of kill to be achieved"
```

```
            Label2.Text = "Rate constant (k)"
            Label3.Text = ""
            ComboBox1.Items.Clear()
            ComboBox1.Items.Add("sec-1")
            ComboBox1.Items.Add("min-1")
            ComboBox1.SelectedIndex = 0
            Button1.Text = "&Calculate time"
    End Sub

    Private Sub Button1_Click(ByVal sender As
        System.Object, ByVal e As
        System.EventArgs) Handles Button1.Click
        Dim N, No, k, t As Double
        No = Val(TextBox1.Text)
        k = Val(TextBox2.Text)
        N = (100 - No) / 100
        t = -(1 / k) * (Math.Log10(N))
        If ComboBox1.SelectedIndex = 0 Then
            Label3.Text = "Contact time = " +
                FormatNumber(t, 2) + " seconds"
        Else
            Label3.Text = "Contact time = " +
                FormatNumber(t, 2) + " minutes"
        End If
    End Sub
End Class
```

Example 3.22 (see Program 3.21 Listing)

1) The rate constant (to base 10) for chlorination of *E. coli* is 0.6 min^{-1} for free chlorine residual. What retention time would be needed to obtain 99.9 percent kill?
2) Write a computer program to compute the retention time that would be needed to obtain a certain percent kill for *E. coli* given rate constant.
3) Verify your program by solving example 3.22a.

Solution

By using Chick's as defined above, then the retention time needed by the disinfectant to attain the desired percentage of kill:

$$T = -\frac{1}{k'}\log\frac{N}{N_o} = -\frac{1}{0.6'}\log 10^{-3} = 5\,minutes$$

194

Example 3.23

1) Given that for chlorine as a disinfectant and a 99 percent kill, Chick's law could be liberalized and written in the form:

$C^{0.86}$ t = 0.24 for *E. coli*
Where:
C = concentration of disinfectant.
T = contact time between disinfectant and *E. coli*.
Find the contact time required by these organisms to achieve a kill of 99 percent for a chlorine concentration of 15 mg/L.

2) Write a computer program to compute contact time required to disinfect a certain type of organisms given percent kill required, and chlorine concentration.

3) Verify your program by solving example 3.23a.

Solution

Using the given equation, then contact time

$$T = \frac{0.24}{15^{0.86}} = 0.02 \, seconds$$

Exercise 3.6

1) Discuss significance of usage of **chlorine** in water and wastewater treatment installations. (SQU, 1992).

2) Groundwater is to be utilized for consumption by rural inhabitants in El-Gezira area. The responsible water engineer suggested using bleaching powder to disinfect the water prior to consumption. Compute the **contact time** required to accomplish 99 percent kill for the disinfection of microorganisms present given that the rate constant (to base 10) equals 0.04 sec^{-1}. (Ans. 50 s).

3) For *E. coli*, evaluate the **contact time** required to achieve a 99 percent kill for chlorine doses of: a) 0.06 g/m^3. B) 0.1 mg/L. c) 6 ppm. And d) 10 g/m^3. (Ans. 1.7, 0.05, 0.03 s).

4) Compare the **contact times** required for chlorine to kill 99.99 percent of certain species of microorganisms in water for free residual chlorine of 0.3 g/m^3, given that the rate constants are 10^{-1} and 10^{-4} seconds^{-1} respectively. (Ans. 9, 283 s).

5) In a water treatment plant the production is 25000 m^3/day. 10 kg/day of chlorine has been used to disinfect the water. Calculate the chlorine **dosages** in mg/ per litre, and the chlorine **demand** of the water if the chlorine residual after 10 minutes contact time is 0.2 mg/L. (Ans. 0.4, 0.2 mg/L).

6) Bleaching powder is used to disinfect 20000 m^3 of water daily. The bleaching powder used contains 35 percent available chlorine. The chlorine required is 0.4 mg/L to maintain a residual of 0.2 mg/L. Compare the amount of **bleaching powder** that is needed to disinfect the water. (Ans. 23 kg/d).

7) A test of chlorine demand on a raw water sample gave the following tabulated results:

Chlorine applied, mg/L	Residual chlorine, mg/L
1	1.0
2	2.0
3	2.9
4	3.5
5	4.2
6	4.1
7	3.1
8	2.0
9	1.1
10	1.4
11	2.1
12	3.4

Plot the chlorine **demand** and find the **breakpoint dosage**. Find also the chlorine **demand** at a dosage of 6, 8 and 12 mg/L.

8) Comment about merits and demerits of **chlorination** of water. For a mixed industrial-residential city of 100,000 people the average daily water consumption is 660 Lpcd. Calculate the number of kg **chlorine** needed per day in a water treatment plant supplying the city. The chlorine demand is 1 mg/L and a free available chlorine concentration of 0.2 mg/L is to be provided. The treatment plant must be capable of operating at the maximum daily flow-rate (Hint: maximum daily rate = 1.5 x average daily rate).

9) The following data was obtained in a chlorination experiment.

Dosage (mg/L)	1.00	2.00	3.00	4.00	5.00	6.00	7.00
Residual (mg/L)	0.80	1.55	1.95	1.25	0.50	0.85	1.95

     i)   Plot the data.

     ii)   What **dosage** is required to provide a free residual of 1 mg/L? (OIU, UNESCOC, 2004).

10) a) What are the requirements for an effective **disinfectant**? (8 marks)

    a)   Which method of disinfection will you recommend for a refugee camp? State your reasons.

    b)   A test of chlorine demand on a raw water sample showed the following tabulated results: (U of K, 2001).

Residual chlorine, mg/L	Chlorine applied, mg/L
1.0	1.0
2.0	2.0
2.9	3.0
3.5	4.0
4.2	5.0
4.1	6.0
3.1	7.0
2.0	8.0
1.1	9.0
1.4	10.0
2.1	11.0
3.4	12.0

- Plot the chlorine-demand **curve**.
- Determine the **break-point dose**.
- What is the chlorine **demand** at a dose of 8.5 mg/L?.

    c)   How much **SO_2** (g) has to be dosed to remove the Cl_2 from a flow of water containing 2 ppm of Cl_2? (U of K, 1988). (Ans. 1.8 mg/L).

11) A solution of hydrofluosilicic acid (H_2SiF_6) which contains 170 g of (H_2SiF_6) per litre is to be used to dose a flow of 200 L/s. the water naturally contains 0.3 mg/L of F^-, but it is required that the treated water have a fluoride content of 1 mg/L calculate the rate of feed of **hydrofluosilicic acid** required. (UAE, 1989). (Ans. 90 L/d s).

12) Outline possible parallel **competitive reactions** when chlorine is added to water. Chloroform, $CHCl_3$, may be prepared in the laboratory by the reaction between chlorine and methane:

$$3Cl_2 + CH_4 \longrightarrow CHCl_3 + 3HCl$$

Calculate the number of grams of **chlorine** that are required to produce 45 grams of $CHCl_3$. (SQU, 1991).

13) Pathogens decrease exponentially when exposed to a disinfectant. The decrease is described by the Chick's law: $Ln(N_t/N_0) = -kt$

Where: N_0 = initial number of pathogens (cells/L)
N_t = number of pathogens after time t (cells/L)
t = duration of disinfection
k = empirical constant (t^{-1})

The following data were obtained in a disinfection experiment dealing with inactivation of poliovirus (Floyd et al., 1978. Environ. Sci. Technol. 12: 1031-1035):

Time (sec)	N/N_0
4	1/13
8	1/158
12	1/2000

Plot $Ln(N/N_0)$ versus time, calculate the k value and find the time required for a 1/5000 reduction of poliovirus (B.Sc. UoD, 2014)

3.7 Water Stabilization

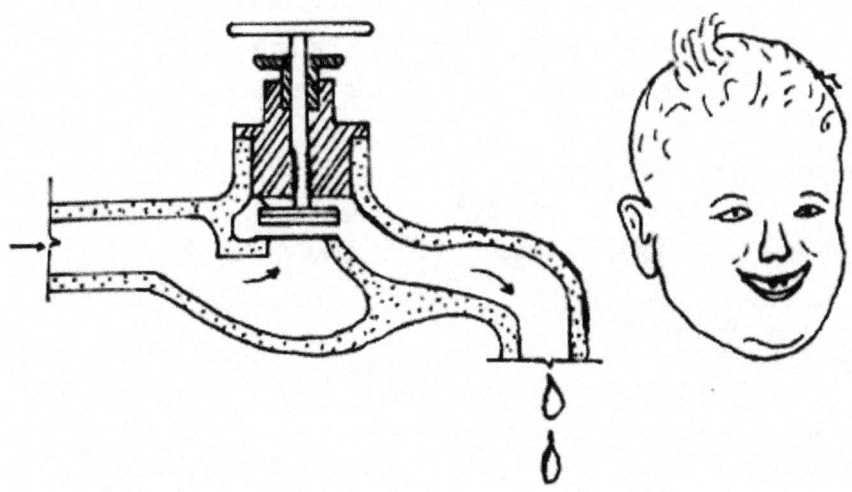

Introduction

Water stabilization is a process that involves chemical and bacteriological operations of significance to water quality when it is being transported to consumers.

Water, in an unstable state, could either be aggressive or precipitative. In the former case there is an excess of carbon dioxide that can affect concrete and even asbestos cement by dissolving calcium carbonate. Corrosiveness of water creates the problem of lead dissolution in mains, etc. And in the latter case water lacks carbon dioxide.

Example 3.24

1. Laboratory tests of a sample of drinking water gave the following composition:

Determination	Value
Calcium	60 mg/L
Bicarbonate	180 mg/L
pH	7.9

Indicate whether this water is aggressive or precipitative?
If it is aggressive how much is the amount of aggressive carbon dioxide?

2. Write a computer program to indicate whether a cerain sample of water is aggressive or precipitative. Let the program also deterimene the amount of aggressive carbon dioxide.

3. Verify your program by solving example 3.21a.

Solution

To use Tillman's curve, the ratio between calcium and bicarbonate concentration should be equal to 1:2. Checking this then,

$$\frac{Ca^{++}}{HCO_3^-}=\frac{60}{40}:\frac{180}{61}=1.5:3=1:2$$

Therefore, Tillman's curve could be used.

From Tillman's curve for HCO_3^- = 180 mg/L and pH of 7.4: carbon dioxide is found to be 12 mg/L. Therefore, this water is aggressive.

The amount of aggressive carbon dioxide could be found by plotting from the point of HCO_3^- = 180 mg/L and pH = 7.4, a parallel line to XB and horizontally it gives a value of carbon dioxide of 7.4

Thus, amount of aggressive CO_2 = 12 – 7.4 = 4.6 mg/L

Example 3.25

Drinking water was found to have the following constituents:

Total salt content	500 mg/L
Bicarbonate	61 mg/L
Total hardness	3.6 mg/L
Magnesium	12 mg/L
pH value	7.5
Temperature	20° C

Indicate whether this water is aggressive or not.

Solution

Bicarbonate hardness = 1 meq/L

Magnesium = 12/24 = 0.5 m mole

Therefore, Calcium = 1.8 – 0.5 = 1.3 m mole

$$\frac{Ca^{++}}{HCO_3^-}=1.3:1$$

Thus Tillman's curve can not be used, but the problem could be solved by resorting to the Langelier-Hoover nomograph.

pHs = (pk$_2$ – pk$_s$) + pCa+ pAlk

Where:

pH$_s$: value of the pH water should when there is equilibrium with solid calcium carbonate.

pCa = - Log (Ca)   (Ca moles/L)

pAlk = - Log alkalinity (alkalinity in equivalents)

pKs = - Log [ca^{++}][CO$_3^-$]

And a saturation index indicated:

pHa – pHs = SI

Where:

pH$_a$ = actual or measured pH

SI = saturation index

When SI is greater than zero, water is said to be precipitative and

When SI is smaller than zero, water is said to be aggressive.

Alkalinity may be found as:

$$alkalinity = 100\left[\frac{1}{2}[HCO_3^-]\right]+CO_3^- +\frac{1}{2}[OH^-]=100\,[\frac{1}{2}x\,1]+0+0$$

= 50 ppm CaCO$_3$

[ca^{++}] = 1.3 x 40 = 52 mg/L

From the Langelier-Hoover graph for TDS of 500 mg/L, calcium of 52 mg/L and alkalinity = 50 ppm, then:

pH$_s$ = 8.3

Therefore, saturation index, SI = 7.5 - 8.3 = -0.8 a value lower than zero and in this case the water is aggressive.

Exercise 3.7

1) A sample of river water was analyzed in the laboratory and the test results indicated the following composition:

Determination	Value
Calcium	80 mg/L
Bicarbonate	244 mg/L
pH	7.36

State whether this water is **aggressive** or not.

If it is aggressive compute the concentration of aggressive **carbon dioxide**. (Ans. Aggressive, 1.4 mg/L).

2) The tabulated results have been obtained by the sanitary engineer when analyzing a sample of water:

Determination	Value
Total salt content	600 mg/L
Bicarbonate	122 mg/L
Calcium	80 mg/L
pH value	7.8
Temperature	30° C

Find whether this water is **aggressive or precipitative**. (Ans. 1 precipitative).

3) A water sample has the following composition: total salts content 500 mg/L, bicarbonate 60 ppm, magnesium 12 ppm, total hardness 3.6 meq/L, pH of 7.5 and the water temperature is 20° C. Indicate whether this water sample is **aggressive or precipitative**.
Define each of the flowing corrosion potential indicators: **Langelier, Ryznar** and **Aggressiveness.** (DU, 2012). (Ans. − 0.8, aggressive).

4) Define the **terms** used in the equation for determining the aggressiveness index:
AI = pH + Log_{10} [Alk]*[HCa] (DU, 2012).

Laboratory analysis results for three water samples gave the results shown in table (1). Using equation of aggressiveness index and classification of aggressiveness index presented in table (2), find which sample of water is more **aggressive** to asbestos-cement pipes. Show your computations and state your reasons. (DU, 2012). (Ans. Sample 1, 9.9).

Table (1): Laboratory analysis results.

Item	Sample (1)	Sample (2)	Sample (3)
pH		7.1	
hydrogen ion concentration, mol/L	$2.1*10^{-6}$		$2.75*10^{-9}$
Alkalinity,	140 mg/L as $CaCO_3$	3 meq/L	207.4 mg HCO_3^-/L
Calcium hardness	120 mg/L as $CaCO_3$	64 mg Ca^{++}/L	2 meq/L

Table (2): Aggressiveness index description.

AI value	Description
Less than 10	Highly aggressive
10 to 11.9	Moderately aggressive
Greater than 12	Non-aggressive

5) Analysis of a water sample revealed the following data:
1) Calcium hardness = 4 mg/L as Ca^{++},
2) Alkalinity = 240 mg/L as $CaCO_3$,
3) Hydrogen ion concentration of $4.7*10^{-7}$ mol/L.

Determine whether this water is **aggressive** to asbestos-cement pipes using the aggressiveness index classification system presented in the following table (DU, 2012). (Ans. 9.7, highly aggressive).

Value	Description
Less than 10	Highly aggressive
10 to 11.9	Moderately aggressive
Greater than 12	Non-aggressive

6) Analysis of a water sample revealed the following data:
- Calcium hardness = 140 mg/L as $CaCO_3$,
- alkalinity = 180 mg/L as $CaCO_3$,
- hydrogen ion concentration = $2.5*10^{-7}$ mol/L.

Using the aggressiveness index (AI = pH + Log_{10} (Alk)*(HCa)) determine whether this water is **aggressive** to asbestos-cement pipes. (DU, 2011). (Ans. 11.9 moderately aggressive).

AI value	Description
Less than 10	Highly aggressive
10 to 11.9	Moderately aggressive
Greater than 12	Non-aggressive

7) What is saturation index (SI) as defined by **Langelier** for the equilibrium of $CO_2 - CaCO_3$? Explain how this index can be used to indicate **aggressiveness or precipitativeness** of water in asbestos cement or concrete pipes.

Drinking water has the following composition(U of K, 1983).

Total salt content	500 mg/L
HCO_3^-	60 mg/L
Mg^{++}	12 mg/L
pH	7.3

Total hardness	3.6 meq/L
Temperature	20°C

Is this water aggressive or precipitative? Why? (Ans. − 0.8 aggressive).

8) The tabulated results below have been obtained by a sanitary engineer while analyzing a sample of water

Total salt content	500 mg/L
Water temperature	25°C
Bicarbonate concentration	213.5 mg/L
Calcium content	80 ppm
Turbidity	150 NTU

Find whether this water is aggressive or precipitative? What solutions do you recommend for this situation? (U of K, 1984). (Ans. 0.45 precipitative).

General Problems

a) Solve the following: (U of K, 1988).

1) Air of atmospheric pressure contains 21% of oxygen by volume. Determine the **saturation concentration** of oxygen in water at a depth of 5m and 30°C temperature. (Ans. 19.3 g/m^3).

2) **Hindered** settling seldom occurs in drinking water practice because: …………………….

3) Filters are to be used for treating water in an amount of 2880 m3/hr. Compute the **number** of slow sand filters required. If rapid sand filters are to be preferred, find their needed number. (Ans. 14, 11).

4) A filter of bed thickness 1.7 m is composed of anthracite with a pore space of 40% and grain size of 1.2 mm. Find the **porosity** after backwashing at 20% expansion. (Ans. 50 %).

5) The lethal dose LD$_{50}$ of cyanide for man is 70 mg. The ADI of cyanide is 0.05 mg/Kg body weight. Estimate the **maximum permissible concentration** in drinking water of the compound in mg/L. (assume a daily intake of 3 liters of water per person, an average weight of 60 kg for a person and a contribution of 10% to the daily intake of the compound by water consumption). (Ans. 0.1 mg/L).

6) **Tillmans'** curve is restricted to waters with the following composition: (Ans. $[Ca^{++}]:\{HCO_3^-\} = 1:2$).

7) Give examples of materials that give the water an **"apparent" color.** (Ans. algae, silt, clay, $Fe(OH)_3$, MnO_2).

8) Possible parallel **competitive reactions** that may occur when chlorine is added to water include: ....

9) The contact time required to achieve a certain kill for a disinfectant microorganism system with a rate constant (to base 10) of $5x10^{-2}$ per second is 80 seconds. Determine the percent **kill**. (Ans. 99.99%).

10) Give examples of **de-chlorinating** agents.

b) A consultant engineer selected a water source to supply a town with the needed demand. The recorded population of the town is 66173 with a population growth rate of 2.5%. The laboratory analysis indicated the following data:

parameter	measurement
Temperature	20° C
pH	6.6
Suspended solids matter	100 mg/L
Carbon dioxide	12 mg/L
Carbonate (CO_3^{--})	0
Bicarbonate (HCO_3^-)	305 mg/L
Calcium (Ca^{++})	72 mg/L
Magnesium (Mg^{++})	48.8 mg/L
Iron (Fe^{++})	3.5 mg/L
Sodium (Na^+)	9.2 mg/L
Sulfate (SO_4^{--})	134.4 mg/L
Chloride (Cl-)	7.1 ppm
Total alkalinity	250 mg $CaCO_3$/L
Ammonia (NH_4^+)	0.5 mg/L
Turbidity	10 NTU
Coliform MPN	100 /100 mL
Oxygen saturation concentration at 20° C	9 mg/L
Dynamic viscosity at 20° C	$1.002x10^{-3}$ Ns/m^2

In order to supply the inhabitants with an amount of 200 L/capita/day of potable and palatable water, the consultant engineer proposed treatment by using screens, cascade aeration, sedimentation, rapid filtration, water conditioning and post chlorination. Further experimental analysis indicated the following:

Cascade aerator can decrease the oxygen deficiency from 90% to 5% of the saturation value.

0.17 mg oxygen are required to oxidize 1 mg Fe^{++}

3.6 mg oxygen are needed to oxidize 1 mg ammonia.

The results of a batch settlement test are as tabulated below:

Sampling depth (m)	Sampling time (hr)	% suspended removed from sample
1	1	39
	3	57
	6	70
2	1	36
	3	44
	8	64
3	1	35
	2	37
	6	49

Total surface area of sedimentation tanks needed is 1000 m².

Sand has been selected as a filter media with the flowing characteristic:

Particle diameter = 0.6mm.

Porosity = 40%

Uniformity coefficient = 1.3

Shape factor = 0.92

For each rapid filter used, the underneath values were taken:

Bed thickness = 1 m

Supernatant water layer = 1.5m

Filtration rate = 15 m³/m²/hr

Chlorine concentration is 20 ppm to achieve 99 % kill for *E.coli* (Chick's law could be modified and written as:

$C^{0.86} t = 0.24$ Where: C = concentration of disinfectant in mg/L, t = contact time between disinfectant and *E. coli* in seconds).

Given that the treatment units are required to be installed for a design period of 20 years, you are asked to compute:

a) Expected **oxygen content** of the treated water.

b) Suspended **solids content** of the effluent from the sedimentation tanks.

c) **Number** of **filters** to be applied.

d) Unit **area** of **filters** to be used.

e) **Head loss** through the **filters**.

    f) **Contact time** required by *E.coli* to achieve a kill of 99% for the given chlorine concentration.

    g) **Carbonate** and **non-carbonate hardness** of the water in mg/L $CaCO_3$.

    h) By constructing a **bar diagram** for the water amount of lime needed to soften the water if lime used is pure.

    i) The **aggressiveness** index of the water and indicate whether this water is aggressive to asbestos-cement pipes.

c) A Daniel cell is discharging at 100 mA, the rate of corrosion of zinc rod anode amounted to 1×10^{-6} mm/s. Given that the anode is 0.5 cm radius, find the **depth of immersion** in the electrolyte to yield the indicated corrosion rate. (Take density of zinc as 7000 kg/m^3 and zinc electrochemical equivalent to be 0.3387 mg/C). (U of K, 1988). (Ans. 15 cm).

d) Fill in the blank spaces:

    1. Spherical discrete particles of O.05mm diameter & specific gravity of 2.6, when settle in water of viscosity 1.003 x 10-6 m^2/s, will have a **settling velocity** of .... m/s. (Ans. 2.17×10^{-3} m/s).

    2. A well with a water depth of 3m, has a diameter of 1.5m. The amount of bleaching powder needed for disinfection at a **dose** of 50 mg/l is .... (Assume amount of available chlorine in bleaching powder to be 36%). (Ans. 736 g).

    3. In a water treatment plant the production is 25000 m^3/day. If 10 kg/day chlorine has been used to disinfect the water, the **chlorine dosage** of the water will be ....mg/l. (Ans. 0.4 mg/l).

    4. In the above problem the chlorine **demand** of the water will be .... mg/l, when the chlorine residual after 10 minutes contact time is 0.2mg/l. (Ans. 0.2 mg/l).

    5. The **ratio** of the molar concentration of OH- ions in a solution at pH of 12 to that at pH of 10 is equal to .... (U of K, 1986). (Ans. 100:1).

e) Write briefly about THREE of the following: (DU, 2012).

    a) Potential sources of water **pollution &/or infection**.

    b) Objectives of water treatment.

c) Types of **diseases** related to water and environmental sanitation.
d) Water **resources** available for public use and affecting factors.
e) Factors affecting **selection** of a water **source**.
f) Objectives of the **legislation**, laws, provisions, guidelines, standards, and orders

Chapter Four

Water supply, storage and distribution

4.1 Water supply
Example 4.1
1) Outline most important factors affecting productivity of a well (B.Sc., DU, 2013).

Solution
Most important factors affecting productivity of a well:
- Lowering of groundwater within aquifer (drawdown aspects).
- Dimensions of aquifer & its lateral extent.
- Ground water storage.
- Transmissivity & specific yield or storage coefficient of aquifer.
- Conditions of flow (steady or unsteady).
- Depth of well.
- Establishment of well & methods of construction, properties & condition.

2) A well of diameter 0.3 m contains water to a depth of 50 m before pumping commences. After completion of pumping the draw-down in a well 20 m away is found to be 5 m, while the draw-down in another well 40 m further away reached 3 m. For a pumping rate of 2500 L/minute, determine:
      i) radius of zero draw-down.
      ii) coefficient of permeability, and
      iii) draw-down in the pumped well (B.Sc., DU, 2013).

3) Using expected well yield estimates as presented in table (4.1), comment about yield of the well as related to its diameter, & suggest a more suitable well diameter. Explain and validate your answer(B.Sc., DU, 2013).

$$Q_o = \frac{\pi k \left(H^2 - h_o^2 \right)}{Ln \dfrac{R}{r_o}}$$

Table (4.1) Expected well yield.

Well diameter, cm	Expected well yield, m^3/d
15	< 500
20	400 to 1000
25	800 to 2000
30	2000 to 3500
35	3000 to 5000
40	4500 to 7000
50	6500 to 10000
60	8500 to 17000

4) Write a computer program to compute radius of zero draw-down, coefficient of permeability, draw-down in the pumped well and yield of the well as related to its diameter given well diameter, water depth before pumping, draw-downs in two observation wells at given distarnces from the pupmped well after completion of pumping and pumping rate.

5) Verify your program by solving example 3.21a.

Solution

- Given: D= 0.3 m, H = 50 m, r_1 = 20 m, x_1 = 5 m, r_2= 40 m, x_2= 3 m, Q_o = 2500

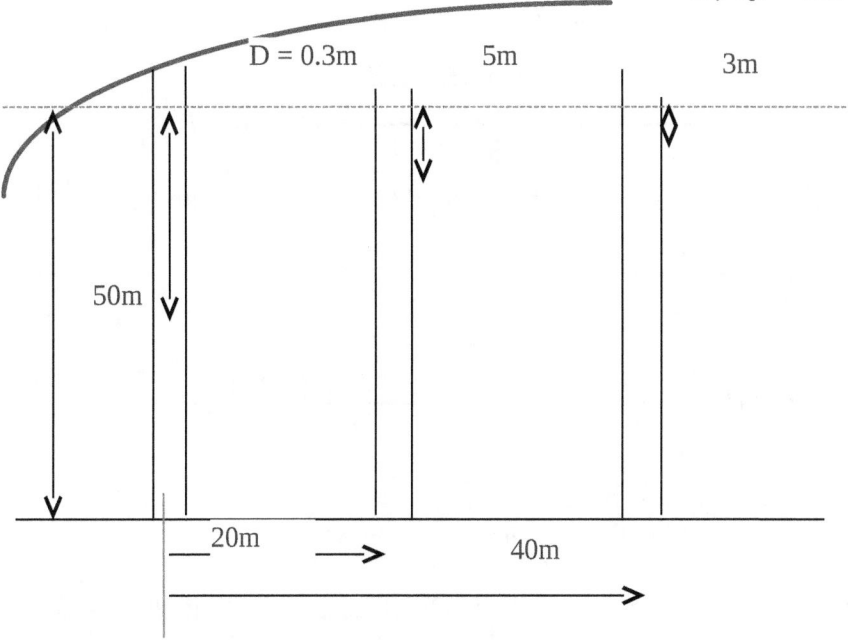

Find $h_1 = h - x_1 = 50 - 5 = 45$ m, and $h_2 = h - x_2 = 50 - 3 = 47$ m.
Use the following equation for both observation wells:

$$Q_o = \frac{\pi k \left(H^2 - h_0^2 \right)}{Ln\dfrac{R}{r}}$$

$$\therefore \left[\frac{\pi k \left(H^2 - h_0^2 \right)}{Ln\dfrac{R}{r_o}} \right]_{istwell} = \left[\frac{\pi k \left(H^2 - h_0^2 \right)}{Ln\dfrac{R}{r_o}} \right]_{2ndwell}$$

By substituting given values into the previous equation, then:

$$\left[\frac{\left(15^2 - 13.7^2 \right)}{Ln\dfrac{R}{40}} \right]_1 = \left[\frac{\left(15^2 - 14.3^2 \right)}{Ln\dfrac{R}{90}} \right]_2$$

- This yields R = 119.7 m.

- Find the permeability coefficient by using the data of one of the wells.

    Thus, for h = 50 m, ho = 45 m, r = 20 m, R = 119.7 m, Q = 2500*10⁻³*60*24 = 3600 m³/day,

$$k = \frac{QLn\frac{R}{r_o}}{\pi\left(H^2 - h_0^2\right)} = \frac{4320\,Ln\frac{242.29}{40}}{\pi\left(15^2 - 13.7^2\right)} = 66.36\,m/d$$

- Depth of the water in the pumped well may be found as:

$$Q_o = \frac{\pi k\left(H^2 - h_o^2\right)}{Ln\frac{R}{r_o}} \quad \text{Or}$$

$$h_0^2 = H^2 - \frac{Q}{\pi k}Ln\frac{R}{r} = 15^2 - \frac{1576.8}{\pi x\,24.9}Ln\frac{254.9}{0.15}$$

    = 75.09
    This yields, h = 27 m
    Determine the draw-down at the well as:
    r_o = h - h_1 = 50 - 27 = 23 m.

- From table for a well diameter of 30 cm, yield is 2000 – 3500 m³/d.

    The amount of water abstracted is Q_o = 2500 L/min = 2500*60*24/1000 = 3600 m³/d. A better design would be selected a well of diameter of 35 cm (giving a yield of 3683 m³/d for computed drawdown sat well (O.k. between 3000 - 5000).

$$Q_o = \frac{\pi \times 4.32\left(50^2 - 27^2\right)}{Ln\frac{119.7}{(0.35/2)}} = 3683 \quad \text{cubic meter per day}$$

Program 4.1 Listing:
Radius of zero draw-down, coefficient of permeability, draw-down in the pumped well and yield of the well as related to its diameter

```
'***************************
'EXAMPLE 4.1: Groundwater
'***************************
Public Class Form1

    Private Sub Form1_Load(ByVal sender As
        System.Object, ByVal e As
        System.EventArgs) Handles MyBase.Load
        Me.Text = "Example 4.1"
        Me.FormBorderStyle =
            Windows.Forms.FormBorderStyle.FixedSingle
        Me.MaximizeBox = False
        Label1.Text = "Well's diameter, d (m)"
        Label2.Text = "Well's depth, H (m)"
        Label3.Text = "r1 (m)"
        Label4.Text = "x1 (m)"
        Label5.Text = "r2 (m)"
        Label6.Text = "x2 (m)"
        Label7.Text = "Qo (L/m)"
        Label8.Text = ""
        Button1.Text = "&Calculate"
    End Sub

    Private Sub Button1_Click(ByVal sender As
        System.Object, ByVal e As
        System.EventArgs) Handles Button1.Click
        Dim diameter, H, r1, x1, r2, x2, Qo, h1, h2
            As Double
        Dim R, a, b, c, d, k, h0 As Double

        diameter = Val(TextBox1.Text)
        H = Val(TextBox2.Text)
        r1 = Val(TextBox3.Text)
        x1 = Val(TextBox4.Text)
        r2 = Val(TextBox5.Text)
        x2 = Val(TextBox6.Text)
        Qo = Val(TextBox7.Text)
        h1 = H - x1
        h2 = H - x2

        'calculate R
        a = (H ^ 2) - (h1 ^ 2)
        b = Math.Log(r1)
```

```
      c = (H ^ 2) - (h2 ^ 2)
      d = Math.Log(r2)
      R = Math.E ^ (((a * d) - (c * b)) / (a - c))
      Label8.Text = "R = "
            + FormatNumber(R, 2) + " m"
      'find permeability coefficient
      Qo = Qo / 1000 * 60 * 24
      k = (Qo * Math.Log(R / r1)) /
         (Math.PI * ((H ^ 2) - (h1 ^ 2)))
      Label8.Text += vbCrLf
      Label8.Text += "k = "
            + FormatNumber(k, 2) + " m/d"
      Label8.Text += vbCrLf
      'find depth of pumped well
      h0 = Math.Sqrt((H ^ 2) - ((Qo / (Math.PI * k))
         * Math.Log(R / (diameter / 2))))
      Label8.Text += "Water depth, h1 = " +
         FormatNumber(h0, 2) + " m"
      Label8.Text += vbCrLf
      Label8.Text += "Drawdown at well = " +
         FormatNumber(H - h0, 2) + " m"
   End Sub
End Class
```

Example 4.2

1) In your opinion, what is the most significant source of groundwater pollution in this country? Explain why is groundwater contamination so difficult to detect and clean up? (6 Marks)

2) A well penetrates into an unconfined aquifer having a saturated depth of 100 m. The discharge is 250 litres per minute at 12 m drawdown. Assuming equilibrium flow conditions and a homogeneous aquifer, estimate the discharge at 18 m drawdown. The distance from the well where the drawdown influence are not appreciable may be taken to be equal for both cases. (UAE, 1990).

3) Write a computer program to compute the discharge at a certain drawdown given unconfined aquifer saturated depth, the discharge at a certain drawdown and that the distance from the well where the drawdown influence are not appreciable may be taken to be equal for both case.

4) Verify your program by solving example 4.2a.

Solution:

- – Pollutants.   - Difficulty.

$$Q = \frac{\pi k \left(H^2 - h_1^2\right)}{\ln \dfrac{R}{r_o}}$$

$$A_1 = 250 = \frac{\pi k \left[100^2 - (100 - 12)^2\right]}{\ln \dfrac{R}{r_o}}$$

$$\frac{\pi k}{\ln \dfrac{R}{r_o}} = \frac{250}{(100 + 88)(100 - 88)} = \frac{250}{188 \times 12}$$

$$Q_2 = \frac{\pi k}{\ln \dfrac{R}{r_o}}\left[100^2 - (100 - 12)^2\right] = \frac{250}{188 \times 12}\left[100^2 - 82^2\right]$$

$$\frac{250}{188 \times 12} \times 182 \times 18 = 363 \, L/min$$

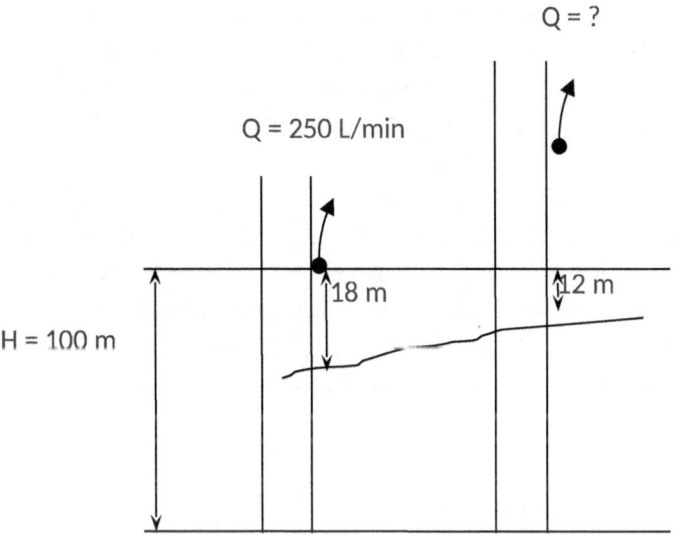

Q = ?

Q = 250 L/min

18 m

12 m

H = 100 m

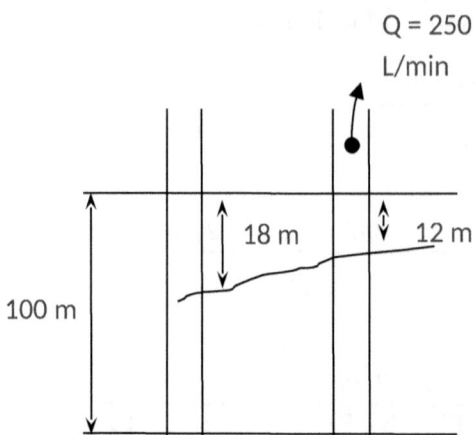

Example 4.3

a) Indicate factors that may influence yield of wells.
b) A 0.3 m diameter well penetrates vertically through an aquifer to an impervious strata which is located 18 m below the static water table. After a long period of pumping at a rate of 1 m^3/min., the drawdown in test holes 14 and 40 m from the pumped well is found to be 2.62 and 1.5 m, respectively.
   i)   Determine coefficient of permeability of the aquifer.
   ii) What is the transmissibility of the aquifer?
   iii) Compute the specific capacity of the pumped well. (UAE, 1990).
c) Write a computer program to compute coefficient of permeability, transmissibility of an aquifer and the specific capacity of the pumped well given well diameter, aquifer depth below the static water table, pumping rate, drawdown in two test holes and their distances from from the pumped well.
d) Verify your program by solving example 4.3a.

Solution:

1) r, Q, H, T, specific yield storage, lateral extent of aquifer, well

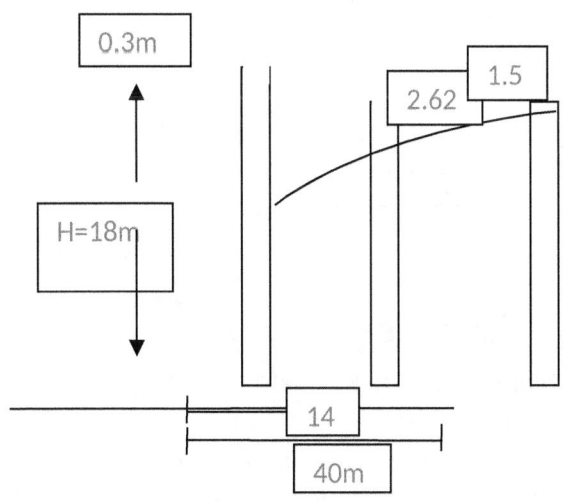

2) $Q = \dfrac{\pi k \left(H^2 - h_1^2 \right)}{\ln \dfrac{R}{r_o}}$

$\dfrac{\pi k \left(18^2 - [18 - 2.62]^2 \right)}{\ln \dfrac{R}{20}} \left(first\ well \right) = \dfrac{\pi k \left(18^2 - [18 - 1.6]^2 \right)}{\ln \dfrac{R}{40}}$

$\ln \dfrac{R}{40} = 0.5917 \ln \dfrac{R}{14}$

$R^{0.40627} = \dfrac{40}{14^{0.5917}} = 8.3926 \rightarrow R = 183.21\ m$

i] $k = \dfrac{Q \ln \dfrac{R}{r}}{\pi \left(H^2 - h^2 \right)} \left(first well \right) = \dfrac{\dfrac{1}{60} \ln \dfrac{133.21}{14}}{\pi \left(18^2 - 15.38^2 \right)} = 1.56 \times 10^{-4}\ m/s$

    $= 13.5$ m/d

ii] $T = kH = 13.5 \times 18 = 243$ m^3/m.d

iii] depth in well

$$h^2 = H^2 - \frac{G}{\pi k} \ln \frac{R}{r} = 18^2 - \frac{1}{60 \, \pi x \, 1.56 \, x \, 10^{-4}} \ln \frac{183.21}{0.15}$$

$= 82.28$

$h = 9.1$ m

drawdown (r) = H −h = 18-9.1 = 8.9m

iv] specific capacity = $\dfrac{G}{r} = \dfrac{1}{60 \, x \, 8.9}$ m²/s

Exercise 4.1

1) Define parameters that may influence **yield** of wells.

A 30 cm well is pumped at the rate of Q m3/minute. At observation wells 15 m and 30 m away the draw downs noted are 75 and 60 cm, respectively. The average thickness of the aquifer at the observation wells is 60 m and the coefficient of its permeability amounts to 26 m/day.

   i) Find the coefficient of **transmissibility** of the aquifer?
   ii) Determine the **rate of pumping** Q.
   iii) Compute **specific capacity** of the well.
   iv) What is the **drawdown** in the pumped well? (UAE, 1989). (Ans. 65 m²/hr, 1.5 m³/min, 0.73 m²/min, 2 m).

2) Define the term "**Transmissibility** of aquifer".

A 30 cm well penetrates 45 m below the static water table. After a long period of pumping at a rate of 1200 Lpm, the drawdown in the wells 20 and 45 m from the pumped well is found to be 3.8 and 2.4 m respectively.

   1. Determine the **transmissibility** of the aquifer.

   2. What is the **drawdown** in the pumped well? (UAE, 1989). (Ans. 7.1 m²/hr, 13.5 m).

3) a) Outline most important factors affecting **productivity** of a well.

b) A well of diameter 0.3 m contains water to a depth of 50 m before pumping commences. After completion of pumping the draw-down in a well 20 m away is found to be 5 m, while the draw-down in another well 40 m further away reached 3 m. For a pumping rate of 2500 L/minute, determine:

      i)   radius of **zero draw-down**.

     ii) coefficient of **permeability**, and

    iii) **draw-down** in the pumped well.

c) Using expected well yield estimates as presented in table (2.1), comment about **yield** of the well as related to its diameter, & suggest a more suitable well diameter. Explain and validate your answer. (Ans. 120 m, 4.3 m/d, 23 m).

$$Q = \frac{\pi k \left(H^2 - h^2 \right)}{\operatorname{Ln} \dfrac{R}{r}}$$

4) Outline main differences between confined and unconfined aquifer.

Define parameters shown in **Theims** equation:

$$S_o = \frac{Q_o}{2\pi kH} \ln \frac{R_o}{r} = \frac{Q_o}{2\pi T} \ln \frac{R_o}{r}$$

A fully-penetrating well, with an outside diameter of 50 cm, discharges a constant 3 m³/min from an aquifer whose coefficient of transmissibility is 0.03 m²/s. The aquifer is in contact with a lake 2 km away & has no other source of supply.

    i.   Estimate the **drawdown** at the well surface. (Take R_0 as twice distance between aquifer & lake).

   ii. Using expected well **yield** estimates as presented in table (1), comment about yield of the well as related to its diameter, & suggest a more suitable well diameter. Explain your answer.

  iii. Give a suggestion for **type** of aquifer **soil** with assuming an aquifer thickness of 20 m. (Ans. 2.6 m, 35 cm, fine gravel).

5) Comment about **Theim's** equation (equilibrium equation) assumptions.

Define parameters shown in Theims equation:

$$S_o = \frac{Q_o}{2\pi kH} \ln \frac{R_o}{r} = \frac{Q_o}{2\pi T} \ln \frac{R_o}{r}$$

A fully-penetrating well, with an outside diameter of 0.3 m, discharges a constant 4 m³/min from an aquifer whose coefficient of transmissibility is 1.4 m²/min. The aquifer is in contact with a lake 1.5 km away & has no other source of supply.

  i.   Estimate the **drawdown** at the well surface. (Take R_0 as twice distance between aquifer & lake).

 ii.   Using expected well **yield** estimates as presented in table (1), comment about yield of the well as related to its diameter, & suggest a more suitable well diameter. Explain your answer.

iii.   Give a suggestion for **type** of aquifer **soil** with assuming an aquifer thickness of 25 m. (Ans. 4.5 m, 40 cm, coarse sand).

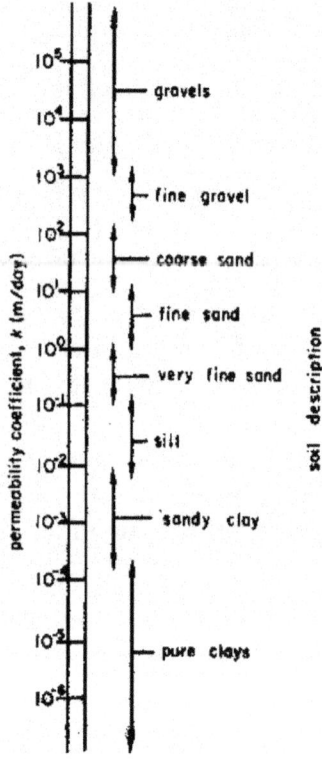

Figure 4.1 Soil descriptions with permeability coefficient

6) Define factors that may influence **yield** of wells.

The static level of water table in an unconfined aquifer was 30 m above the underlying impermeable stratum. A 150 mm diameter well, penetrating the aquifer to its full depth, was pumped at the rate of 20 litres per second. After several weeks of pumping, the drawdown in observation wells 20 m and 50 m from the well were 3.5 m and 2 m respectively, and the observed drawdowns were increasing very slowly.

   a) Assuming equilibrium conditions, estimate the **hydraulic conductivity** and **transmissivity** of the aquifer.
   b) Estimate the **drawdown** just outside the pumped well.
   c) What will be the **yield** of a 300 mm diameter well which will produce the same drawdowns just outside the well and at the 50 m distance observation well in (2)? What would be the **drawdown** at a nearer observation well? (Ans. 7.1*10⁻⁵ m/s, 185 m²/d, 2m).

7) Define terms used in the equation used to estimate rate of constant pumping from a well penetrating an unconfined aquifer

$$Q_o = \frac{\pi k \left(H^2 - h_o^2 \right)}{Ln \dfrac{R}{r_o}}$$

A well of diameter 30 cm penetrates an aquifer, water depth in it 15 meters before pumping. When pumping is being done at a rate of 3000 liters per minute, the drawdowns in two observation wells 40 and 90 meters away from the well are found to be 1.3 and 0.7 meters, respectively.

   i.   Find radius of **zero draw down**.
   ii.   Determine the coefficient of **permeability**, and
   iii. Compute **drawdown** in the well.
   iv.   Give a suggestion for **type** of aquifer **soil** for the estimated permeability.
   v.   Comment on **yield** of well compared to its diameter. Give suggestions for improvement? (Ans. 242 m, 66 m/d, 6.3 m, coarse sand, 4322 m³/d).

221

8) Define factor that may influence **yield** of wells.

The static level of water table in an unconfined aquifer was 33.5m above the underlying impermeable stratum. A 150mm diameter well, penetrating the aquifer to its full depth, was pumped at the rate of 25 litres per second. After several weeks of pumping, the drawdown in observation wells 20m and 50m from the well were 3.55m and 2.27m respectively, and the observed drawdowns were increasing very slowly.

1. Assuming equilibrium conditions, estimate the **hydraulic conductivity** and **transmissivity** of the aquifer.
2. Estimate the **drawdown** just outside the pumped well.
3. What will be the **yield** of a 300mm diameter well which will produce the same drawdowns just outside the well and at the 50m distance observation well in (ii)? What would be the **drawdown** at a nearer observation well? (Ans. 6.4 m/d, 216 m²/d, 13 m, 2.3 m).

9) a] What is the difference between a water table well and artesian well? (Draw sketches). b] a 30.0 cm well serves a community of 1100 capita with a water consumption rate of 250 l/day. Under steady-state conditions the drawdown in the well was 1.83 m. The well penetrates to an impermeable stratum 32 m below the water table of a homogeneous-isotropic unconfined aquifer. Compute the change in the **discharge** in l/s for a well drawdown of 1.83 m if the diameter of the well was: a) 20.0 cm; and b) 50 cm. Assume that the radius of influence in all cases is 760 m. (SQU, 1991).

10) A community of 15,000 capita and fire demand of 35 l/s for six hours is to be served by a 50 cm diameter well. The well is constructed in a confined aquifer with a uniform thickness of 15 m and hydraulic conductivity of 100 m/d. Two observation wells are installed at radial distance of 50 m and 150 m. The drawdowns in the wells are 1.7 m and 1.3 m respectively. Find:

a) the **discharge** of the well.
b) water **consumption** (l/c/d).
c) the **power** needed to lift the water to the

ground surface if the original piezometer level is 20 m below the ground surface. (SQU, 1991).

4.2 Water Storage
Example 4.4

1) Outline benefits of a mass curve (flow duration curve or Rippl diagram).
2) A water reservoir is designed to collect water from the adjacent catchment basin & to regulate water use across an average regular flow of 230 cubic meters per minute. The table below shows the monthly records of the stream flow. Find amount of **storage** needed to keep up with regular consumption assuming no loss of water (B.Sc., DU, 2013).

Monthly records of stream flow.

Month	Volume of water, million cubic meter
January	8
February	65
March	45
April	35
May	25
June	12
July	2
August	3
September	9
October	45
November	67
December	77

3) Write a computer program to compute amount of **storage** needed in water reservoir to keep up with regular consumption given average regular flow and monthly records of the stream flow.
4) Verify your program by solving example 4.4a.

Solution

- Data: regular consumption of 230 m³/min, & data of monthly water flow.
- Find cumulative total flow as shown in the following table:
- Draw mass curve for data by plotting cumulative flow values as a variable with time.

Month	Volume of water (million cubic meter)	Cumulative Volume of water (million cubic meter)
1	8	8
2	65	73
3	45	118
4	35	153
5	25	178
6	12	190
7	2	192
8	3	195
9	9	204
10	45	249
11	67	316
12	77	393

- Find the value of the annual use rate (for the month of December) =
  230 (m³/min) × 60 (minutes/hour) × 24 (hours/day) × 365 (day/year) = 120.89×10^6 m³/year = 121 million m³/year.
- Draw draft line of uniform use from the point of origin to the point (a) on the mass curve.
- Draw a line parallel to the draft line from the point where the reservoir is full (b), & then find the value of minimum required storage for the reservoir to keep pace with consumption = 20×10^6 m³.

Exercise 4.2

1) Outline benefits of a mass curve (flow duration curve or **Rippl** diagram).

   A water reservoir is designed to collect water from the adjacent catchment basin & to regulate water use across an average regular flow of 230 cubic meters per minute. Table (5) shows the monthly records of the stream flow. Find amount of **storage** needed to keep up with regular consumption assuming no loss of water. (Ans.121 Mm^3/year).

Table (5) Monthly records of stream flow.

Month	Volume of water, million cubic meter
January	8
February	65
March	45
April	35
May	25
June	12
July	2
August	3
September	9
October	45
November	67
December	77

2) Define **mass curve** "Rippl diagram". What information does it provide?

The peak water consumption on the day of maximum water usage as follows:

Time	L/s	Time	L/s
Midnight	220	13	640
1	210	14	630
2	180	15	640
3	140	16	640
4	130	17	670
5	120	18	740
6	200	19	920
7	350	20	840
8	500	21	500
9	600	22	320
10	640	23	280
11	700	Midnight	220
Noon	660		

 i.   Calculate hourly cumulative consumption values.

 ii. Plot (i) to a mass diagram **curve**.

 iii. What is the constant 24 hour **pumping rate**.

iv.   Compute required **storage** capacity to equalize demand over the 24 hour period.

4.3 Water distribution
Example 4.5

a)The following expression is used in analysis of pipe network: h = k*Qn. Define the terms used in the expression and show unit of measurement of each term.

b)Water enters the single loop shown in figure (4) at point A at the rate of 400 L/s and is delivered at B, C and D at the rate of 150, 100 and 150 L/s. All pipes are 0.6 m in diameter with a friction coefficient of 0.0312 and their lengths are AB and CD 150 m, BC 300 m AND DA 240 m. Determine the flow through each pipe and the pressures at B, C and D if that at A is 105 kN/m². (Hint assume flow along pipe AB = 260 L/s) (SQU, B.Sc., 1995).

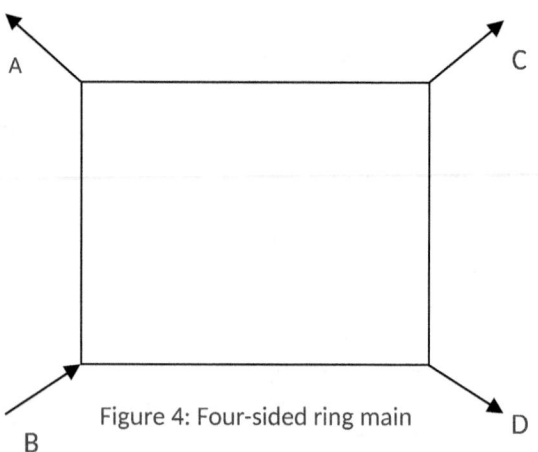

Figure 4: Four-sided ring main

3) Write a computer program to compute flow through each pipe in a loop and the pressures at nodes given pipe characteristics and relevant information.

4) Verify your program by solving example 4.5a.

Solution

1) H = head loss, m
   K = resistance coefficient
   Q = flow, m^3/s
   N = factor depending on equation

$$k = f \frac{L}{2g\left(\frac{\pi}{4}\right)^2 D^5} = 0.0312 \frac{L}{2*9.81\left(\frac{\pi}{4}\right)^2 0.6^5} = 0.03315\, L$$

$$h = f \frac{L}{D}\frac{v^2}{2g} = f\frac{L}{2}g\frac{Q^2}{\left(\frac{\pi}{4}\right)^2 D^5}$$

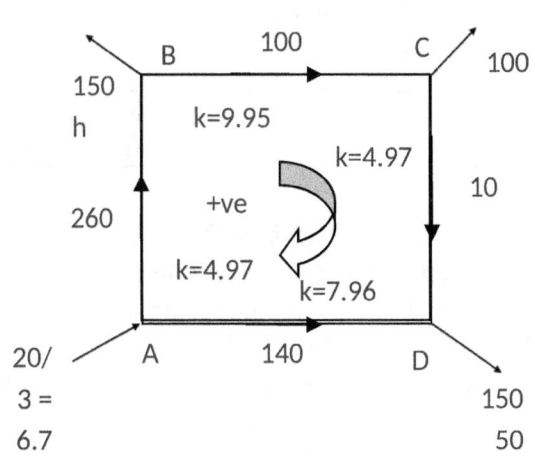

pipe	Q assumed	h = kQ²	h/Q	δQ
AB	260	335972	1292.2	=
BC	110	120395	1094.5	-3008.48/
CD	10	497	49.7	(2*1322)
DA	-140	-156016	-1114.4	= -114
	Sum	300848	1322	

pipe	Q assumed	$h = kQ^2$	h/Q	δQ
AB	146	106252.51	726.69	=-
BC	-4	-142.55	37.66	460101.86/
CD	-104	-53533.49	515.81	(2*3300.29)
DA	-254	-512678.34	2020.13	= 70
	Sum	-460101.86	3300.29	

Pipe	Q assumed	$h = kQ^2$	h/Q	P
AB	216	0.232 m	2.3 kN/m^2	P_A =15 kN/m^2
BC	66	0.043	0.42	P_B =15 – 2.3 (losses AB) =102.7
CD	-34	-0.006	- 0.06	$P_c = P_B$ – losses in BC =102.7 – 0.42 =102.3
DA	-184	-0.269	2.6	$P_D = P_C$ – losses in CD =102.3 – (-0.06) = 102.4

Exercise 4.3

1) Differentiate between in-series and in-parallel connection.

A pipe network consists of two loops as shown in figure. The network contains three pipes ACB, ADB and AEB having diameters of 250, 200 and 300mm, and lengths of 500, 200 and 100m respectively. Each pipe has a friction factor of 0.002. Point A is at a height of 16m, while point B is at a height of 15m. Water flowing through the network is at a temperature of 20oC, and at an hourly rate of 1500 m3. The pressure at point A is 100 kPa. Using Darcy-Weisbach equation, find:

    i) The rate of **flow** in each pipe.

    ii) The **pressure** at point B. (SQU, B.Sc., 1995).

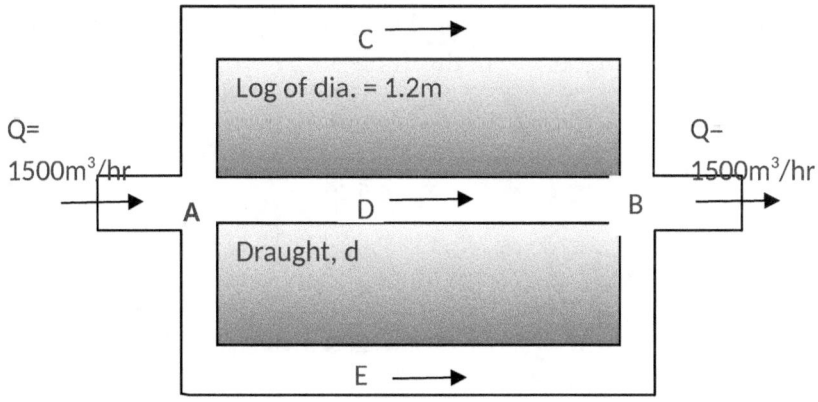

2) Calculate **head loss** in a 500 m long CI pipe of diameter 0.4m carrying a discharge of 0.2 m³/s. The kinematic viscosity of water is $1*10^{-6}$ m²/s. The roughness of CI pipe is $2.5*10^{-4}$ m. Gravity acceleration constant is 9.81 m/s². (DU, 2012). (Ans. 2.9 m).

3) What are the Advantages and Disadvantages of **Looped** Water Distribution Systems? (10 marks)

4) What are the required resources and data for the design of water supply system (6 marks).

5) What can we do in the following cases?
   1- **The minimum pressure head is negative** (i.e., the pressure is below the atmospheric pressure).
   2- The **maximum pressure head** is **too high**, :
   3- **If** the maximum velocity constraint is violated marginally,

6) Find the pipe discharges for the Gravity-Sustained Distribution Main with the data presented in the table. (DU, 2011).

Pipe i	Elevation z_i (m)	Length L_i (m)	Demand Discharge q_i (m³/s)	Pipe Discharge Q_i (m3/s)
0	100			
1	92	1500	0.01	0.065
2	94	200	0.015	0.055
3	88	1000	0.02	0.04
4	85	1500	0.01	0.02
5	87	500	0.01	0.01

7) Find the pipe discharges for the network presented in the figure with the data presented in the Table.

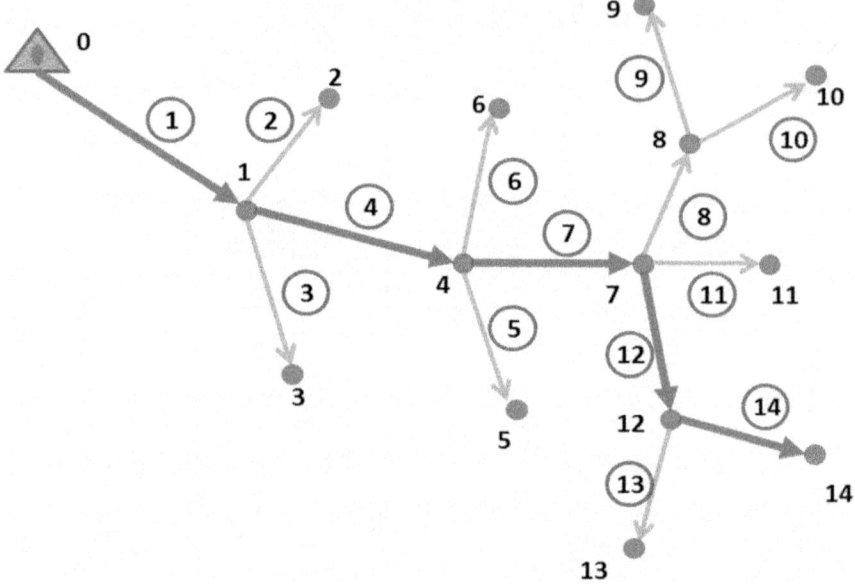

8) Estimated **Nodal Discharges** for the water distribution network presented in the figure for the data presented in the Table. peak discharge factor $\theta_p = 2.5$, rate of water supply $\omega = 400$ liters/capita/day (L/c/d). (DU, 2011).

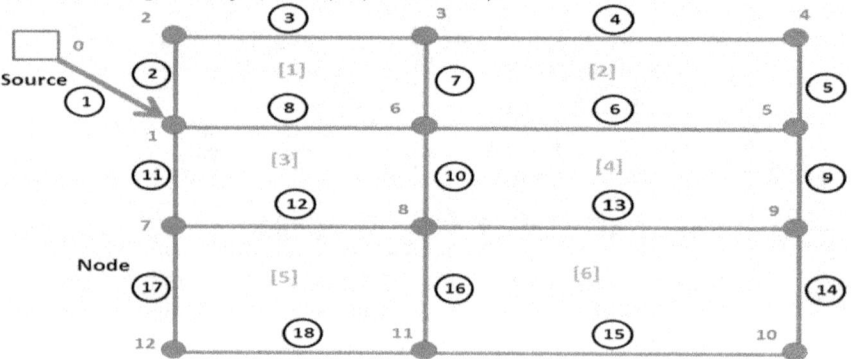

9) **Analyze** a looped pipe network as shown in the Figure for pipe discharges using Hardy Cross method. Assume a constant friction factor f = 0.02 for all pipes in the network. Assume the initial value of the discharge in pipe 1 is 0.25 m³/s and in pipe 6 is 0.1, m³/s. (DU, 2011).

$$k = \frac{8fL}{\pi^2 qD^5}$$

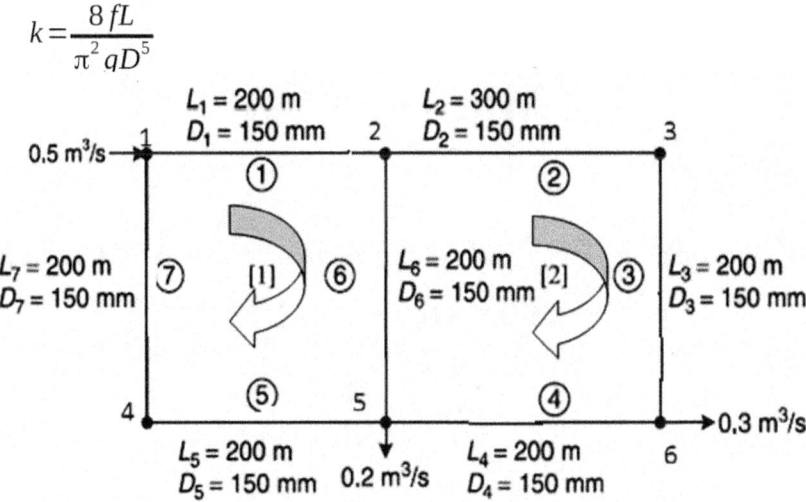

10) For the square loop shown, find the discharge in all the pipes. All pipes are 1 km long, and 300 mm in diameter with a friction factor of 0.0163. Assume that minor losses can be neglected. (B.Sc. UoD, 2014) (Ans. **45.8, 25.8, -14.2, -54.2 L/s**)

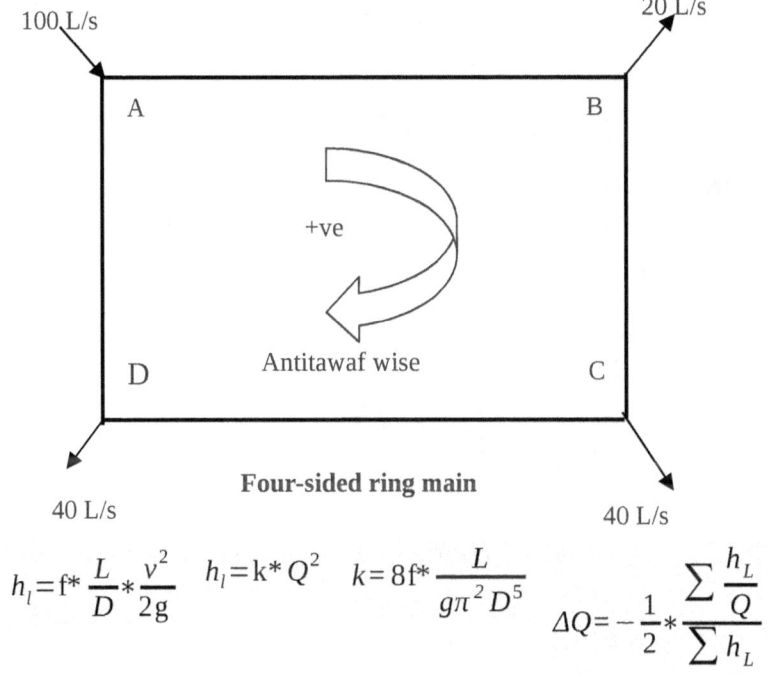

$$h_l = f * \frac{L}{D} * \frac{v^2}{2g} \qquad h_l = k * Q^2 \qquad k = 8f * \frac{L}{g\pi^2 D^5} \qquad \Delta Q = -\frac{1}{2} * \frac{\sum \frac{h_L}{Q}}{\sum h_L}$$

Chapter Five

Wastewater Volumes, Collection and Transportation

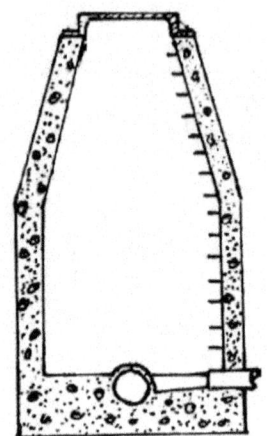

5.1 Population estimates
Example 5.1

1. Show how to use the following equations to estimate population growth for inter-censal and post-censal estimates:

$$P_m = P_e + (P_l - P_e)\left(\frac{t_m - t_e}{t_l - t_e}\right)$$

$$P_m = P_l + (P_l - P_e)\left(\frac{t_m - t_l}{t_l - t_e}\right)$$

$$LogP_m = LogP_e + (LogP_l - LogP_e)\frac{t_m - t_e}{t_l - t_e}$$

$$LogP_m = LogP_l + (LogP_l - Log\,P_e)\frac{t_m - t_l}{t_l - t_e}$$

2. A city recorded a population of 620,000 in its earlier decennial census and 700,000 in its later one. Assuming a census date of first April, estimate the mid-year (First July) populations for the fifth intercensal year and for the ninth postcensal year by geometric increase. Determine your design population. Justify your selection. (B.Sc. UoD, 2012)
3. Write a computer program to compute design population for a city given its population in its earlier and later decennial census, census date, mid-year timing.
4. Verify your program by solving example 5.1a.

Solution

1. Given: P_l = 620,000 ,P_e =700,000
2. Inter censal Geometric growth estimates after the fifth year
    $t_m - t_e$ = 5.25 year, then: $t_l - t_e$ = 10 years

    $$\frac{t_m - t_e}{t_l - t_e} = \frac{5.25}{10} = 0.525$$

    Arithmetic growth

    $$P_m = P_e + \left(P_l - P_e\right)\frac{t_m - t_e}{t_l - t_e} = 620000 + \left(700000 - 620000\right) x \, 0.525$$

    = 662000
    Geometric growth

    $$LogP_m = LogP_e + \left(LogP_l - LogP_e\right)\frac{t_m - t_e}{t_l - t_e}$$

    $Log\,620000 + \left(Log\,700000 - Log\,620000\right)0.525 = 5.82$
    $Pm = 660787$

3. Post censal Geometric growth estimates for the 9th year
    $t_m - t_l$ = 9.25 year, then $t_l - t_e$ = 10 years

    $$\frac{t_m - t_l}{t_l - t_e} = \frac{9.25}{10} = 0.925$$

    Arithmetic growth

    $$P_m = P_l + \left(P_l - P_e\right)\frac{t_m - t_l}{t_l - t_e} = 700000 + \left(700000 - 620000\right) x \, 0.925$$

$= 774000$

Geometric growth

$$\mathrm{LogP}_m = \mathrm{LogP}_l + \left(\mathrm{LogP}_l - \mathrm{Log}\, P_e\right) \frac{t_m - t_l}{t_l - t_e}$$

$¿ \mathrm{Log}700000 + \left(\mathrm{Log}700000 - \mathrm{Log}620000\right) * 0.925 = 5.894$

$P_m = 783162$

Select highest value for design purposes. Thus design population, P = 783,162

Program 5.1 Listing:
Design population for a city

```
'*********************************
'EXAMPLE 5.1: Population growth
'*********************************
Imports System.Math

Public Class Form1

    Private Sub Form1_Load(ByVal sender As
        System.Object, ByVal e As
        System.EventArgs) Handles MyBase.Load
        Me.Text = "Example 5.1"
        Me.FormBorderStyle =
            Windows.Forms.FormBorderStyle.FixedSingle
        Me.MaximizeBox = False
        Label1.Text = "P1 = "
        Label2.Text = "Pe = "
        Label3.Text = "Estimate population after (yrs)"
        Label4.Text = ""
        Button1.Text = "&Estimate"
        RadioButton1.Text = "Intercensal"
        RadioButton2.Text = "Postcensal"
        RadioButton1.Checked = True
    End Sub

    Private Sub Button1_Click(ByVal sender As
        System.Object, ByVal e As
        System.EventArgs) Handles Button1.Click
        Dim P1, Pe, t, Pm1, Pm2, k As Double
        P1 = Val(TextBox1.Text)
        Pe = Val(TextBox2.Text)
        t = Val(TextBox3.Text)
```

```
    k = (t + 0.25) / 10
    'estimate population by arithmetic growth
    If RadioButton1.Checked Then Pm1 =
        P1 + ((Pe - P1) * k) _
        Else Pm1 = Pe + ((Pe - P1) * k)
    Label4.Text = "Population estimate after "
        + t.ToString + " yrs:"
    Label4.Text += vbCrLf
    Label4.Text += "By arithmetic growth: "
    Label4.Text += Pm1.ToString
    'estimate population by geometric growth
    If RadioButton1.Checked Then _
        Pm2 = 10 ^ (Log10(P1) + ((Log10(Pe) -
            Log10(P1)) * k)) _
        Else Pm2 = 10 ^ (Log10(Pe) + ((Log10(Pe) -
            Log10(P1)) * k))
    Label4.Text += vbCrLf
    Label4.Text += "By geometric growth: "
    Label4.Text += Pm2.ToString

    Label4.Text += vbCrLf
    Label4.Text += "Select higher value for design,
        so population P="
    If Pm1 > Pm2 Then
        Label4.Text += Pm1.ToString
    Else
        Label4.Text += Pm2.ToString
    End If
    End Sub
End Class
```

Example 5.2

Determine amount of water (in daily cubic meters) to be supplied to the city for the estimated population given that the responsible authority avails 600 L/c/d of treated water. (B.Sc. UoD, 2012)

Solution

Amount of water = P*q = 783,162*600 = 469,897,200 L/d = 469,897 m^3/d

Program 5.2 Listing:

```
'*****************************
'EXAMPLE 5.2: Water demand
'*****************************
Public Class Form1

    Private Sub Form1_Load(ByVal sender As
        System.Object, ByVal e As
        System.EventArgs) Handles MyBase.Load
        Me.Text = "EXAMPLE 5.2: Water demand"
        Label1.Text = "Population, P ="
        Label2.Text = "Available water, q (L/c/d) ="
        Label3.Text = ""
        Button1.Text = "&Estimate"
    End Sub

    Private Sub Button1_Click(ByVal sender As
        System.Object, ByVal e As
        System.EventArgs) Handles Button1.Click
        Dim P, q, w As Double
        P = Val(TextBox1.Text)
        q = Val(TextBox2.Text)
        w = P * q
        Label3.Text = "Amount of water needed = " _
            + FormatNumber(w, 2) + " L/d"
    End Sub
End Class
```

Exercise 5.1

1) Define the terms used in the following equations to estimate geometric population growth for inter-censal and post-censal estimates:

$$LogP_m = LogP_e + \left(LogP_l - LogP_e\right)\frac{t_m - t_e}{t_l - t_e}$$

$$LogP_m = LogP_l + \left(LogP_l - Log\,P_e\right)\frac{t_m - t_l}{t_l - t_e}$$

A city recorded a population of 310000 in its earlier decennial census and 370000 in its later one. Estimate the mid-year (1 July) **populations**

    i.   for the fifth intercensal year.

236

ii.   for the ninth postcensal year by geometric increase. Assume a census date of 1 April. (DU, 2012). (Ans. 340176, 435792).

Solution (see Program 5.1 Listing)

1.   Given: $P_l = 210000$,$P_e = 150000$
2.   Inter censal Geometric growth estimates after the fifth year
     $t_m - t_e = 5.25$ year, then: $t_l - t_e = 10$ years

$$\frac{t_m - t_e}{t_l - t_e} = \frac{5.25}{10} = 0.525$$

$$LogP_m = LogP_e + \left(LogP_l - LogP_e\right)\frac{t_m - t_e}{t_l - t_e}$$

$Log\,310000 + \left(Log\,370000 - Log\,310000\right)0.525, P_m =$
$Pm = 340176$

- Post censal Geometric growth estimates for the 9th year
  $t_m - t_l = 9.25$ year, then $t_l - t_e = 10$ years

$$\frac{t_m - t_l}{t_l - t_e} = \frac{9.25}{10} = 0.925$$

$$LogP_m = LogP_l + \left(LogP_l - Log\,P_e\right)\frac{t_m - t_l}{t_l - t_e}$$

$¿Log370000 + \left(Log370000 - Log310000\right)0.925$
$P_m = 435792$

2) Population figures are shown for three communities in the accompanying table. All are residential with similar characteristics. By graphical comparison with other cities, predict the **population** of community B in 1980. (UAE, 1989).

City	1900	1910	1920	1930	1940	1950	1960
A	53,388	58,744	84,706	141,122	160,842	190,384	206,984
B	21,748	36,420	76,002	86,984	116,972	130,496	136,294
C	63,862	86,718	101,350	133,310	163,618	206,314	212,686

3) Find the future **fire demand** for a town having a population of 20 thousand inhabitants with a population growth rate of 2.5% for a design period of 15 years. (DU, 2012). (Ans. 19.7 m^3/min).

$$Q = 3860 \sqrt{P}(1 - 0.01 \sqrt{P})$$

Solution

Given: POP = 20000, growth rate = 2.5%

Population after 15 years = $20000(1-0.025)^{15}$ = 28966 = 29 in thousands.

Determine the fire demand as:

$Q = 3860(P)^{1/2}[(1 - 0.01(P^{1/2})] = 3860*29^{1/2}*(1 + 0.01*29^{1/2})$. = 19.7 m^3/min

4) Show how to use the following equations to predict population growth for a city during a certain period of time.

$$P_m = P_l + (P_l - P_e)\frac{t_m - t_l}{t_l - t_e} = 19500 + (195000 - 125000)0.925$$

$$= 259750$$

$$P_m = P_l + (P_l - P_e)\left(\frac{t_m - t_l}{t_l - t_e}\right)$$

$$LogP_m = LogP_e + (LogP_l - LogP_e)\left(\frac{t_m - t_e}{t_l - t_e}\right)$$

$$LogP_m = LogP_l + (LogP_l - LogP_e)\left(\frac{t_m - t_l}{t_l - t_e}\right)$$

A city recorded a population of 125000 in its earlier decennial census and 195000 in its later one. Estimate the mid-year (1 June) populations

a. For the sixth **intercensal** year.

b. For the ninth **postcensal** year by arithmetic increase. Assume a census date of 1 March.

c. Comment on your results. (DU, 2012). (Ans. 168750, 259750 capita).

Solution (see Program 5.1 Listing)

Given: $P_1 = 195000$, $P_e = 125000$

Arithmetic growth

Inter censal estimates for the sixth year

$t_m - t_e = 6.25$ year, then: $t_1 - t_e = 10$ years

$$\frac{t_m - t_e}{t_1 - t_e} = \frac{6.25}{10} = 0.625$$

$$P_m = P_e + \left(P_1 - P_e\right)\frac{t_m - t_e}{t_1 - t_e} = 125000 + \left(195000 - 125000\right)0.625$$

$$= 168750$$

Post censal estimates for the 9th year

$t_m - t_1 = 9.25$ year, then $t_1 - t_e = 10$ years

$$\frac{t_m - t_1}{t_1 - t_e} = \frac{9.25}{10} = 0.925$$

$$P_m = P_1 + \left(P_1 - P_e\right)\frac{t_m - t_1}{t_1 - t_e} = 19500 + \left(195000 - 125000\right)0.925$$

$$= 259750$$

5) Indicate how to use the following equations to estimate arithmetic and geometric population growth for inter-censal and post-censal estimates:

$$P_m = P_e + \left(P_1 - P_e\right)\frac{t_m - t_e}{t_1 - t_e}$$

$$Log P_m = Log P_e + \left(Log P_1 - Log P_e\right)\frac{t_m - t_e}{t_1 - t_e}$$

$$P_m = P_1 + \left(P_1 - P_e\right)\frac{t_m - t_1}{t_1 - t_c}$$

$$Log P_m = Log P_1 + \left(Log P_1 - Log P_e\right)\frac{t_m - t_1}{t_1 - t_e}$$

A city recorded a population of 150000 in its earlier decennial census and 210000 in its later one. Estimate the mid-year (1 July) populations

i.   for the fifth **intercensal** year.

239

ii.   for the ninth **postcensal** year by arithmetic increase and geometric increase. Assume a census date of 1 April. (DU, 2011). (Ans. 181500, 178982, 265500, 286674 capita).

Solution (see Program 5.1 Listing)

1.   Given: $P_l = 210000$, $P_e = 150000$

2.   Inter censal estimates after the fifth year

$t_m - t_e = 5.25$ year, then: $t_l - t_e = 10$ years

$$\frac{t_m - t_e}{t_l - t_e} = \frac{5.25}{10} = 0.525$$

Arithmetic growth

$$P_m = P_e + \left(P_l - P_e \right) \frac{t_m - t_e}{t_l - t_e} = 150000 + \left(210000 - 150000\right)0.525$$

$\quad = 181500$

Geometric growth

$$LogP_m = LogP_e + \left(LogP_l - LogP_e \right) \frac{t_m - t_e}{t_l - t_e}$$

$Log\,150000 + \left(Log\,210000 - Log\,150000 \right)0.525, P_m =$

$Pm = 178982$

N.B. Geometric growth estimates lesser for inter censal estimates.

i.   Post censal estimates for the 9th year

$\_t_m - t_l = 9.25$ year, then $t_l - t_e = 10$ years

Arithmetic growth

$$\frac{t_m - t_l}{t_l - t_e} = \frac{9.25}{10} = 0.925$$

$$P_m = P_l + \left(P_l - P_e \right) \frac{t_m - t_l}{t_l - t_e} = 210000 + \left(210000 - 150000\right)0.925$$

$\quad = 265500$

Geometric growth

$$LogP_m = LogP_l + \left(LogP_l - Log\,P_e\right)\frac{t_m - t_l}{t_l\ t_e}$$

$\iota Log210000 + \left(Log210000 - Log150000\right)0.925$

$P_m = 286674$

N.B. Geometric growth estimates are higher for post censal population estimates

6) If the "typical value" for Saudi domestic water consumption is 220 L/cap-d, calculate the actual and the equivalent 2008 **population** if a town has a record of the following volumes consumed over that entire year: (B.Sc., UoD, 2013)

Category	m^3
Domestic	x 10^6 8.03
Commercial	x 10^6 4.62
Industries	x 10^6 5.07
Public	x 10^5 3.55
Fire	x 10^4 7.61
Losses	x 10^5 6.82

7) **Population** statistics are given in the table below for a town which for various reasons has been limited to a saturation population of approximately 135,000. Using graphical approximation as well as a method based on calculations, estimate the population in the year 2030 (B.Sc., UoD, 2013)

Year	1910	1930	1950	1970	1990	2010
Population in thousands	60	82	100	115	124	128

5.2 Wastewater Volumes, Collection and Transportation

5.2.A. Dry Weather Flow (DWF)

This is defined as "the average daily flow rate of sewage on days when the rainfall has not exceeded 2.5 mm during the previous 24 hours".

DWF = PxQ + I_r + T_w − E (l/d)

Where:

P = population served by the sewer

I_r = average, dry weather, infiltration into sewer due to:

- Poor joints
- Pervious materials. (usually 0 – 30 % of DWF) l/d

Q = average daily water usage (l/c.d)

T_w = average trade waste discharge (l/d)

E = evaporation, in hot climates I may reach 30 – 50 % of water consumption

Example 5.2.1

1) A sanitary sewer is to serve an area having an estimated future population of 4000. The average sewage flow is 0.4 m^3/d.capita. Daily infiltration in the area is estimated to be 80 m3/km length of sewer. Assuming the total length of sewer is 4 km, determine the DWF.

2) Write a computer program to compute the DWF given future population, average sewage flow, daily infiltration in the area and total length of sewer.

3) Verify your program by solving example 5.2.1.1.

Solution

Infiltration = 80 (m^3/km length.d) x 4 km = 320 m^3/d

DWF = 0.4 x 400 x 320 = <u>1920 m^3/d</u>

Exercise 5.2.A

1) a) A sanitary sewer is to be designed to serve an area having a population of 80 persons per hectare. The average daily sewage flow is 0.25 m^3/c, and the estimated infiltration is 37.5 m^3 per km of sewer per day. The total length of sewer is 4 km, while the area is 25 ha. Compute the **DWF**. (Ans. 650 m^3/d).

2) Write a computer program to compute the **DWF** given population per hectare, average daily sewage flow, infiltration, total length of sewer, and area.

3) Verify your program by solving example 1a.

Solution

DWF = PxQ + I_r + T_w − E = 80*25*0.25 + 37.5*4 = 650 m^3/d

Program 5.2.A.1 Listing:
DWF

```
'**************************
'EXERCISE5.2.A.1: DWF I
'**************************
Public Class Form1

    Private Sub Form1_Load(ByVal sender As
        System.Object, ByVal e As
        System.EventArgs) Handles MyBase.Load
        Me.Text = "Exercise5.2.A.1: DWF I"
        Me.MaximizeBox = False
        Me.FormBorderStyle =
            Windows.Forms.FormBorderStyle.FixedSingle
        Label1.Text = "Population (per hectare)"
        Label2.Text = "Avg. sewage flow/d (m3/c)"
        Label3.Text = "Infiltration (m3/km/d)"
        Label4.Text = "Sewer length (km)"
        Label5.Text = "Area (ha)"
        Label6.Text = ""
        Button1.Text = "&Calculate DWF"
    End Sub

    Private Sub Button1_Click(ByVal sender As
        System.Object, ByVal e As
        System.EventArgs) Handles Button1.Click
        Dim P, F, I, L, A, DWF As Double
        P = Val(TextBox1.Text)
        F = Val(TextBox2.Text)
        I = Val(TextBox3.Text)
        L = Val(TextBox4.Text)
        A = Val(TextBox5.Text)

        DWF = (P * A * F) + (I * L)
        Label6.Text = "DWF = " +
            FormatNumber(DWF, 2) + " m3/d"
    End Sub
End Class
```

4) For a city having a population of 20000 and a growth rate of 2%, the average daily water consumption is 175 l/c. The area of

the city is 1200 ha. It is expected to return 70 % of water consumed to the sewerage system. Compute the sewer's **DWF** if infiltration amounts to 30 m^3/d. (Ans. 268 m^3/d).

5) Write a computer program to compute sewer's DWF given population, growth rate, average daily water consumption, area, percent water consumed returned to the sewerage system and infiltration.

6) Verify your program by solving example d.

Solution

Given: P = 20,000, r = 0.02, Q = 17 L/c.d

Qw =0.7*17*20000/1000 = 238 m^3/d

I_r = 30 m^3/d

DWF = 238 + 30 = 268 m^3/d

Program 5.2.A.2 Listing:
Sewer's DWF

```
'**************************
'EXERCISE5.2.A.2: DWF II
'**************************
Public Class Form1

    Private Sub Form1_Load(ByVal sender As
        System.Object, ByVal e As
        System.EventArgs) Handles MyBase.Load
        Me.Text = "Exercise5.2.A.2: DWF II"
        Me.MaximizeBox = False
        Me.FormBorderStyle =
            Windows.Forms.FormBorderStyle.FixedSingle
        Label1.Text = "Population "
        Label2.Text = "Growth rate (%)"
        Label3.Text =
            "Avg. daily water consumption (L/c.d)"
        Label4.Text = "Percent. water return (%)"
        Label5.Text = "Infiltration (m3/d)"
        Label6.Text = "Area (ha)"
        Label7.Text = ""
        Button1.Text = "&Calculate DWF"
    End Sub

    Private Sub Button1_Click(ByVal sender As
        System.Object, ByVal e As
        System.EventArgs) Handles Button1.Click
```

```
        Dim P, r, Q, L, Ir, A, Qw, DWF As Double
        P = Val(TextBox1.Text)
        r = Val(TextBox2.Text)
        Q = Val(TextBox3.Text)
        L = Val(TextBox4.Text)
        Ir = Val(TextBox5.Text)
        A = Val(TextBox6.Text)

        r /= 100
        Qw = (L / 100) * Q * P / 1000
        DWF = Qw + Ir
        Label7.Text = "DWF = " +
            FormatNumber(DWF, 2) + " m3/d"
    End Sub
End Class
```

5.2.B. Population Equivalent (PE)

The PE of a sewage is an expression that denotes some characteristics (e.g. flow, BOD, COD, SS... etc) as related to domestic sewage. Therefore, the PE of any sewage would indicate the number of persons responsible for production of a sewage that would have the same characteristics as the standard sewage. The assumption is that the average person excretes a BOD_5^{20} of 0.06 kg/d.

The PE has been used as a technique for evaluating and assessing industrial treatment costs.

Example 5.3

1) Compute the PE of a certain factory that produces 10^6 l of wastewater each day, knowing that the 5-day BOD of the generated waste is 300 mg/l.
2) Write a computer program to compute PE of a certain factory given daily wastewater productionand 5-day BOD of the generated waste.
3) Verify your program by solving example 5.3.1.

Solution

PE = BOD_5 of waste x flow rate / BOD_s of standard sewage

   = 300×10^{-3} g/l.d x 10^6 l/0.06×10^3 g/d $\underline{= 5000}$

245

Program 5.3 Listing:
PE of a certain factory

```
'*************************
'EXAMPLE 5.3: PE
'*************************
Public Class Form1
    Const BODs = 0.06 * 1000   'BODs of standard sewage

    Private Sub Form1_Load(ByVal sender As
        System.Object, ByVal e As
        System.EventArgs) Handles MyBase.Load
        Me.Text = "Example5.3: PE"
        Me.FormBorderStyle =
            Windows.Forms.FormBorderStyle.FixedSingle
        Me.MaximizeBox = False
        Label1.Text = "Waste flow rate (L/d)"
        Label2.Text = "5-day BOD (mg/L)"
        Label3.Text = ""
        Button1.Text = "&Calculate PE"
    End Sub

    Private Sub Button1_Click(ByVal sender As
        System.Object, ByVal e As
        System.EventArgs) Handles Button1.Click
        Dim Qw, BOD5, PE As Double
        Qw = Val(TextBox1.Text)
        BOD5 = Val(TextBox2.Text)
        PE = (Qw * BOD5) / BODs
        PE /= 1000
        Label3.Text = "PE = " +
            FormatNumber(PE, 2) + " mg/L"
    End Sub
End Class
```

Exercise 5.2.B. (see Program 5.3 Listing)

1) A milk processing plant produces wastewater in an amount of 120000 l/d with an average 5-day BOD of 1500 mg/l. If this plant disposes off its wastewater in the neighboring sewer, calculate the added load on the city's treatment works, in terms of **PE**. (Ans. 3,000).

Solution

Q_w = 120,000 L/d, BOD = 1500 mg/L

PE = BOD_5 of waste x flow rate / BOD_s of standard sewage = (1500 mg/l/1*12000 L/d)/(0.06x10^6 mg/d) = 3,000

2) The sewage flow from a town is 100000 l/d. If the average BOD_5^{20} of the sewage is 250 mg/l, compute the **PE**. (Ans. 417).

Solution

Q_w = 100,000 L/d, BOD_5 = 250 mg/L

PE = BOD_5 of waste x flow rate / BOD_s of standard sewage = (250 g/m^3/*100m^3/d)/(0.06x10^3 g/d) = 417

3) An industrial plant generates daily 1 m^3 of sewage that have a 5-daym3/d) BOD of 350 ppm. Compute the **PE** of the sewage of the plant. (Ans. 6).

Solution

Q_w = 1 m^3/d, BOD_5 = 350 mg/L

PE = (350 g/m^3/*1 m^3/d)/(0.06x10^3 g/d) = 6

4) The average sewage flow of a certain city is 720 m^3/d. If the average BOD5 of the generated waste is 275 ppm, calculate the **PE** of the waste. (Ans. 3,300).

Solution

Q_w = 720 m^3/d, BOD_5 = 275 g/m^3

PE = (275*720)/(60) =3300

5) An industry discharges 1500 kg of 5-day BOD each day. Find the **PE** of the industrial waste discharge. (Ans. 25,000).

Solution

BOD_5 = 1500 kg/d

PE = (1500 kg/d)/(0.06 kg/d) =25,000

6) Determine maximum daily, average daily and minimum daily per capita wastewater **flow** for a city with population of 40,000 given the following information: Annual average water consumption = 400 L/capita/d; dry weather flow, DWF = 80

percent of water consumption; and minimum flow amounts to around 30 percent of average flow.

7) An area with a population of 30,000 is to be served by a 6 kilometers long sanitary sewer. The average daily per capita wastewater flow is 500 m^3. The daily infiltration in the area is estimated to be 50 m^3 per kilometer length of sewer. Determine the dry weather flow, **DWF**. Estimate **DWF** for a population growth rate of 3 percent in 50 years period. (SAS, 2008).

8) An area with a population of 20000 is to be served by a sanitary sewer. The average per capita wastewater flow is 20 m^3 per hour. The daily infiltration in the area is estimated to be 40 m^3 per kilometer length of sewer. For a total sewer length of 5 kilometers, compute the dry weather flow, **DWF**.

9) Determine the different per capita wastewater **flows** (i.e. maximum daily, average daily and minimum daily flow) for a population of 30000 using the following data: Annual average water consumption = 300 L/capita/d; DWF = 90 percent of water consumption; and minimum flow = 40 percent of average flow. (SAS, 2007).

5.3 Wastewater Collection and Transportation
5.3.1 Storm Sewage

The primary source of storm flow is precipitation. The Rational method (Lloyd Davies Method) is used to estimate the rate of runoff.

Q = 0.278 C I A

Where:

Q = Maximum rate of runoff (m^3/s)

C = coefficient representing the ratio of runoff to rainfall (or the fraction of the incident rainfall that appears as surface flow). It depends on type and character of surface (0 less than C less than 1)

I = average intensity of rainfall, for the time of maximum rainfall of a given frequency of occurrence having a duration equal to the time required for the entire drainage area to contribute to flow (mm/hr)

A = drainage area (km^2).

Table (5.1): Runoff Coefficient for the Rational Method

Type of surface	Coefficient, C
Gardens	0.0 – 0.1
Lawns depending on surface slope and character of subsoil	0.1 – 0.25
Unimproved area, parks, cemeteries, playgrounds	0.2 – 0.3
Suburb, single family areas	0.3 – 0.5
Town, apartment – dewelling areas	0.5 – 0.7
Town centre (Urban area)	0.7 – 0.9
Paved driveways and walks	0.75 – 0.85
Water tight roofs, streets	0.7 – 0.95
Paved streets	0.8 – 0.9
Asphaltic cement streets	0.85 – 0.9

Example 5.4

1) Design the storm sewer in the figure below by using the following data:
   - Flowing full velocity in sewer = 0.75 m/s
   - Roughness coefficient (n) = 0.013

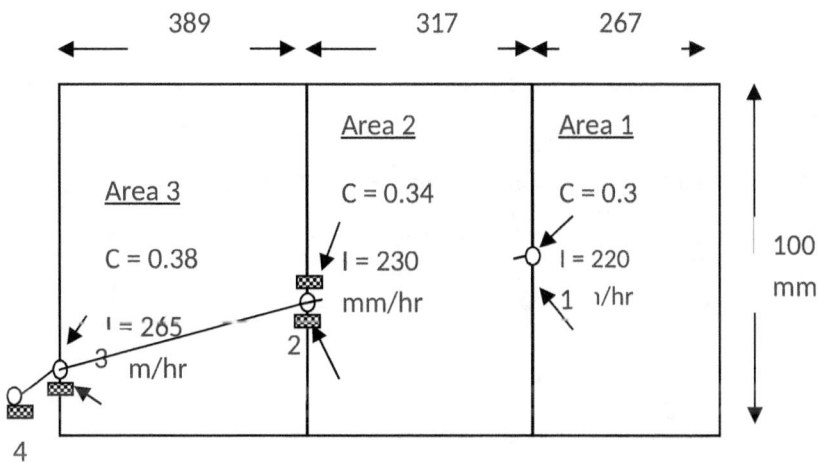

2) Write a computer program to design a storm sewer given full velocity in sewer, roughness coefficient and area properies.

3) Verify your program by solving example 5.4.1.

Solution

Area 1 = 100x267 = 0.0267 km^2

Area 2 = 100x317 = 0.0317 km^2

Area 3 = 100x389 = 0.0389 km^2

Flow from manhole 1 to manhole 2

Q_1 = 0.278 x 0.3 x 220 x 0.0267 = 0.49 m^3/s

From Manning's nomograph use a sewer of diameter 915 mm at a slope of 0.78 %. This gives a discharge of 0.53 m^3/s

Flow from manhole 2 to manhole 3

Q_2 = Q_1 + 0.278 x 0.34 x 230 x 0.0317 = 0.49 + 0.69 = 1.1 m^3/s

From Manning's nomograph use a sewer of diameter 1.37 m at a slope of 4.5 %. This gives a discharge of 1.18 m^3/s

Flow from manhole 3 to manhole 4

Q_3 = Q_1 + Q_2 + 0.278 x 0.38 x 265 x 0.0389 = 2.19 m^3/s

From Manning's nomograph use a sewer of diameter 1.98 m at a slope of 2.5 %. This gives a discharge of 2.34 m^3/s.

Program 5.4 Listing:

Design a storm sewer

```
'*********************
'Example5.4: Runoff
'*********************
Public Class Form1

    Private Sub Form1_Load(ByVal sender As
        System.Object, ByVal e As
        System.EventArgs) Handles MyBase.Load
        Me.Text = "Example5.4: Runoff"
        Me.FormBorderStyle =
            Windows.Forms.FormBorderStyle.FixedSingle
        Me.MaximizeBox = False
        Label1.Text = "Enter data for all areas below:"
        Label2.Text = ""
        Button1.Text = "&Calculate Runoff"
        DataGridView1.Columns.Clear()
        DataGridView1.Columns.Add("CCol",
            "Coefficient, C")
        DataGridView1.Columns.Add("ICol",
            "Intensity, I (mm/hr)")
        DataGridView1.Columns.Add("ACol",
```

```vb
                "Area, A (km2)")
        DataGridView1.Columns.Add("QCol",
            "Runoff, Q (m3/s)")
        DataGridView1.Columns(3).ReadOnly = True
    End Sub


    Private Sub Button1_Click(ByVal sender As
        System.Object, ByVal e As
        System.EventArgs) Handles Button1.Click
        If DataGridView1.RowCount <= 1 Then
            MsgBox("Enter at least one reading!",
                    vbOKOnly)
            Exit Sub
        End If

        Dim C, I, A, Q, QT As Double
        QT = 0
        'last row is the "Add new" row, so ignore
        'it by subtracting 2
        For j = 0 To DataGridView1.RowCount - 2
            C = Val(DataGridView1.Rows(j).
                Cells(0).Value)
            I = Val(DataGridView1.Rows(j).
                Cells(1).Value)
            A = Val(DataGridView1.Rows(j).
                Cells(2).Value)
            Q = QT + 0.278 * C * I * A
            QT = Q
            DataGridView1.Rows(j).Cells(3).Value =
                FormatNumber(Q, 2)
        Next
        Label2.Text = "Use Manning's nomograph to
                determine diameter and slope"
    End Sub
End Class
```

251

Exercise 5.3.1

Size of sewer

1) Given the drainage area in the figure below and the indicated information, determine the **size** and **gradient** of sewers.

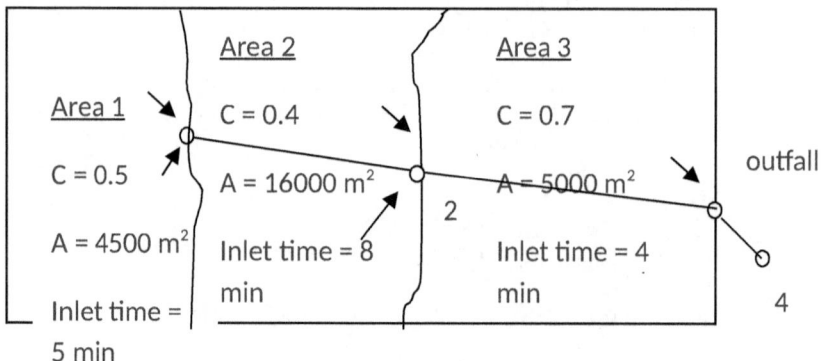

1. Length of sewer from manhole 1 to manhole 2 = 100 m
2. Length of sewer from manhole 2 to manhole 3 = 150 m
3. Average velocity of flow in sewers = 0.75 m/s
4. Rainfall intensity duration I = 750/(t + 10) mm/hr, t is the duration in minutes (concentration time is the sum of inlet time and the flow time). (Ans. 685 mm, 0.001; 760 mm, 0.0009; 1228 mm, 0.00045).

Solution

Flow from manhole 1 to manhole 2

Flow time from manhole 1 to 2 = 100m/0.75*60m/min = 2.2 min

Concentration time = inlet time + flow time = 5 + 2.2 = 7.2 min

Thus, rainfall intensity = 750/(7.2 + 10) = 436 mm/hr

Flow Q_1 = 0.278CIA = 0.278 x 0.5 x 436 x 4500/10^6 = 0.27 m³/s = 16.4 m³/min

From Manning's nomograph use a sewer of diameter 685 mm at a slope of 0.001. This gives a discharge of 0.28 m³/s

Flow from manhole 2 to manhole 3

Time of concentration from manhole 2 to manhole 3 = 8 + 150/0.75*60 = 11.3 min

Rainfall intensity = 750/(11.3 + 10) = 35.2 mm/hr

$Q_2 = Q_1 + 0.278 \times 0.4 \times 35.2 \times 16000/10^6 = 0.27 + 0.063 = 0.333$ m³/s = 19.96 m³/min

From Manning's nomograph for Q = 19.96 m³/min & v = 0.75 m/s, take a sewer of diameter 760 mm at a slope of 0.0009. This gives a discharge of 0.346 m³/s

Flow from manhole 3 to manhole 4

Concentration time = 4 min

Rainfall intensity = 750/(4 + 10) = 53.57 mm/hr

$Q_3 = Q_2 + 0.278 \times 0.7 \times 53.57 \times 50000/10^6 = 0.27 + 0.063 + 0.271 = 0.85$ m³/s = 51.3 m³/min

From Manning's nomograph take a sewer of diameter 1228 mm at a slope of 0.00045. This gives a discharge of 0.88 m³/s.

Determine the **size** of sewer below manhole number 3 for the shown drainage area. (Ans. 685 mm, 0.001; 915 mm, 0.00065; 1228 mm, 0.00048).

2)

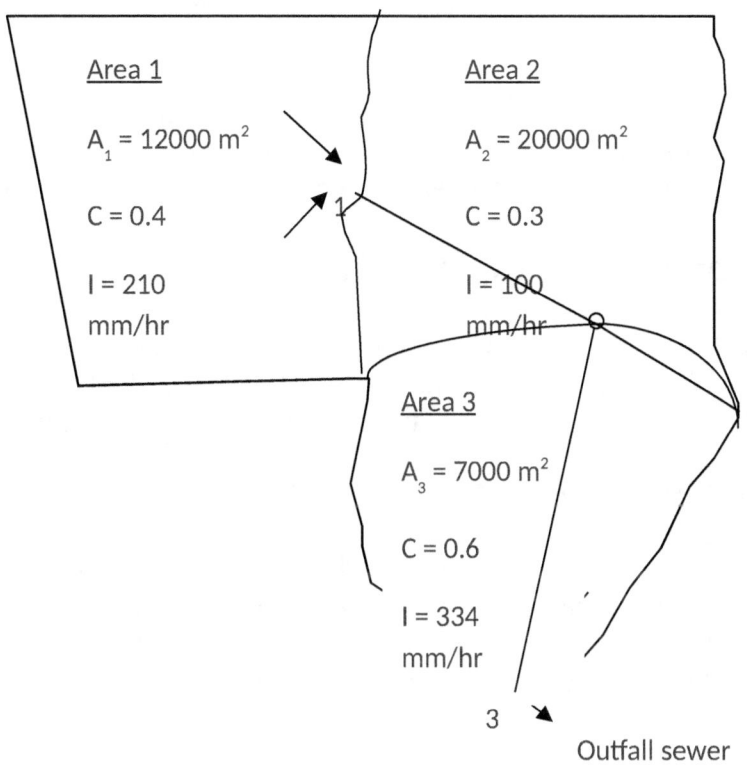

Solution

Flow from manhole 1 to manhole 2

$Q_1 = 0.278CIA = 0.278 \times 0.4 \times 210 \times 12000/10^6 = 0.28$ m^3/s = 16.8 m^3/min

From Manning's nomograph use a sewer of diameter 685 mm at a slope of 0.001. This gives a discharge of 0.28 m^3/s

Flow from manhole 2 to manhole 3

$Q_2 = Q_1 + 0.278 \times 0.3 \times 100 \times 20000/10^6 = 0.28 + 0.17 = 0.45$ m^3/s = 26.8 m^3/min

From Manning's nomograph take a sewer of diameter 915 mm at a slope of 0.00065. This gives a discharge of 0.48 m^3/s

Flow from manhole 3 to manhole 4

$Q_3 = Q_2 + 0.278 \times 0.6 \times 334 \times 70000/10^6 = 0.85$ m^3/s = 0.28+ 0.17 + 0.39 = 0.84 m^3/s = 50.4 m^3/min

From Manning's nomograph take a sewer of diameter 1228 mm at a slope of 0.00048. This gives a discharge of 0.91m^3/s.

3) **Design** a circular storm sewer that would serve an area of 0.4 km^2 with a runoff coefficient of 0.6 and an average ground slope of 1 in 400. The design intensity rainfall is 10 cm/hr. (Ans. 1830 mm).

Solution (see Example 5.4 Listing)

A = 0.4 km^2, C = 0.6, I = 100 mm/hr

Q = 0.278 x 0.6 x 100 x 0.35 = 5.8 m^3/s = 350.3 m3/min

From Manning's nomograph take a sewer of diameter 1830 mm and for a slope of 1 in 400 it gives a discharge of 6 m^3/s at a velocity of flow of

$v = (1/0.013)*(1.83/4)^{2/3}*(2.5*10^{-3})^{1/2} = 2.29$ m/s

4) a) A circular storm sewer is needed to serve an area of 800 ha, the average ground slope is 1 in 6000. Compute the **size** of the sewer at point of discharge knowing that C = 0.5, the time of concentration (t) = 21 minutes, and I = 750/(t + 10). (Ans. 5180 mm).

b) Write a computer program to compute the **size** of the sewer at point of discharge given roughness coefficient, time of concentration, area served, sewer diameter, and slope.

c) Verify your program by solving example 5.4.a.

Solution

A = 800 ha = 8 km², C = 0.5, slope = 1/5000 = 0.0002

I = 750/(21 + 10) = 24.9 mm/hr

Q = 0.278 x 0.5 x 24.19 x 8 = 26.9 m³/s = 1614 m³/min

From Manning's nomograph take a sewer of diameter 5.18 m at the specified slope which gives a discharge of 27.3 m³/s.

Program 5.3.1.4 Listing:

Size of the sewer at point of discharge

```
'***************************
'EXERCISE 5.3.1.4: RUNOFF II
'***************************
Public Class Form1

    Private Sub Form1_Load(ByVal sender As
        System.Object, ByVal e As
        System.EventArgs) Handles MyBase.Load
        Me.Text = "Exercise 5.3.1.4: Runoff II"
        Me.MaximizeBox = False
        Me.FormBorderStyle =
            Windows.Forms.FormBorderStyle.FixedSingle
        Label1.Text = "Area (ha)"
        Label2.Text = "Slope"
        Label3.Text = "Time (min)"
        Label4.Text = "Coefficient, C"
        Label5.Text = ""
        Button1.Text = "&Calculate"
    End Sub

    Private Sub Button1_Click(ByVal sender As
        System.Object, ByVal e As
        System.EventArgs) Handles Button1.Click
        Dim A, S, T, C, I, Q As Double
        A = Val(TextBox1.Text)
        S = Val(TextBox2.Text)
        T = Val(TextBox3.Text)
        C = Val(TextBox4.Text)
        A /= 100  'convert to km2
        I = 750 / (T + 10)
        Q = 0.278 * C * I * A
```

255

```
      Label5.Text = "Runoff Q = " +
         FormatNumber(Q, 2) + " m3/s"
      Label5.Text += vbCrLf + "Use Manning's
         nomograph to determine diameter"
   End Sub
End Class
```

d) Determine the **design flow** and **size** of sewers needed to serve the area indicated in the figure shown below.
   - Duration of storm = 25 minutes
   - Rainfall intensity I = 1000/(t + 20), where t is the duration in minutes. (Ans. 380 mm, 0.0023; 685 mm, 0.001; 107 mm, 0.00055).

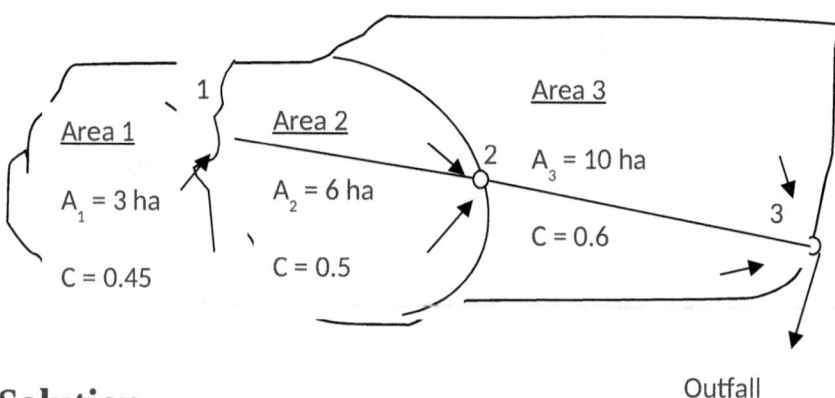

Outfall

Solution

$Q = I(\Sigma CA)$

$A_1 = 3ha = 0.03 \text{ km}^2$, $A_2 = 0.06 \text{ km}^2$, $A_1 = 0.1 \text{ km}^2$

$I = 750/(25 + 10) = 22.22 \text{ mm/hr}$

$Q = 0.278 \times 22.22 \times (0.45*0.03 + 0.5*0.06 + 0.6*0.1) = 0.64 \text{ m}^3/s$

Flow from manhole 1 to manhole 2

$Q_1 = 0.278 \times 0.45 \times 22.22 \times 0.03 = 0.083 \text{ m}^3/s = 5 \text{ m}^3/min$

From Manning's nomograph take a sewer of diameter 380 mm at a slope of 0.0023. This gives a discharge of 0.087 m³/s

Flow from manhole 2 to manhole 3

$Q_2 = Q_1 + 0.278 \times 0.5 \times 22.22 \times 0.06 = 0.083 + 0.278 \times 0.5 \times 22.22 \times 0.06 = 0.268 \text{ m}^3/s = 16.16 \text{ m}^3/min$

From Manning's nomograph take a sewer of diameter 685 mm at a slope of 0.001. This gives a discharge of 0.277 m³/s

Flow from manhole 3 to manhole 4
Q_3 = 0.268 + 0.278 x 0.6 x 22.22 x 0.1 = 0.64 m³/s = 38.3 m³/min
From Manning's nomograph take a sewer of diameter 107 mm at a slope of 0.00055. This gives a discharge of 0.67 m³/s.

Number of people to be served by sewer

e) Determine the maximum **number of people** that can be served by a sewer 255 mm in diameter laid at a minimum grade. (Use a design flow of 1.5 m3/person.d and a flowing full velocity of 0.75 m/s). (Ans. 2207 persons).

Solution

Q = vA = 0.75*π*(0.255)²/4 = 0.038 m³/s = 3309.4 m3/d
Maximum number of people = 3390.4 m³/d/1.5 m³/c.d =

f) Compute design **population** that can be served by a 200 mm sanitary sewer laid at a slope of 0.4 percent. Assume that the design flow per person is 0.017 L/s (1500 L/person.d). (UAE, 1990). (Ans. 1236 persons).

g) g Write a computer program to compute design **population** that can be served by a sanitary sewer laid given its diameter, slope, and design flow per person.

h) Verify your program by solving example f.

Solution

D = 200 mm, s = 0.4/100 = 0.004, n = 0.013

$$Q_f = \frac{1}{0.013}\left(\frac{0.2}{4}\right)^{\frac{2}{3}}\sqrt{0.004} \times \frac{\pi}{4}(0.2)^2 = 0.021\, m^3/s = 21\, L/s$$

Q_f = 0.21×10³ = 21 L/s

$$Population = \frac{21\, L/s}{0.017\, L/c/d} = 1236\, persons$$

Program 5.3.1.7 Listing:
Design population that can be served by a sanitary sewer

```vb
'*****************************
'EXERCISE 5.3.1.7
'*****************************
Public Class Form1
    Const n = 0.013

    Private Sub Form1_Load(ByVal sender As
        System.Object, ByVal e As
        System.EventArgs) Handles MyBase.Load
        Me.Text = "Exercise 5.3.1.7"
        Me.MaximizeBox = False
        Me.FormBorderStyle =
            Windows.Forms.FormBorderStyle.FixedSingle
        Label1.Text = "Sewer diameter (mm)"
        Label2.Text = "Slope (%)"
        Label3.Text = "Design flow/person (L/s)"
        Label4.Text = ""
        Button1.Text = "&Calculate"
    End Sub

    Private Sub Button1_Click(ByVal sender As
        System.Object, ByVal e As
        System.EventArgs) Handles Button1.Click
        Dim D, S, I, Qf, P As Double
        D = Val(TextBox1.Text)
        S = Val(TextBox2.Text)
        I = Val(TextBox3.Text)
        D /= 1000   'convert to m
        S /= 100
        Qf = (1 / n) * ((D / 4) ^ (2 / 3)) *
            (Math.Sqrt(S)) * (Math.PI / 4) * (D ^ 2)
        Qf *= 1000
        'we use Math.Ceiling() function to round
        'a number up to
        'the next near integer, e.g. 3.7 -> 3
        P = Math.Ceiling(Qf) / I
        Label4.Text = "Design population that can be
            served = " + _
            Math.Ceiling(P).ToString + " persons"
    End Sub
End Class
```

Velocity of flow in sewer

i) Outline major classes of **sewer** systems. List merits and demerits of each system. A circular sewer 1520 mm (60 in) diameter is laid on a slope of 0.00036. Determine the **velocity** of flow and **depth** of flow when the minimum flow of 1020 m^3/hr (10 ft^3/s) occurs. (UAE, 1990). (Ans. 0.77 m/s, 532 mm).

Solution

    1) Separate, Combined, Psuedo-separate
    2) i] D = 1520 mm, s = 0.00036, n = 0.013
       (assumed)

$$v = \frac{1}{n} R_h^{\frac{2}{3}} S^{1/2} = \frac{1}{0.013} \left(\frac{1.52}{4}\right)^{\frac{2}{3}} \sqrt{0.00036} = 0.766 \, m/s$$

$$Q_f = v . \frac{\pi}{4} D^2 = 0.766 \times \frac{\pi}{4} (1.52)^2 = 1.39 \, m^3/s$$
$$= 83.4 \, m^3/min$$

or, from numograph: $v = \dfrac{85}{\dfrac{\pi}{4}(1.52)^2} = 0.78 \, m/s$

and Q_f = 85 m^3/min

  ii] $\dfrac{q}{Q_f} = \dfrac{1020 \, m^3/hr}{1.39 \, m^3/s} = \dfrac{0.283}{1.39} = 0.2$

from chart d/D = 0.35
∴d = 0.35×1520 = 532 mm

5.3.2 Sanitary sewer

Merits of sewers systems include:

- Collection and transportation of sewage from the point of production to the treatment plant or final disposal places.
- Prevention of development of septic conditions.
- Prevention of malfunctions and ill consequences that endanger public health.

Factors to be considered in the design stage include:
- Availability of sufficient space.
- Wastewater flow patterns and characteristics.
- Expected effects on treatment plant.
- Availability of spare parts and required power.
- Effects on receiving bodies of water.
- Environmental and climatic conditions prevailing in the area.
- Economic aspects.
- Required topographical investigations.
- Master plan and future developments in the area.
- Political interventions.
- Existing utilities in the area.

Example 5.5
1) For a concrete sewer of diameter 1.68 m and a slope of 0.0009 compute the capacity when flowing full, and the velocity. Take the frictional coefficient n to be 0.013.
2) Write a computer program to compute capacity and velocity of a sewer when flowing full, sewer of diameter and type, and frictional coefficient.
3) Verify your program by solving example 5.5.1.

Solution
Manning's formula could be used to compute the velocity since it was found that it satisfies experimental results, and it proved to be an easy formula in calculations.

$$v = \frac{Rh^{2/3} S^{21/2}}{n}$$

Where:
V = average velocity in the pipe (m/s)
R_h = hydraulic radius (m) = area of flow/wetted perimeter = A/W_p
S = slope of the sewer (m/m)
N = roughness coefficient.

In this case Rh = $\pi D^2/4\pi D$ = D/4 = 1.68/4 = 0.42 m

Therefore, $v = (0.42)^{2/3}(0.0009)^{1/2}/0.013 = 1.29$ m/s

Capacity $Q = v.A = 1.29 \times \pi (1.69)^2/4 = 2.86$ m^3/s

Otherwise, the nomograph based on Manning's formula may be used.

Program 5.5 Listing:
Capacity and velocity of a sewer

```
'*********************************
'EXAMPLE 5.5: Manning's Formula
'*********************************
Public Class Form1

    Private Sub Form1_Load(ByVal sender As
        System.Object, ByVal e As
        System.EventArgs) Handles MyBase.Load
        Me.Text = "Example 5.5: Manning's Formula"
        Me.MaximizeBox = False
        Me.FormBorderStyle =
            Windows.Forms.FormBorderStyle.FixedSingle
        Label1.Text = "Sewer diameter (m)"
        Label2.Text = "Slope"
        Label3.Text = "Frictional coefficient, n"
        Label4.Text = ""
        TextBox3.Text = "0.013"
        Button1.Text = "&Calculate"
    End Sub

    Private Sub Button1_Click(ByVal sender As
        System.Object, ByVal e As
        System.EventArgs) Handles Button1.Click
        Dim D, S, n, v, Q As Double
        D = Val(TextBox1.Text)
        S = Val(TextBox2.Text)
        n = Val(TextBox3.Text)
        v = ((D / 4) ^ (2 / 3)) * (Math.Sqrt(S)) / n
        Q = v * Math.PI * (D ^ 2) / 4
        Label4.Text = "Velocity (v) = "
            + FormatNumber(v, 2) + " m/s"
        Label4.Text += vbCrLf
        Label4.Text += "Capacity (Q) = " +
            FormatNumber(Q, 2) + " m3/s"
    End Sub
End Class
```

Example 5.6

1) A 460 mm diameter sewer with n = 0.013 is laid on a grade of 0.02. Determine:
   a) capacity when flowing half full.
   b) The velocity when the depth of flow is 115 mm.
2) Write a computer program to compute capacity of a sewer when flowing half full and velocity for a ceryain depth of flow given sewer diameter value of n and gradient at which sewer is laid.
3) Verify your program by solving example 5.6.1.

Solution

Full capacity, Q = v.A

$$Q = v.\ A = 0.312 \frac{D^{8/3} S^{21/2}}{n} = 0.312 \frac{0.46^{8/3} 0.02^{21/2}}{0.013} = 0.43\ m^3/s$$

a)capacity when flowing half full = 0.43/2 = 0.215 m³/s

b) Central angle: cos ϕ/2 = 1 − 2d/D = 1 − 2x0.115/0.46 = 0.5

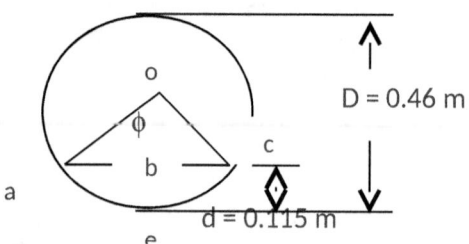

Thus, ϕ = 120°

Wetted perimeter, wp = πDϕ/360

Area = (D²/4)(πϕ/360 −((sinϕ)/2)

Hydraulic radius, Rh = (D/4)(1 − 360sinϕ/2πϕ) = (0.46/4)(1 − 360 sin120/2πx120 = 0.067 m

Thus, velocity = (0.067)²/³ x (0.02)¹/²/0.013 = 1.79 m/s

Program 5.6 Listing:

Capacity of sewer when flowing half full and velocity for a ceryain depth of flow

```
'********************************
'EXAMPLE 5.6: Manning's Formula II
'********************************
```

```vbnet
Imports System.Math

Public Class Form1

    Private Sub Form1_Load(ByVal sender As
        System.Object, ByVal e As
        System.EventArgs) Handles MyBase.Load
        Me.Text = "Example 5.6: Manning's Formula II"
        Me.MaximizeBox = False
        Me.FormBorderStyle =
            Windows.Forms.FormBorderStyle.FixedSingle
        Label1.Text = "Sewer diameter (mm)"
        Label2.Text = "Slope"
        Label3.Text = "Frictional coefficient, n"
        Label4.Text =
            "Find velocity at what depth (mm)?"
        Label5.Text = ""
        TextBox3.Text = "0.013"
        Button1.Text = "&Calculate"
        Me.Height = 180
    End Sub

    Private Sub Button1_Click(ByVal sender As
        System.Object, ByVal e As
        System.EventArgs) Handles Button1.Click
        Dim D, d2, S, n, v, Q As Double
        Dim angle, Rh As Double

        Me.Height = 256

        D = Val(TextBox1.Text)
        S = Val(TextBox2.Text)
        n = Val(TextBox3.Text)
        d2 = Val(TextBox4.Text)
        D /= 1000
        d2 /= 1000
        'find v and Q at full flow
        v = ((D / 4) ^ (2 / 3)) * (Sqrt(S)) / n
        Q = v * PI * (D ^ 2) / 4
        Label5.Text = "At full flow:" + vbCrLf
        Label5.Text += "Velocity (v) = "
            + FormatNumber(v, 2) + " m/s"
        Label5.Text += vbCrLf
        Label5.Text += "Capacity (Q) = "
            + FormatNumber(Q, 2) + " m3/s"
        'find the central angle from its cosine
        Dim tmp = 1 - (2 * d2 / D)
            'this is the value of cosine angle/2
        angle = Acos(tmp)
            'find the angle in radians
```

263

```
        angle = angle * 180 / PI    'convert to degrees
        angle *= 2                  'calculate the angle
        'find hydraulic radius
        Rh = (D / 4) *
            (1 - ((360 * Sin(angle / 180 * PI)) /
                (2 * PI * angle)))
        'find v at new depth
        v = (Rh ^ (2 / 3)) * (Sqrt(S)) / n
        Label5.Text += vbCrLf
        Label5.Text += "Velocity at new depth = " +
            FormatNumber(v, 2) + " m/s"
    End Sub
End Class
```

Example 5.7

A sewer of diameter 200 mm is to flow at a depth of 30 % of full depth on a grade that enables self-cleaning, similar to that at full depth at a velocity of 0.75 m/s. Find the needed grades, associated velocities and rates of flow at full depth and 30 % depth. (Assume n = 0.013)

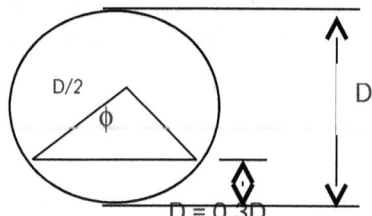

Solution

$$v_{full} = \frac{\left[\dfrac{D}{4}\right]^{2/3} S^{21/2}}{n}$$

$$0.75 = \frac{\left[\dfrac{0.2}{4}\right]^{2/3} S^{21/2}}{0.013}$$

This yields a needed grade at full depth of 5.2 %o

$$Q_f = \frac{\pi D^2 v_f}{4} = \frac{\pi 0.2^2 \times 0.75}{4} = 0.024 \, m^3/s$$

Central angle = ϕ = 2 cos^{-1}0.2D/0.5D = 132.8°

Ratio of areas at partial and full flow $(A_p/A_f) = \phi/360 - (\sin\phi)/2\pi = 132.8/360 - \sin 132.8/2\pi = 0.252$

From the hydraulic elements diagram for ratio of areas of 0.252
$v_p/v_f = 0.776$

The velocity at a depth of 30% = 0.776x0.75 = 0.582 m/s

$Q_p/Q_f = 0.196$, this gives a discharge at a depth of 30% of
$Q_p = 0.196 \times 0.024 = 0.0046$ m^3/s

Example 5.8

1. A sewer is designed to carry a maximum flow of 2 m^3/min, and a minimum flow of 0.674 m^3/min. If the allowable self-cleaning velocity is 0.75 m/s in both cases, find:
    - Standard size and slope of sewer needed.
    - Depth of flow (assume n = 0.013).

2. Write a computer program to compute standard size and slope of a sewer and depth of flow needed given maximum flow to be carried by sewer, minimum flow and allowable self-cleaning velocity.

3. Verify your program by solving example 5.8.1.

Solution

Ratio of flows:

$Q_p/Q_f = 0.674/2 = 0.337$, for this value the ratio of areas could be determined from the hydraulic element diagram as equal to (A_p/A_f) 0.374.

Area at partial flow $A_p = Q_p/v = 0.015$ m^2.

Therefore, $A_f = 0.015/0.374 = 0.041 = \pi D^2/4$

This yields a diameter, D of 228 mm

The hydraulic radius, Rh = D/4 = 0.228/4 = 0.057 m

From diagram of partially full pipe (or from table 5.2)
$(r_h)_p/(R_h)_f = 0.857$

Then, $(r_h)_p = 0.857 \times 0.057 = 0.49$ m

Thus, slope (S) = 5.3‰

b) From hydraulic element diagram $d_p/D_f = 0.4$

Thus depth of flow, $d_p = 0.4 \times 228 = 91$ mm

Program 5.8 Listing:
Standard size, slope of a sewer and depth of flow

```
'***************
'EXAMPLE 5.8
'***************
Imports System.Math

Public Class Form1
    Dim dpDfRatio(19) As Double
    Dim rhDfRatio(19) As Double
    Dim ApAfRatio(19) As Double
    Dim rhRhRatio(19) As Double
    Dim vpvfRatio(19) As Double
    Dim QpQfRatio(19) As Double

    Private Sub Form1_Load(ByVal sender As
        System.Object, ByVal e As
        System.EventArgs) Handles MyBase.Load
        Me.Text = "Example 5.8"
        Me.MaximizeBox = False
        Me.FormBorderStyle =
            Windows.Forms.FormBorderStyle.FixedSingle
        Label1.Text = "Max. flow (m3/min)"
        Label2.Text = "Min. flow (m3/min)"
        Label3.Text = "Self-cleaning velocity (m/s)"
        Label4.Text = "Frictional coefficient, n"
        Label5.Text = ""
        TextBox4.Text = "0.013"
        Button1.Text = "&Calculate"
        init_Ratios_Table()
        Me.Height = 193
    End Sub

    Private Sub init_Ratios_Table()
        'this sub will initialize the ratio table
        'with data from table 5.2.
        dpDfRatio(0) = 0.1
        dpDfRatio(1) = 0.15
        dpDfRatio(2) = 0.2
        dpDfRatio(3) = 0.25
        dpDfRatio(4) = 0.3
        dpDfRatio(5) = 0.35
        dpDfRatio(6) = 0.4
        dpDfRatio(7) = 0.45
        dpDfRatio(8) = 0.5
        dpDfRatio(9) = 0.55
        dpDfRatio(10) = 0.6
        dpDfRatio(11) = 0.65
```

```
dpDfRatio(12) = 0.7
dpDfRatio(13) = 0.75
dpDfRatio(14) = 0.8
dpDfRatio(15) = 0.85
dpDfRatio(16) = 0.9
dpDfRatio(17) = 0.95
dpDfRatio(18) = 1.0

rhDfRatio(0) = 0.064
rhDfRatio(1) = 0.093
rhDfRatio(2) = 0.121
rhDfRatio(3) = 0.147
rhDfRatio(4) = 0.171
rhDfRatio(5) = 0.193
rhDfRatio(6) = 0.214
rhDfRatio(7) = 0.233
rhDfRatio(8) = 0.25
rhDfRatio(9) = 0.265
rhDfRatio(10) = 0.278
rhDfRatio(11) = 0.288
rhDfRatio(12) = 0.296
rhDfRatio(13) = 0.302
rhDfRatio(14) = 0.304
rhDfRatio(15) = 0.303
rhDfRatio(16) = 0.298
rhDfRatio(17) = 0.286
rhDfRatio(18) = 0.25

ApAfRatio(0) = 0.052
ApAfRatio(1) = 0.094
ApAfRatio(2) = 0.142
ApAfRatio(3) = 0.196
ApAfRatio(4) = 0.252
ApAfRatio(5) = 0.312
ApAfRatio(6) = 0.374
ApAfRatio(7) = 0.436
ApAfRatio(8) = 0.5
ApAfRatio(9) = 0.564
ApAfRatio(10) = 0.626
ApAfRatio(11) = 0.688
ApAfRatio(12) = 0.748
ApAfRatio(13) = 0.804
ApAfRatio(14) = 0.858
ApAfRatio(15) = 0.906
ApAfRatio(16) = 0.948
ApAfRatio(17) = 0.981
ApAfRatio(18) = 1.0

rhRhRatio(0) = 0.254
rhRhRatio(1) = 0.372
```

```
rhRhRatio(2) = 0.482
rhRhRatio(3) = 0.587
rhRhRatio(4) = 0.684
rhRhRatio(5) = 0.774
rhRhRatio(6) = 0.857
rhRhRatio(7) = 0.932
rhRhRatio(8) = 1.0
rhRhRatio(9) = 1.06
rhRhRatio(10) = 1.111
rhRhRatio(11) = 1.153
rhRhRatio(12) = 1.185
rhRhRatio(13) = 1.207
rhRhRatio(14) = 1.217
rhRhRatio(15) = 1.213
rhRhRatio(16) = 1.192
rhRhRatio(17) = 1.146
rhRhRatio(18) = 1.0

vpvfRatio(0) = 0.401
vpvfRatio(1) = 0.517
vpvfRatio(2) = 0.615
vpvfRatio(3) = 0.701
vpvfRatio(4) = 0.776
vpvfRatio(5) = 0.843
vpvfRatio(6) = 0.902
vpvfRatio(7) = 0.954
vpvfRatio(8) = 1.0
vpvfRatio(9) = 1.04
vpvfRatio(10) = 1.073
vpvfRatio(11) = 1.1
vpvfRatio(12) = 1.12
vpvfRatio(13) = 1.134
vpvfRatio(14) = 1.14
vpvfRatio(15) = 1.137
vpvfRatio(16) = 1.124
vpvfRatio(17) = 1.095
vpvfRatio(18) = 1.0

QpQfRatio(0) = 0.021
QpQfRatio(1) = 0.049
QpQfRatio(2) = 0.087
QpQfRatio(3) = 0.137
QpQfRatio(4) = 0.196
QpQfRatio(5) = 0.263
QpQfRatio(6) = 0.337
QpQfRatio(7) = 0.416
QpQfRatio(8) = 0.5
QpQfRatio(9) = 0.586
QpQfRatio(10) = 0.672
QpQfRatio(11) = 0.756
```

```
      QpQfRatio(12) = 0.838
      QpQfRatio(13) = 0.911
      QpQfRatio(14) = 0.978
      QpQfRatio(15) = 1.03
      QpQfRatio(16) = 1.066
      QpQfRatio(17) = 1.074
      QpQfRatio(18) = 1.0
End Sub

Private Sub Button1_Click(ByVal sender As
    System.Object, ByVal e As
    System.EventArgs) Handles Button1.Click
    Dim Qp, Qf, v, n, S, index As Double
    Dim QpQf, vp As Double

    Me.Height = 256

    Qf = Val(TextBox1.Text)     'Qp
    Qp = Val(TextBox2.Text)     'Qf
    v = Val(TextBox3.Text)
    n = Val(TextBox4.Text)
    vp = Val(TextBox5.Text)
    QpQf = Qp / Qf

    'find the index into the table
    For i = 0 To 18
        If QpQf = QpQfRatio(i) Then
            index = i
            Exit For
        End If
    Next
    'find value of Ap/Af from table
    Dim Ap, Af, D, Rh, _rh, dp As Double
    Ap = (Qp / 60) / v
    Af = Ap / ApAfRatio(index)
    D = Sqrt(Af * 4 / PI)
    'D /= 1000 'convert to m
    Rh = D / 4
    Label5.Text = "Hydraulic radius Rh = " +
        FormatNumber(Rh, 3) + " m"
    Label5.Text += vbCrLf
    _rh = rhRhRatio(index) * Rh * 10
    S = Sqrt((v * n) / (_rh ^ (2 / 3)))
    Label5.Text += "rh = " + FormatNumber(_rh, 3)
        + " m"
    Label5.Text += vbCrLf
    Label5.Text += "Slope S = " +
        FormatNumber(S, 3) + "%"
    Label5.Text += vbCrLf
    dp = dpDfRatio(index) * D * 1000
```

```
        Label5.Text += "Depth of flow dp = " +
            FormatNumber(dp, 3) + " mm"
    End Sub
End Class
```

Table (5.2) Hydraulic element of a Circular sewer Running Partly Full

Depth ratio (d_p/D_f)	Hydraulic radius/dia. (r_h/D_f)	Area ratio (A_p/A_f)	Hydraulic radius ratio (r_h/R_h)	For same n	
				Velocity ratio (v_p/v_f)	Discharge ratio (Q_p/Q_f)
0.10	0.064	0.052	0.254	0.401	0.021
0.15	0.093	0.094	0.372	0.517	0.049
0.20	0.121	0.142	0.482	0.615	0.087
0.25	0.147	0.196	0.587	0.701	0.137
0.30	0.171	0.252	0.684	0.776	0.196
0.35	0.193	0.312	0.774	0.843	0.263
0.40	0.214	0.374	0.857	0.902	0.337
0.45	0.233	0.436	0.932	0.954	0.416
0.50	0.250	0.500	1.000	1.000	0.500
0.55	0.265	0.564	1.060	1.040	0.586
0.60	0.278	0.626	1.111	1.023	0.672
0.65	0.288	0.688	1.153	1.100	0.756
0.70	0.296	0.748	1.185	1.120	0.838
0.75	0.302	0.804	1.207	1.134	0.911
0.80	0.304	0.858	1.217	1.140	0.978
0.85	0.303	0.906	1.213	1.137	1.030
0.90	0.298	0.948	1.192	1.124	1.066

| 0.95 | 0.286 | 0.981 | 1.146 | 1.095 | 1.074 |
| 1.00 | 0.250 | 1.000 | 1.000 | 1.000 | 1.000 |

p = partial flow
f = full flow

Table (5.3) Values of (n) in Manning's Formula

Surface Description	n
Plastic pipe, well planed timber evenly laid, asphalted cast iron	0.009
Smooth metal, neat cement	0.010
Asbestos pipes	0.011
Cast iron pipe of ordinary roughness, unplanned timber	0.012
Good concrete, riveted steel pipes, verified clay pipe, brick work (well laid)	0.013
Concrete	0.014
Cast iron pipe, average brick work	0.015
Rough brick, tuberculated iron pipe	0.017
Smooth earth or firm gravel	0.018 – 0.020
Natural channels	0.025 – 0.0035
Ditches and rivers in good order, some stones and weeds	0.030
Ditches and rivers with rough bottom and much vegetation	0.040

Hydraulic Elements Diagram of Circular Sewers (7)

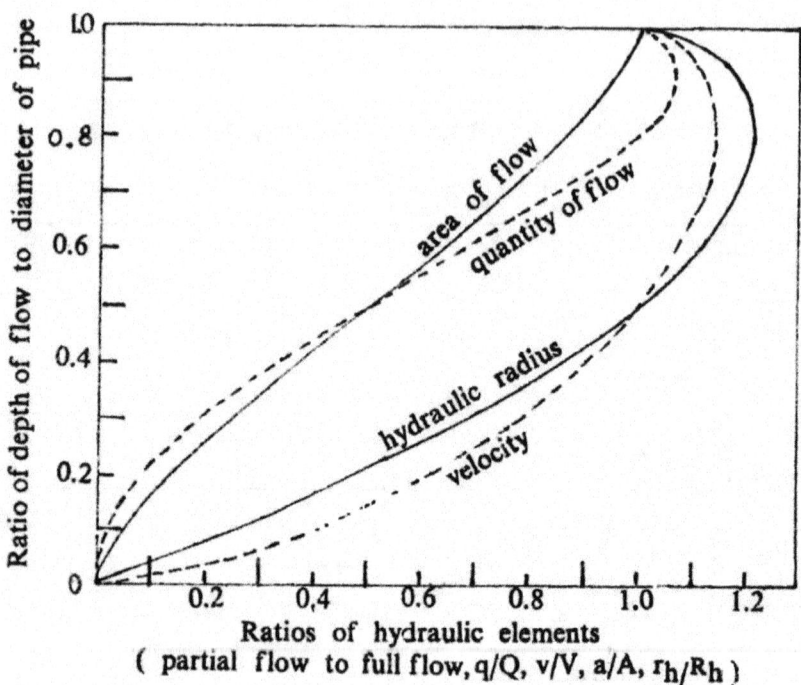

Fig. 5.1 Hydraulic elements diagram of circular sewers.

Theoretical Problems 5.3.2

1) State the merits and demerits of combined and separate sewer systems. What system would

Exercise 5.3.2

Velocity & depth of flow

1) A circular sewer 1.37 m in diameter is laid on a slope of 0.0035. Find the **velocity** and **depth** of flow when a minimum flow of 15 m^3/min occurs. (Take n to be equal to 0.013). (Ans. 1.3m/s 395 mm).

Solution

$$Q_f = vA = \frac{1}{n}\left[\frac{D}{4}\right]^{2/3} S^{1/2} \pi \frac{D^2}{4}$$

$$= 23.976 \, D^{8/3} \, S^{1/2} = 23.976 \, (1.37)^{8/3} \, (0.0035)^{1/2}$$

$$= 3.28 \, m^3/s$$

$v_f = 2.23$ m/s
Ratio of flows $= Q_p/Q_f = (15/60)/(3/28) = 0.076$
From graph of hydraulic element diagram: $(v_p/v_f = 0.59$
Thus, $v_p = 0.59*2.23 = 1.3$ m/s
$a_p = Q_p/v_p = 15/1.3*60 = 0.192$
Therefore, $d = (4*0.192/\pi)^{1/2} = 0.495$ m

2) A sewer of diameter 1.52 m is to carry 85 m^3/min of sewage when flowing full. The sewer is laid on a grade of 0.003 and n = 0.013. Find the **velocity** and **depth** of sewage at a dry weather flow of 1341 m^3/hr. (Ans. 1.86m/s 532 mm).

Solution

$d_f = 1.52$m, $Q_f = 85$ m^3/min, s $= 0.003$, $Q_p = 1314$ m^3/hr $= 22.35$ m^3/min
$Q_p/Q_f = 22.35/85 = 0.263$
From graph, $v_p/v_f = 0.843$

$$v_f = \frac{1}{n}\left[\frac{D}{4}\right]^{2/3}_h S^{1/2} = \frac{1}{0.013}\left[\frac{1.52}{4}\right]^{2/3}(0.003)^{1/2} = 2.21 \, m/s$$

Thus, $v_p = 0.843*2.21 = 1.86$ m/s
$d/D = 0.35$, thus, $d = 0.35*1.52 = 0.532$ m

3) Given that the velocity of flow in a sewer is 0.75 m/s and n = 0.013, compute ratios of **depths** and **velocities** of flow in sewers discharging:
   1) One fourth full flow.
   2) $1/10^{th}$ full flow. (Ans. 0.6 0.5 m/s).

Solution

$v_f = 0.75$m/s, n $= 0.013$
$Q_p/Q_f = ¼ = 0.25$

From graph, d/D = 0.34, v_p/v_f = 0.83
Given v_f = 0.75, thus, v_p = 0.83*0.75 = 0.6 m/s
For Q_p/Q_f = 0.1
From graph, d/D = 0.21, v_p/v_f = 0.62
Given v_f = 0.62*0.75 = 0.5 m/s

4) A sewer of diameter 530 mm when flowing full carries 6.5 m^3/min, while the minimum flow is 1/10 of the maximum flow. Find the **depth** and **velocity** at maximum flow (n = 0.012). (Ans. 0.3m/s 111 mm).

Solution
Q_p/Q_f = 0.1
From graph, d/D = 0.21, d = 0.21*530 = 111 mm
v_p/v_f = 0.62
thus, v_p = 0.62*(6.5/60)/(π*0.53²/4) = 0.3 m/s

5) A circular sewer is expected to carry a flow of 0.07 m^3/s when flowing half full. If the slope is 6 ‰, determine the **size** of the sewer and **velocity** of flow (n = 0.013). (Ans. 380 mm, 3.4 m/s).

Solution
Q_f = 0.07*2 = 0.1 m^3/s, s = 0.006
From graph, d/D = 0.21, d = 0.21*530 = 111 mm
v_p/v_f = 0.62
thus, v_p = 0.62*(6.5/60)/(π*0.53²/4) = 0.3 m/s

$$Q_f = \frac{1}{n}[D]^{8/3} S^{1/2} \frac{\pi}{4^{5/3}} = \frac{1}{n} D^{8/3} S^{1/2} \frac{\pi}{4^{5/3}}$$

$$(D)^{8/3} = 0.075$$

Then D = 379 mm, take a diameter of 380 mm

$$v = \frac{1}{0.012}(0.38)^{2/3} 0.006^{1/2} = 3.39 m/s$$

6) A concrete sewer discharges 1600 m^3/min at a gradient of 0.0009. The channel is trapezoidal in section with a bottom width of 9 m and the slope is in the ratio of 3 horizontal to 1 vertical. Determine the **depth** of flow given that n = 0.013. (Ans. 1.07 m).

274

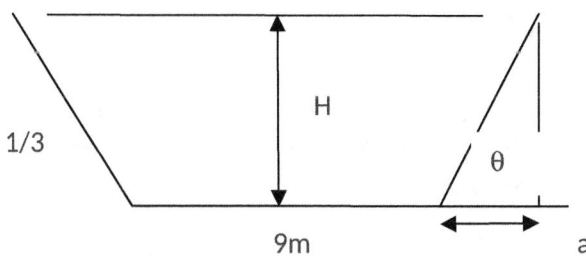

Solution

$Q = 1600$ m3/min $= 26.67$ m^3//s, s $= 0.0009$, n $= 0.013$

Let depth of channel to be $= H$

$\tan(\theta) = 1/3 = H/a$, hen a $= 3H$

Upper long side $= 9 + 3H + 3H = 9 + 6H$

Area $= (9 + 9 + 6H)*H)/2 = 9H + 3H^2$

$Wetted\ perimeter = 9 + 2\sqrt{H^2 + (3H)^2} = 9 + 2\sqrt{10H^2} = 9 + 6.32H$

Hydraulic radius, Rh $= A/wp$

$$hydraulic\ radius = \frac{9H + 3H^2}{9 + 6.32H}$$

From Manning's formula

$$Q = \frac{1}{n}R_h^{2/3}S^{1/2}A = \frac{1}{n}\left[\frac{A}{W_p}\right]_h^{2/3}S^{1/2}A = \frac{1}{n}\frac{A^{5/3}S^{1/2}}{W_p^{2/3}}$$

Then

$$26.67 = \frac{1}{0.013}\frac{(9H + 3H^2)^{5/3}(0.0009)^{1/2}}{(9 + 6.32H)^{2/3}}$$

Assuming values for H, then by trial and error

H (assumed)	LHS of equation
1.5	49.67
1.2	432.8
0.9	19.5
1	23.53
1.1	27.98

1.08	27.05
1.07	26.6

Therefore, H = 1.07 m

7) A sewer of diameter of 450 mm is laid on a slope of 0.0025. Compute the **depth** of flow equivalent to a velocity 0.6 m/s. (Assume n = 0.013). (Ans. 100 m).

Solution

N = 0.013, D = 0.45m, s = 0.0025

$$v=\frac{1}{n}R_h^{2/3}S^{1/2}=\frac{1}{0.013}\left[\frac{0.45}{4}\right]_h^{2/3}(0.0025)^{1/2}=0.9\,m/s$$

$v_p/v_f = 0.6/0.9 = 0.67$
From graph, d/D = 0.23
Thus, d = 0.23*0.45 = 100 mm

8) Determine the **size** of a circular sewer needed to convey 9 m³/min of sewage when the maximum allowable velocity and gradient are 3 m/s and 0.003 respectively. (Ans. 250 mm).

Solution

Q = 9 m³/min = 0.15 m³/s, v = 3 m/s, s = 0.003
Q = vA
Then 0.15/3 = πD²/4
Therefore, D = 0.25 m
Or:

$$Q=0.312\,D^{8/3}S^{1/2}$$

$$v=\frac{1}{n}[\frac{D}{4}]^{2/3}S^{1/2}=\frac{0.397}{n}0.013(D)^{2/3}(S)^{1/2}$$

$$\frac{0.15}{3}=\frac{0.312}{0.397}(D)^2$$

Then, D = 0.25 m

9) Deduce the formula for the **depth** of water in a circular conduit for maximum discharge. Find the depth for maximum discharge in a circular brick sewer 1.5 m in diameter. (Ans. 1.4 m).

276

Solution

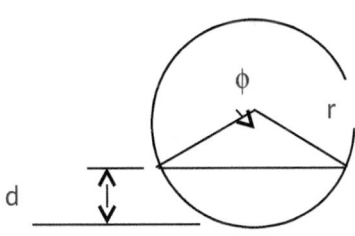

$$Area\ A = r^2\varphi - r^2\cos\varphi\sin\varphi = r^2\left[\varphi - \frac{\sin 2\varphi}{2}\right]$$

Wp = 2rϕ

$$Q = CA\frac{1}{n}\sqrt{R_h S} = C\frac{1}{n}\sqrt{\frac{A^{3S}}{W_p}}$$

For max. Q, A^3/w_p is ma.

$$\frac{dA^3/W_p}{d\varphi} = 0 = \left[\frac{3A^2}{W_p}\right]\frac{dA}{d\varphi} - \left[\frac{A^3}{W_p}\right]\frac{dW_p}{d\varphi}$$

Therefore,

$$\frac{3W_p dA}{d\varphi} = 0 = A\frac{dW_p}{d\varphi}$$

$$3(2r\varphi)(1 - \cos 2\varphi)r^2 = r^2\left[\varphi - \frac{\sin 2\varphi}{2}\right]2r$$

$$3\varphi - 3\varphi\cos 2\varphi = \varphi - \frac{\sin 2\varphi}{2}$$
$$4\varphi - 6\varphi\cos 2\varphi = -\sin 2\varphi$$

Therefore, ϕ = 154°
D = r(1 – cosf) = r(1 + 0.9) = 1.9r
Or, depth = 0.95diameter of circular conduit
Then, depth = 0.95*1.5 = 1.4 m

10) A 750 mm sewer pipe is placed on a slope of 4 mm/m. Taking n
    = 0.013,
    i) At what **depth** of flow is the velocity equal 0.6 m/s?

ii) If the depth of flow is 450 mm, what is the **velocity** and discharge? (UAE, 1989). (Ans. 75 mm, 1.7 m/s).

Solution

i)

$$v = \frac{1}{n} R_h^{2/3} S^{1/2} =$$

$$\frac{1}{n}\left[\frac{D}{4}\right]^{2/3} S^{1/2} = \frac{1}{0.013}\left(\frac{0.75}{4}\right)^{2/3}\left(\frac{4}{100}\right)^{1/2} = 1.594\,m/s$$

$$\frac{v}{v_f} = \frac{0.6}{1.594} = 0.3764$$

from hydraulic element diagram

$$\frac{d}{d_f} = 0.1 \rightarrow d = 0.1\,d_f = 0.1\ x\ 750 = 75\,mm$$

ii) $$\frac{d}{d_f} = \frac{450}{750} = 0.6 \rightarrow \frac{v}{v_f} = 1.073$$

$$\therefore v = 1.073 \times 1.594 = 1.71 \text{ m/s, and}$$

$$\frac{q}{q_f} = 0.672 \rightarrow q = 0.672\left[1.594\ x\ \frac{\pi}{4}(0.75)^2\right]$$

Or, q = 0.672×0.704 = 0.473 m³/s

11) What are the restrictions on the **velocity** of flow in sewer conduits?

A 400 mm sanitary sewer is laid on a slope of 2 units per 1000 units. Find the **velocity** and **discharge** if the sewer is flowing half full. Assuming Manning's n = 0.013

Outline factors involved in evaluation and selection of materials needed for sewer construction. (UAE, 1989). (Ans. 0.74 m/s, 0.05 m³/s).

Solution:

1. i) Storm sewers: min allowable velocity 0.75 – 0.9 m/s.

ii) Sanitary sewers: Vmin = 0.61 – 0.91 m/s

- Sufficient to prevent deposition of suspended materials.
- Required to transport sand and grit.

2. d = 400 mm,   s = 2×10⁻³,   n = 0.013

from chart for

$$\frac{d_P}{D_f}=0.5, \frac{Q_P}{Q_f}=0.5, \frac{V_P}{V_f}=1$$

from chart: $V_f = 0.74$ m/s $Q_f = 0.093$ m³/s
$V_p = V_f = 0.74$ m/s
$Q_p = Q_f/2 = 0.047$ m³/s

or:

$$v_f=\frac{1}{n}R_h^{2/3} S^{1/2}=\frac{1}{0.013}\left[\frac{0.4}{4}\right]^{2/3}(2 \times 10^{-3})^{1/2}=0.741 m/s$$
$$Q_f=0.741 \times \frac{\pi}{4}(0.4)^2=0.093 m^3/s$$

$Q_p = 0.047$ m³/s

3. Use of wastewater, scour conditions, installation requirements, corrosion conditions, flow requirements, infiltration, cost effectiveness, physical properties of pipes, handling requirements.

12) A sewer of diameter 685 mm is laid on a slope of 1 in 500. Find the **depth** of flow equivalent to a velocity of 0.55 m/s. take n = 0.013 (UAE, 1989). (Ans. 103 mm).

Solution

n = 0.013,   D = 0.685 m,   S = 0.002

$$v_f=\frac{1}{1.06}\left(\frac{0.685}{4}\right)^{1/3}(0.002)^{1/2}=1.06 m/s$$

$$\frac{v_p}{v_f}=\frac{0.55}{1.06}=0.519$$

From graph d/D = 0.15
Therefore: d = 0.15×685 = 102.75 mm

13) Outline major advantages and disadvantages of sewers. (UAE, 1989). A circular storm sewer is needed to serve an

area of 600 ha. The average ground slope is 1 in 2500. Using Lloys Daries method, determine the **size** of the sewer at point of discharge, using the following information:

C = 0.44, Time of concentration = 20 minutes,

$$I=\frac{750}{(t+10)}(mm/hr) ,\qquad\qquad Q = 0.278 \text{ CIA (Ans. 3.96 m).}$$

Solution

b] A = 600 ha = 6 km², C = 0.44, S = 1/2500 = 4x10⁻⁴,

$$I=\frac{750}{(20+10)}=25\,mm/hr$$

Q = 0.278 C.I.A = 0.278×0.44×25×6 = 18.348 m³/s = 1100.88 m³/min

From nomograph take sewer diameter 3.96 m at specified slope and this yields a flow of 18.82 m³/s = 1129.3 m³/min.

14) What are the benefits of a **sewerage** system? For a sewer of diameter D, flowing partially full at a depth d, and velocity of sewer when flowing full v_f, prove any one of the following hydraulic elements:

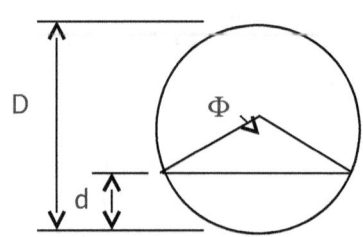

$$\frac{(rH)_p}{D}=\frac{1}{4}-\frac{360}{\pi\varphi}\left[\frac{1}{2}-\frac{d}{D}\sqrt{\frac{d}{D}-[\frac{d}{D}]^2}\right]$$

$$\frac{(rH)_p}{D}=\frac{1}{4}-\frac{360}{\pi\varphi}\left[\frac{1}{2}-\frac{d}{D}\sqrt{\frac{d}{D}-[\frac{d}{D}]^2}\right]$$

$$\frac{(rH)_p}{(rH)_f}=1-1440\,\pi\,\varphi\left[\frac{1}{2}-\frac{d}{D}\sqrt{\frac{d}{D}-[\frac{d}{D}]^2}\right]$$

$$\frac{v_p}{v_f}=\left[\frac{(rH)_p}{(rH)_f}\right]^{2/3}=\left[1-\frac{1440}{\pi\varphi}\right]\left[\frac{1}{2}-\frac{d}{D}\sqrt{\frac{d}{D}-\left[\frac{d}{D}\right]^2}\right]^{2/3}$$

$$\frac{Q_p}{Q_f}=\frac{v_pA_p}{v_fA_f}=$$

$$\left[1-\frac{1440}{\pi\phi}\right]\left[\frac{1}{2}-\frac{d}{D}\sqrt{\frac{d}{D}-\left[\frac{d}{D}\right]^2}\right]^{2/3}\left[\frac{\phi}{360}-4\pi\left[\frac{1}{2}-\frac{d}{D}\right]\sqrt{\left[\frac{d}{D}-\left[\frac{d}{D}\right]^2\right]}\right]$$

A sewer of diameter 150 mm is to flow at a depth of 40 percent on a grade that enables self-cleaning, similar to that at full depth at a velocity of 0.85 m/s. find the needed **grades**, associated **velocities**, and **flow rates** at full depth and at 40 percent depth. Take Manning's coefficient of friction to be equal to 0.013. (SUST, 2001). (Ans. $9.7*10^{-3}$, 0.015 m3/s; 0.76 m/s).

Solution

D= 150 mm= 0.15 m, d =0.4D, v_f =0.85 m/s, n= 0.013

- Find slope as

$$I=\sqrt{\frac{v_fn}{rH^{2/3}}}=\sqrt{\frac{0.85\,x\,0.013}{\left(\frac{0.15}{25}\right)^{2/3}}}=9.7\,x\,10^{-3}$$

- Compute $Q_f = (\pi/4)\,D^2\,v_f= (\pi/4)\,(0.15)^2x0.85 = 0.015$ m³/s
- Find central angle

$$\Phi=2\,cose^{-1}\left(\frac{\frac{D}{2}-d}{\frac{D}{2}}\right)$$

$= 2 \cos^{-1}(1-(2d/D)) = 2 \cos^{-1}(1- 2x0.4)$
$= 156.93°$

- Find ratio of areas at partial and full flow

$$\frac{A_p}{A_f}=\frac{\Phi}{360}=\sin\frac{\Phi}{2\pi}=\frac{156.93}{360}-\sin\frac{156.93}{2\pi}=0.37$$

- Determine $A_p = (\pi/4)\,(0.15)^2x0.37 = 6.5x10^{-3}$ m²
Or:

$$A_p = \frac{D^2}{4}\left(\frac{\Phi\pi}{360} - \frac{\sin\Phi}{2}\right) = \frac{0.15^2}{4}\left(\frac{\pi x\,156.93}{360} - \frac{\sin 156.93}{2}\right)$$

$$= 6.6 \times 10^{-3} m^2$$

- Determine hydraulic radius as:

$$rH = \frac{D}{4}\left(1 - \frac{360\sin\Phi}{2\pi\Phi}\right) = \frac{0.15}{4}\left(1 - \frac{360\sin 156.93}{2\pi\,156.93}\right) = 0.032$$

$$\therefore \quad v_p = 0.032^{2/3}\frac{(9.7 \times 10^{-3})^{1/2}}{0.013} = 0.76\,m/s$$

$$a_p = v_p\,A_p = 0.76 x\,0.37 =$$

15) A sewer line is to be laid to serve a community of 200 persons/ha in a district of area 20 ha. The average water supply is 175 L/c/d. the available slope is 1 in 60. Using manning's formula (with n = 0.013), select a suitable **diameter** of sewer to carry the peak discharge, flowing half full in the section. (UAE, 1991).

16) A rough brick sewer of diameter 1.52 m has a slope of 0.0008 with n = 0.017. Determine the **capacity** when flowing full and the **velocity** of flow. (Ans. 95 m³/min, 0.87 m/s).

Solution

$$Q = 0.312*(1.52)^{8/3}(0.0008)^{1/2}/0.017 = 1.58\,m^3/s = 95\,m^3/min$$

$$v = \frac{1}{n}R_h^{2/3}S^{1/2} = \frac{1}{0.017}\left[\frac{1.52}{4}\right]_h^{2/3}(0.0008)^{1/2} = 0.87\,m/s$$

Flow rate

17) For 255 mm diameter concrete sewer laid to a slope of 1 in 400, compute the **flow rate** and **velocity** when running full. (Ans. 1.9 m³/min, 0.61 m/s).

Solution

$$Q = \frac{0.312*(0.225)^{8/3}(0.0025)^{1/2}}{0.013} = 0.031\,m^3/s = 1.9\,m^3/min$$

$$v=\frac{1}{n}R_h^{2/3}S^{1/2}=\frac{1}{0.013}\left[\frac{0.225}{4}\right]_{Jh}^{2/3}(0.0025)^{1/2}=0.61\,m/s$$

18) A circular sewer is designed to give a velocity of 0.75 m/s and a discharge of 6 m³/min. Find the **velocity** and **discharge** when the sewer is flowing at 20% full depth. (Assume n = 0.013). (Ans. 0.55 m³/min, 0.46 m/s).

Solution

d/D = 0.2, v_p/v_f = 0.615
v_p = 0.615*0.75 = 0.46 m/s
Q_p/Q_f = 0.087, Q_p = 0.087*6 = 0.522 m³/min

19) A brick lined sewer has a semi-circular bottom and vertical side wall 0.6 m apart. If the slope is 1 in 1000, determine the **discharge** when the maximum depth of water is 0.9 m. (Assume n = 0.017). (Ans. 17 m³/min).

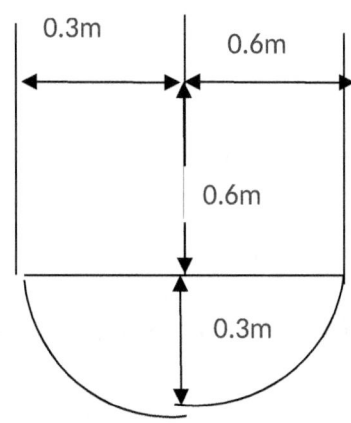

Solution

S = 1/1000
A = π(.3)²/8 + 0.6*0.6 = 0.395 m²
w_p = π(0.3)/2 + 0.6 + 0.6 = 1.67 m
R_h = A/w_p = 0.395/1.67 = 0.237 m

$$v=\frac{1}{n}R_h^{2/3}S^{1/2}=\frac{1}{0.017}[0.237]_h^{2/3}(0.001)^{1/2}=0.712\,m/s$$

Q = vA = 0.712*0.395 = 0.28 m³/s = 17 m³/min

Slope of sewer

20) A sewer, made of average brick work, of diameter 255 mm carries flow at a velocity of 0.9 m/s. Find the minimum **slope** at full depth given n = 0.015. (Ans. 7.2‰).

Solution

$$v = \frac{1}{n}\left[\frac{D}{4}\right]^{2/3} S^{1/2}$$

$$0.9 = \frac{1}{0.015}\left[\frac{0.255}{4}\right]^{2/3} S^{1/2}$$

Therefore, s = 7.2 ‰

21) Compute the **gradient** at which a circular sewer 760 mm in diameter, should be laid in order to obtain a minimum velocity of 1.22 m/s when flowing full or half full (n = 0.103). (Ans. 2.3‰).

Solution

$$v = \frac{1}{n}\left[\frac{D}{4}\right]^{2/3} S^{1/2}$$

$$1.22 = \frac{1}{0.013}\left[\frac{0.76}{4}\right]^{2/3} S^{1/2}$$

Therefore, s = 2.3 ‰

$Q_f = v_f A = 1.22*\pi*(0.76)^2/4$

$Q_{half-full} = v_f A = 1.22*\pi*(0.76)^2/4*2$

$v_{half-full} = (1.22*\pi*(0.76)^2/4*2)/(*\pi*(0.76)^2/8) = 1.22$

Thus, s = 2.3‰

22) A circular sewer is expected to carry 0.08 m³/s when flowing full while the grade adopted is the minimum allowable one. If n = 0.012, determine the **size** and **grade** that is to be used. (Ans. 380 mm, 1.9‰).

Solution

$$Q_f = 0.08 = \frac{1}{n}\left[\frac{D}{4}\right]^{2/3} S^{1/2}\frac{\pi D^2}{4}$$

$$0.08 = \frac{1}{0.012}[D]^{8/3} S^{1/2}\frac{\pi}{4^{5/3}}$$

284

$$v=\frac{1}{n}\left[\frac{D}{4}\right]^{2/3}S^{1/2}=0.75$$

A = Q_f/v = 0.08/0.75 = 0.1067 = $\pi D^2/4$

Thus, D = 0.369 m, use diameter of 380 mm

Then, $s^{1/2}$ = 0.012*0.75/$(0.38/4)^{2/3}$ = 0.043

S = 1.9 ‰

23) A circular sewer 1.52 m in diameter is to discharge 1 m³/s when flowing half full. If the inner surface is finished with concrete, on what **slope** should the line be laid to insure uniform flow? (Assume n = 0.012). (Ans. 0.62‰).

Solution

Q = 01 m³/s, n = 0.012, D = 1.52 m

Rh = $(\pi D^2/4)/(\pi D/2)$ = D/4= 1.52/4 = 0.38 m

$$Q=\frac{1}{n}R_h^{2/3}S^{1/2}A/2\,half\,flow$$

$$Q=\frac{1}{n}R_h^{2/3}S^{1/2}A/2=\frac{1}{0.012}0.38^{2/3}S^{1/2}\frac{\pi*1.52^2}{8}=39.666\sqrt{s}$$

S = 0.6 ‰

Design population

24) Outline major factors to be considered in design of sewers. Compute the **design population** that can be served by a 300 mm sanitary sewer laid at a slope of 0.35 percent. Assume manning coefficient to be equal to 0.012 and that the daily design flow per person is 2000 litre per person. (UAE, 1989). (Ans. 2679).

Solution

a)

Topography – Surface, Subsurface – Existing system – Water supply – Public services – Structures – Town plan – Industries – Population – Metrological and Hydrological town history – Political information – Finance

b) $\frac{1}{n}AR_h^{2/3}S^{1/2}$ D = 300 mm,   s = 0.4/100

$$A=\frac{\pi}{4}D^2=\frac{\pi}{4}(0.3)^2=0.0707\,m^2$$

$$R_h = \frac{A}{W_p} = \frac{\pi}{4} x \frac{D^2}{\pi D} = \frac{D}{4} = \frac{0.3}{4} = 0.075 m$$

$$s = 3.5 \times 10^{-3}$$

$$Population = \frac{62}{2000/3600 \, x \, 24} = 2679 \, persons$$

$$Q = \frac{1}{0.012} x \, 0.0707 \, x \, (0.075)^{2/3} (3.5 \, x \, 10^{-3})^{1/2} = 0.062 \, m^3/s$$

$$= 62 \, L/s$$

Time of concentration

25) A circular storm sewer of diameter 1500 mm is to serve an area of 30 ha, the average ground slope is 1 in 5000. Taking coefficient c = 0.45, compute **time** of **concentration**, given that average intensity of rainfall: (UAE, 1989). (Ans. 18 min).

$$I = \frac{750}{(t+10)} (mm/hr)$$

d) A = 30 ha = 0.3km², $s = \frac{1}{5000} = 2 \, x \, 10^{-4}$

from monograph for D = 1500 mm, and s = 2×10⁻⁴ Q = 1 m³/s

[or,

$$Q = \frac{1}{n} R_h^{2/3} S^{1/2} \quad A = \frac{1}{n} \left(\frac{D}{4}\right)^{2/3} S^{1/2} \frac{\pi}{4} D^2 =$$

$$\frac{1}{0.013} \left(\frac{1.5}{4}\right)^{2/3} (2x \, 10^{-4})^{2 x 10^{-4}} \frac{\pi}{4} (1.5)^2 = 1 m^3/s$$

]

Q = 1000 L/s = 1 m³/s = 0.278 C.I.A = 0.278×45I×0.3

Or, I = 26.65 = $\frac{750}{t+10}$, $t = \frac{750}{26.65} - 10 = 18 \, min$

Wastewater flows

26) Outline major factors affecting quantity and flow patterns of wastewater in a community.

Compute various per capita wastewater **flows** (maximum daily, average daily, ultimate (peak) and minimum daily flow) for a population of 40,000 using:
- Annual average water consumption of 175 L/capita/d.
- DWF amounting to 80 percent of consumption.
- Minimum flow equals 40 percent of average flow. (UAE, 1989). (Ans. 245, 357, 392, 56 L/c).

Solution:

- Population, water use, water demand, water consumption, industrial & commercial requirements, expansion of service, groundwater, topography of area.
- – annual average daily water consumption = 175 L/c

  DWF (wastewater flow) = 80% water consumption

  $$= 0.8 \times 175 = 140 \text{ L/c}$$
- Factor for max daily, or max 24 hr (from figure) = 1.7 for P > 40

  $\therefore$ max daily flow = $140 \times 1.7 = 245$ L/c
- Design average daily max flow considering 16 hr activity = $238 \times 24/16 = 357$ L/c
- Peak design (ultimate flow)

  Peak factor from figure (for P = 40) = 2.8

  $\therefore$ ultimate flow (peak) = $140 \times 2.8 = 392$ L/c
- Min. daily flow = 40% average flow = $0.4 \times 140 = 56$ L/c

Sulfide buildup sewer

27) Illustrate effects of sulfides in a sanitary sewer.

Outline sulfide control method.

A 600 mm (24″) diameter pipe sanitary sewer has been designed to carry 0.02 m³/s at a slope of 0.1%. Taking wetted perimeter as 1.99 ft (0.6 m), and surface width of flow 1.79 ft (0.54 m), give an indicator of the likelihood of

sulfide buildup in the sewer, knowing that the effective biochemical oxygen demand is 460 mg/L. (UAE, 1990). (Ans.18,192).

Solution:

a) - impede proper treatment.
- Public complaint (odour).
- H_2S (corrosive, odour, toxic)

b) - Velocity to transport solids.
- Chemical treatment (Cl_2, hypochlorite, Ferrous sulfate, H_2O_2)
- Dissolving of air.
- Addition of NaOH.
- Injecting air.
- Ventilation.

c) d = 600 mm, Q = 0.02 m³/s = 0.7 cfs,   s   =
0.1/100,
P = wetted = 1.99´, b = width = 1.79´,   BOD
= 460 mg/L

$$Z = \frac{BOD_s}{s^{0.5} Q^{0.33}} \frac{P}{b} = \frac{460}{(0.1/100)^{0.5}(0.02)^{0.33}} \times \frac{1.99}{1.79} = 18,192$$

Z	Sulfide condition
< 5000	Sulfide rarely generated
500 ≤ Z ≤ 10,000	Marginal condition
> 10,000	Sulfide generation common

Since Z > 10,000 ∴ sulfide generation common

0.2	0.11

Vector control

28) a] How can you increase the velocity of flow in an open channel? Give reasons justifying needed increase of velocity from a vector control point of view.

b] A village located in a valley gets flooded after heavy rains due to its rather low land and enclosed topography. A stream is about 10 km away and has a high-water level about 8 m below the lowest point of the valley. The watershed area contributing runoff to the valley is 75 ha. Protection is desired against a rainfall of 13 cm/hr, for duration of one and a half hours. **Design** the drainage ditch. [Take roughness coefficient to be equal to 0.025, runoff coefficient for small agricultural water sheds in rolling country side to be 0.4, and assume that the local anopheline vectors would require approximately 10–14 days to develop from egg to the adult stage.]. (SQU, 1991).

Theoretical problems

1) Discuss importance of recording **population** in a certain municipality.

2) "Water requirements for **fire-fighting**, are little compared to public consumption, but the rate of consumption is very high during the period of fire quenching". Discuss this statement. (DU, 2012).

Solution

Factors for the spread and dispersal of fire include:

- Type of material from which structure is constructed.
- Type of fuel used for cooking.
- Condition of electrical wiring.
- Presence of industries using or producing burning materials.
- Industries related to explosive materials.
- Minimum number of main taps required for fire-fighting.
- Size of each hydrant.
- Number of flaring fires.

3) The objectives of the development of water supply **legislation** and laws are suitable for adoption in any place and anytime by anyone working to achieve safety and confidence of its citizens. Discuss factors affecting implementation of standards. (DU, 2012).

Solution

1. Culture, habits, opinions, beliefs, taboos and local habits and customs,
2. Local sustainable development, and general objectives assigned to them,
3. Socio-economic status of citizens.
4. Type of diet and eating patterns
5. Availability, reliability and accessibility of the needed amounts of water.
6. Existence of efficient and suitably run treatment plants..etc WTP operation
7. Methods used for disposal of human, animal and municipal waste
8. Industries established within the country and their national impact, & industrial expansions and growth, regulations governing disposal and discharge of industrial and agricultural effluent along surrounding environs
9. Annexation and extension of services and amenities
10. Climate
11. Degree of self-help and community participation,
12. Existing technology,
13. Elements of training.
14. Legislation and binding standards protecting the environment may.
15. Local and central, laboratories, centres of research and development, units of information, and exchange of technology and expertise.
16. Qualified managerial device.

4) "Water with **corrosive** characteristics can cause problems in distribution network and residential plumbing system. How can you generally, classify problems encountered? (DU, 2012).

Solution

Generally, problems encountered may be classified according to the following categories:

a) Health problems: these problems originate from dissolution of certain substances into water from transmission and distribution pipelines or plumbing system.

b) Economic problems: these problems result from reduced service life of materials through its deterioration within transmission, distribution, and plumbing system.

c) Aesthetic problems: these problems are due to dissolution of certain metallic substances into the water.

5) Health is defined by the WHO as "A state of complete physical, mental & social well-being & not merely the absence of disease or infirmity." Discuss major groups of **diseases** related to water and environmental sanitation from an environmental engineering prospective. (DU, 2012).

Solution

1) Water borne diseases (Water quality related diseases).
2) Water washed diseases (Water quantity and accessibility related).
3) Water contact diseases (Water-based, Body - of - water related) diseases.
4) Water - related / Insect-vector carrier (Water site related) Diseases.
5) Sanitation - related diseases (Fecal polluted soil related).

6) Write briefly about three of the following:
- Appropriate wastewater **discharge** techniques.
- Methods of evaluating wastewater **flows** and governing inter-relationships.
- **Sources** of wastewater and associate environmental impacts.
- **Corrosion** of sewers and abatement means?

7) Write briefly about three of the following:
- Problems associated with improper wastewater **discharges**.

291

- Inter-relationships between averages, maximum, minimum and peak wastewater **flows**? Illustrate – with sketches – their daily and seasonal variations.
- Suitable means of estimating wastewater **flows** in a rural area.
- **Sources** and fate of municipal wastewater.
- Problems encountered with **corrosive** water and the most important corrosion potential indicators?

Chapter Six

Wastewater Treatment

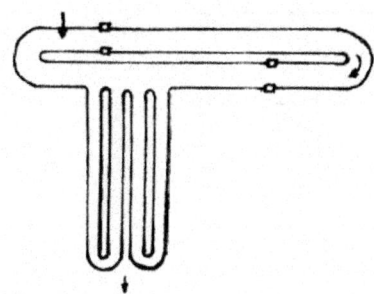

6.1 Grit Chambers

The disadvantages of grit chambers include:
- Promotion of excessive wear of mechanical parts and impellers of sludge pumps.
- Clogging of transmission pipes by deposition.
- Accumulation in sludge holding tanks, sludge digestion units and inverted siphons.

The ideal section grit channel is the best velocity controlled grit chamber and it is parabolic in cross-sectional area. Therefore, the surface area of the channel is directly proportional to the flow through it. This indicates that overflow rate is constant. The width of the channel at any point above invert for a settlement velocity of 30 cm/s is given by:

$$B = 4.92 \frac{Q_{max}}{H_{max}}$$

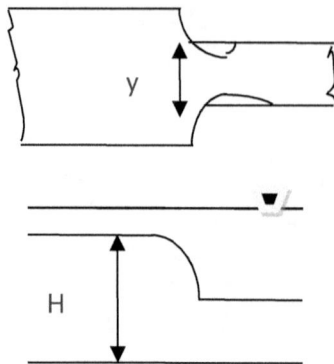

Where:
B = width of the channel (m)
Q_{max} = maximum flow rate (m³/s)
H_{max} = depth of flow (m)

The velocity through the channel of about 30 cm/s is to allow deposition of grit particles and resist settlement of organic matter. Velocity control may be by a flow control device such as a standing-flume, vertical throat, proportional flow weir.
For standing wave flumes:
$Q = 1.71 \ y \ H^{1.5} = 1.14 \ B \ H^{1.5}$
Where:
y = throat width

$$y = \frac{2}{3} B (m)$$

H = depth of flow, Q, through the flume (m)

Example 6.1

1) Design the grit chamber that consists of four mechanically cleaned channels, each designed to carry a maximum flow of 780 l/s for a maximum depth of 0.8 m.
2) Write a computer program to design a grit chamber given number of mechanically cleaned channels, maximum flow to be carried and maximum depth.
3) Verify your program by solving example 6.1.1.

294

Solution

Width of chamber = 4.92x0.78/0.8 = 4.8 m

If four channels are used, width of each channel will be 1.2 m.

Use a standing wave flume in channels for maintaining the velocity. The throat width

Y = (2/3)B = 2x1.2/3 = 0.8 m

Take the length of channel = 18 x 0.8 = 12.4 m

Program 6.1 Listing:

Design a grit chamber

```
'******************
'EXAMPLE 6.1
'******************
Public Class Form1

    Private Sub Form1_Load(ByVal sender As
        System.Object, ByVal e As
        System.EventArgs) Handles MyBase.Load
        Me.Text = "Example 6.1"
        Label1.Text = "Max. flow (1/s)"
        Label2.Text = "Max. depth (m)"
        Label3.Text = "No. of channels"
        Label4.Text = ""
        Button1.Text = "&Calculate"
    End Sub

    Private Sub Button1_Click(ByVal sender As
        System.Object, ByVal e As
        System.EventArgs) Handles Button1.Click
        Dim Q, d, ch, w, Y As Double
        Q = Val(TextBox1.Text)
        d = Val(TextBox2.Text)
        ch = Val(TextBox3.Text)
        Q /= 1000

        w = 4.92 * Q / d
        w /= ch
            'divide total width by number of channels
        Y = (2 / 3) * w

        Label4.Text = "Using " + ch.ToString
            + ": each channel width = " _
            + FormatNumber(w, 2) + " m"
        Label4.Text += vbCrLf + "Throat width Y = " +
            FormatNumber(Y, 2) + " m"
```

```
        End Sub
End Class
```

Table (6.1) Grit Chamber Typical Design Parameters

Parameter	Value
Detention time (s)	60
Horizontal design velocity (cm/s)	30
Equivalent diameter of grit removal (mm)	0.3
Specific gravity of particles captured	2.65
Length of chamber (m)	$18H_{max}$

6.2 Sedimentation

Settlement techniques are adopted to:

1. Get rid of smaller particles of mineral origin.
2. Remove lighter but, settleable organic solids.
3. Extract organic sewage particles prior to discharge.

The factors that influence the settlement of sewage particles include:

1. Sewage rheological properties.
2. Specific gravity of particles.
3. Frictional resistance forces.
4. Velocity of flow through sedimentation basin.
5. Factors that produce erratic tank behavior.

Example 6.2

1) Calculate the settling velocity of spherical discrete particles in water at $15°$ C if their relative density is 1.42 and their diameter is 0.1 mm.
2) Write a computer program to calculate the settling velocity of spherical discrete particles given settling fluid temperature, relative density and diameter of particles.
3) Verify your program by solving example 6.2.1.

Solution

Stoke's law states that:

$$V_{st} = \frac{gxD^2(s.g-1)}{18v}$$

Where:

v_{st} = Settling velocity, m/s.

g = Gravitational acceleration, m/s^2.

D = Diameter of spherical particle, m.

s.g.= Specific gravity .

v = Kinematic viscosity, m^2/s

$$v_{st} = \frac{9.81 \, x \left(0.1 \, x \, 10^{-3}\right)^2 \left(1.42 - 1\right)}{18 \, x \, 1.141 \, x \, 10^{-6}} = 2 \, x \, 10^{-3} \, m/s$$

Check Reynolds number

Re = $v_{st}.D/v$ = 2x10^{-3}x0.1x10^{-3}/1.141x10^{-6} = 0.18 (less than 0.5, OK.).

Program 6.2 Listing:

Settling velocity of spherical discrete particles

```
'********************************
'EXAMPLE 6.2: Settling Velocity
'********************************
Public Class Form1
    Dim rho(21) 'dynamic viscosity of water
    Const g = 9.81     'gravitational acceleration

    Private Sub Form1_Load(ByVal sender As
        System.Object, ByVal e As
        System.EventArgs) Handles MyBase.Load
        Me.Text = "Example 6.2: Settling Velocity"
        Me.MaximizeBox = False
        Me.FormBorderStyle =
            Windows.Forms.FormBorderStyle.FixedSingle
        Label1.Text = "Temp (C)"
        Label2.Text = "Relative density "
        Label3.Text = "Diameter (mm)"
        Label4.Text = ""
        Button1.Text = "&Calculate"

        rho(0) = 1.793
        rho(1) = 1.519
        rho(2) = 1.308
        rho(3) = 1.14
        rho(4) = 1.005
        rho(5) = 0.894
        rho(6) = 0.801
        rho(7) = 0.723
        rho(8) = 0.656
        rho(9) = 0.599
```

297

```
         rho(10) = 0.549
         rho(11) = 0.506
         rho(12) = 0.469
         rho(13) = 0.436
         rho(14) = 0.406
         rho(15) = 0.308
         rho(16) = 0.357
         rho(17) = 0.336
         rho(18) = 0.317
         rho(19) = 0.299
         rho(20) = 0.284
    End Sub

    Private Sub Button1_Click(ByVal sender As
         System.Object, ByVal e As
         System.EventArgs) Handles Button1.Click
         Dim v, T, d, sg, wrho, Re As Double
         T = Val(TextBox1.Text)
         sg = Val(TextBox2.Text)
         d = Val(TextBox3.Text)
         d /= 1000 'convert to m
         'find rho from table
         If T = 0 Then wrho = rho(0) _
               Else wrho = rho(T / 5)

         v = (g * (d ^ 2) * (sg - 1)) /
            (18 * wrho * (10 ^ -6))
         Label4.Text = "Settling velocity = "
            + v.ToString + " m/s"
         'check Reynold's number
         Re = (v * d) / (wrho * (10 ^ -6))
         Label4.Text += vbCrLf
         Label4.Text += "Reynold's number = "
            + Re.ToString
         If Re < 0.5 Then Label4.Text += "(OK)"
    End Sub
End Class
```

Example 6.3

a) A settling column test performed on a discrete particle suspension
   revealed the following results:

Sampling depth (m)	Sampling time (hr)	% suspended solids removed from sample
1	1	40
1	2	57
1	6	71

2	1	37
2	2	45
2	8	64
3	1	35
3	2	38
3	5	50

Plot the cumulative batch settlement curve and determine the total theoretical removal for a horizontal settlement basin with surface area of one hundred meter square when incoming flow rate is 2.7 m^3/min.

b) Write a computer program to plot the cumulative batch settlement curve and determine the total theoretical removal for a horizontal settlement basin given its surface area, incoming flow rate and settling column test results performed on a discrete particle suspension.

c) Verify your program by solving example 6.3.1.

Solution

From the given data the settling velocities could be found for each depth and sampling time

Settling velocity = sampling depth/sampling time

Sampling depth (m)	Sampling time (sec)	Settling velocity (mm/s)	% SS with settling velocity less than that stated
1	3600	0.378	60
1	10800	0.093	43
1	21600	0.046	29
2	3600	0.556	63
2	10800	0.185	55
2	28800	0.069	36
3	3600	0.833	65
3	7200	0.417	62
3	21600	0.139	50

Using this data, then the cumulative batch settlement curve could be plotted as shown in figure.

For a flow of 2.7 m^3/min, the settling velocity

$V_{st} = Q/A = 2.7/60 \times 100 = 0.45$ mm/s

299

From the graph for the specified surface loading,
$x_o = 63\%$
Total removal

$$X_T = 100 - x_o + \frac{1}{V_{st}} \int_0^{x_o} v_{st} \cdot dx$$

To find the integral vst.dx the following procedure may be used:

Interval	dx	V_{st}	$V_{st}.dx$
0 - 10	10	0.01	0.1
10 - 20	10	0.02	0.2
20 - 30	10	0.035	0.35
30 - 40	10	0.06	0.6
40 - 50	10	0.10	1.00
50 - 60	10	0.18	1.80
60 - 63	3	0.35	1.05
Σ			5.10

Thus, $X_T = 100 - 63 + 5.1 \times 10^{-3}/0.45 \times 10^{-3} = 48.3\%$

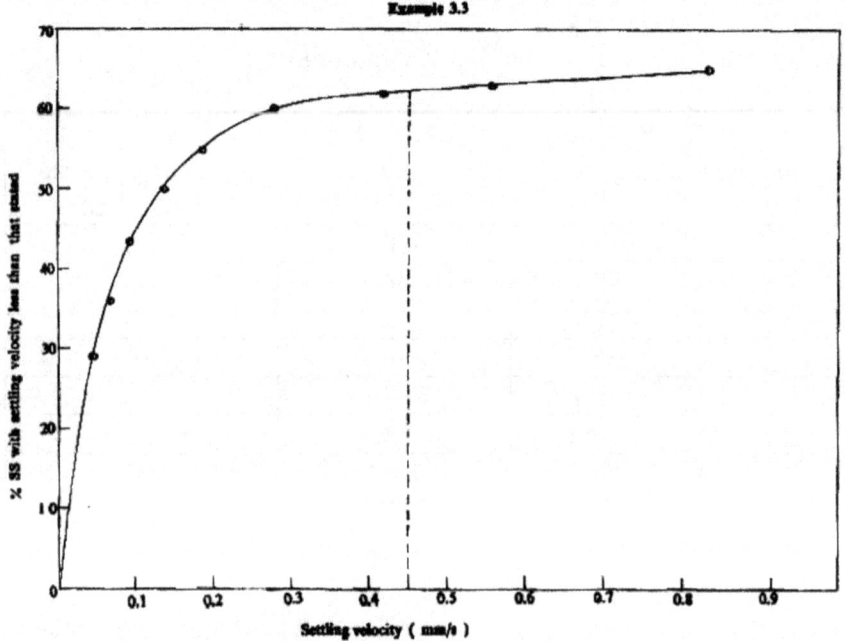

Example 3.3

Exercise 6.1
Discrete settling removal efficiency

1. Spherical particle with a diameter of 0.6 mm and a specific gravity of 2.65 is settling in water at temperature of 22 degree C. Assume **Type 1 settling** and calculate the settling velocity of the particle. (B.Sc., UoD, 2013)

2. The results from a settling column test performed on a municipal sewage from a certain town were as presented below. The experiments were conducted on a sample of sewage which is sieved through a 1.2 mm mesh to remove gross solids that would block valves. 50 l of the sample weighed 1.352 g after drying.

Sampling time (min)	Experimental results for amount of SS remaining in the sample (mg/l)			
	Sampling depth (cm)			
	100	200	300	400
0	230	230	230	230
60	150	150	159	157
180	118	143	147	152
360	90	115	134	145

The sewage treatment work incorporates two primary settlement tanks each of a diameter of 25 m. The sewage flow to the tanks is 0.3 m³/s. Find the percentage expected **SS removal** at the works. (Ans. 53 %).

3. In a laboratory batch settlement test of a flocculating material in a settling column with three sampling points, the following results were obtained:

Depth (m)	Suspended solids (SS) at set time (mg/l)				
	0 min	15 min	30 min	45 min	60 min
1	120	91	65	50	42
1.5	120	94	74	65	51
2	120	95	77	67	56

Determine the **removal efficiency** of solids in a horizontal flow tank at 35 min retention time at a depth of 2 m. Compute the removal at a depth of 3 m. (Ans. 58, 51 %).

4. A settling column test was conducted on a discrete particle suspension. The following results were recorded for a depth of 1.8 m:

Sampling time (sec)	480	600	1200	2400	3600	4800
% SS removal from sample	51	53	63	81	95	98

Find the theoretical **removal of solids** from this suspension in a horizontal flow tank with surface overflow rate of 216 $m^3/m^2.d$. (Ans. 73 %).

5. The following data were obtained from a batch settlement test on a discrete suspension:

Sampling depth (m)	Sampling time (hr)	% suspended solids remaining in the sample
All depths	0	100
1	1	36
1	2	6
1	4	2
2	1	69
2	2	35
3	1	72
3	2	64
4	1.5	71
4	2	68
4	4.5	29

Plot the **removal of suspended solids** versus overflow rate curve for a horizontal flow tank for the theoretical rang of given settling velocities. Compute the overall **removal** for a surface loading of 0.3 mm/s. (Ans. 88 %).

Settling velocity

6. Calculate the **settling velocity** of an organic suspension with a relative density of 1.001 and a diameter of 0.5 mm in water, given that the dynamic viscosity is 1.1×10^{-3} Ns/m^2. Assume that the particles are spherical in shape. (Ans. 0.12 mm/s).

7. Discrete spherical particles of diameter d are to settle in an ideal sedimentation basin. The settling velocity of the particles approaches 5 mm/s and their relative density is 2.6. If the temperature remains constant at 20° C, find the **diameter**, d, of the particles. (Ans. 0.08 mm).

8. Why are indicator organisms used to evaluate the sanitary quality of water?

A wastewater treatment plant has two primary clarifiers, each 20 m in diameter with a 2 m side-water depth. For a flow of 12900 m^3/d, calculate the **overflow rate** and the **detention**

time. (SQU, 1991). (Ans. 24 mm/s, 140 min). (Ans. 0.24 mm/s, 140 min).

Discrete settling removal efficiency

9. In a laboratory batch settlement analysis of a flocculating material, the following results were obtained for percentage of suspended solids removed:

Sampling depth (cm)	Sampling time (min)			
	15	30	45	60
100	29	55	70	78
150	26	46	55	69
200	22	43	53	64

Compute the **removal efficiency** of solids in a horizontal tank at 35 min. at a depth of 2 m. (Ans. 64%).

6.3 Biological Kinetics (Activated sludge)

Biological reaction kinetics are important in the process of wastewater treatment since they influence the size of treatment facility that would be adopted.

Substrate Limiting Growth

The proper control of the environmental conditions that affect the rate of growth of microorganisms answers the highest degree of waste stabilization.

Example 6.4

a) In a fully mixed continuous activated culture with partial feedback of cells, the recycling ratio of the process is 2. The values of the half velocity constant and maximum growth rate of microorganisms are 1 mg/L and 1 per hour, respectively. If the dilution rate is set at a value of 0.8 per hour, compute the value of the growth limiting substrate concentration in reactor and effluent.

b) Write a computer program to compute the value of the growth limiting substrate concentration in fully mixed continuous reactor and effluent with partial feedback of cells given recycling ratio of process, values of the half velocity constant, maximum growth rate of microorganisms and dilution rate.

c) Verify your program by solving example 6.4.1.

Solution

The rate of limiting substrate could be calculated by Monod equation:

$$\mu = \mu_{max} \frac{s^*}{\left[k_s + s^*\right]}$$

Where:

μ = Growth rate of microorganisms, per day = dilution rate (DR)/recycling ratio (R).

μ_{max} = Maximum growth rate of microorganisms, per day.

k_s = Half velocity constant (substrate concentration (mg/l) at half of the maximum growth rate).

S*   = Growth limiting substrate concentration in solution (mg/l).

Given: R = 2, DR = 0.8/hr, K_s = 1 mg/l, μ_{max} = 1/hr.

Therefore, μ = DR/R_u = 0.8/2 = 0.4

Substitute in Monod equation as:

0.4 = 1*s*/(1 + s*)

Therefore, s* = 0.67 mg/L.

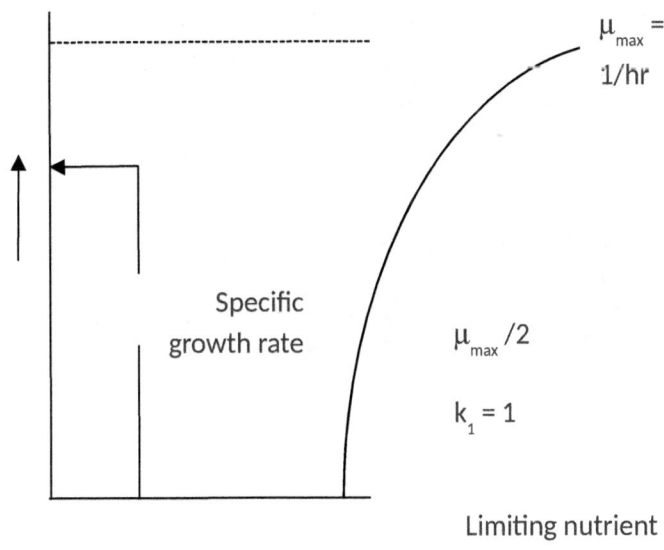

Program 6.4 Listing:

Growth limiting substrate concentration in fully mixed continuous reactor

```
'********************************
'EXAMPLE 6.4: MONOD EQUATION
'********************************
Public Class Form1

    Private Sub Form1_Load(ByVal sender As
        System.Object, ByVal e As
        System.EventArgs) Handles MyBase.Load
        Me.Text = "Example 6.4: Monod Equation"
        Me.MaximizeBox = False
        Me.FormBorderStyle =
            Windows.Forms.FormBorderStyle.FixedSingle
        Label1.Text = "Dilution rate (/hr)"
        Label2.Text = "Recycling ratio"
        Label3.Text = "Maximum growth rate (/hr)"
        Label4.Text = "Half velocity constant (mg/l)"
        Label5.Text = ""
        Button1.Text = "&Calculate"
    End Sub
```

```
Private Sub Button1_Click(ByVal sender As
    System.Object, ByVal e As
    System.EventArgs) Handles Button1.Click
    Dim mu, muMax, ks, S, R, DR As Double
    DR = Val(TextBox1.Text)
    R = Val(TextBox2.Text)
    muMax = Val(TextBox3.Text)
    ks = Val(TextBox4.Text)

    mu = DR / R
    'substitute in Monod's equation
    S = (mu * ks) / (muMax - mu)
    Label5.Text = "S* = " + FormatNumber(S, 2)
        + " mg/L"
    End Sub
End Class
```

The Aeration Tank
Sludge Loading Ratio (SLR), F/M ratio, Organic Loading

It is defined as the mass of BOD$_5$ applied to an activated sludge plant divided by the mass of MLSS in the aeration tank.

SLR = mass of BOD5 input to aeration basin (kg/d)/mass of MLSS in aeration basin (kg)

= BOD$_5$ of sewage (mg/l) x flow of sewage (m^3/d)/MLSS (mg/l) x tank volume (m^3)

$$\frac{F}{M} = SLR = \frac{W}{MLSS * V} = \frac{L_i \, x \, Q_i}{MLSS * V} = \frac{L_i}{MLSS * \tau} \, (per \, day)$$

Table (6.2) SLR for Certain Treatment Units

Unit	Reasonable value of SLR, /d
Conventional plants	0.3 to 0.35
Extended aeration	0.05 to 0.2
Step aeration	0.2 to 0.5

Operation at low F/M ratio results in a high degree of organic matter removal, good settleability of activated sludge and efficient BOD removal.

Volumetric Organic Loading Rate (VOL) is defined as:

$$VOL = \frac{Q * L_i}{V}$$

Therefore,

$$SLR = \frac{VOL}{MLSS}$$

Example 6.5

1) An activated sludge plant has been selected for treating a settled sewage at a daily flow rate of 3000 m^3. The 5-day BOD of the waste is 200 mg/l. The plant consists of two aeration tanks each of 4 m depth, 5 m width and 25 m length. The MLVSS concentration is 2100 mg/l. Compute:
   a) Detention time.
   b) Volumetric organic loading rate.
   c) Sludge loading ratio.
2) Write a computer program to compute detention time, volumetric organic loading rate, and sludge loading ratio for an activated sludge plant treating a settled sewage given daily flow rate, 5-day BOD of the waste, number of aeration tanks and their dimensions and MLVSS concentration.
3) Verify your program by solving example 6.5.1.

Solution

a) Detention time, t = V/Q
Where:
V = aeration volume = 4 x 5 x 25 x 2 = 1000 m^3
Q = rate of sewage flow = 3000 m^3/d
Thus, t = 1000/3000 = 0.33 days = 8 hr
b) Volumetric organic loading rate = (mass BOD load/d)/(aeration volume m^3) = (200x10^{-3} kg/m^{3}x3000 m^3/d)/1000 m^3 = 0.6 kg BOD/m^3.d

c)

$$SLR = \frac{(massBODload/d)}{(MLSSinaerationtankxtankvolume)} = \frac{200 \times 10^{-3} \times 3000\,kg/d}{2100 \times 10^{-3} \times 1000}$$
$$= 0.28\,kgBOD/kgMLSS.d$$

Program 6.5 Listing:
Detention time, volumetric organic loading rate, and sludge loading ratio for an activated sludge plant

```
'*******************
'EXAMPLE 6.5
'*******************
Public Class Form1

    Private Sub Form1_Load(ByVal sender As
        System.Object, ByVal e As
        System.EventArgs) Handles MyBase.Load
        Me.Text = "Example 6.5"
        Me.MaximizeBox = False
        Me.FormBorderStyle =
            Windows.Forms.FormBorderStyle.FixedSingle
        Label1.Text = "Sewage daily flowrate (m3)"
        Label2.Text = "5-day BOD (mg/l)"
        Label3.Text = "MLVSS (mg/l)"
        Label4.Text = "No. of tanks"
        Label5.Text = "Depth of each tank (m)"
        Label6.Text = "Length of each tank (m)"
        Label7.Text = "Width of each tank (m)"
        Label8.Text = ""
        Button1.Text = "&Calculate"
    End Sub

    Private Sub Button1_Click(ByVal sender As
        System.Object, ByVal e As
        System.EventArgs) Handles Button1.Click
        Dim MLVSS, BOD5, t, V, Q As Double
        Dim w, d, l, n As Double
        Dim loadRate, SLR As Double

        Q = Val(TextBox1.Text)
        BOD5 = Val(TextBox2.Text)
        MLVSS = Val(TextBox3.Text)
        n = Val(TextBox4.Text)
        d = Val(TextBox5.Text)
        l = Val(TextBox6.Text)
        w = Val(TextBox7.Text)
```

```
        V = w * l * d * n
        t = V / Q
        Label8.Text = "Detention time = "
            + FormatNumber(t, 2) + " days"
        Label8.Text += vbCrLf

        loadRate = (Q * BOD5 / 1000) / V
        Label8.Text +=
            "Volumetric organic loading rate = "
        Label8.Text += FormatNumber(loadRate, 2)
            + " kg BOD/m3.d"
        Label8.Text += vbCrLf

        SLR = (Q * BOD5 / 1000) / (MLVSS * V / 1000)
        Label8.Text += "Sludge loading ratio = "
        Label8.Text += FormatNumber(SLR, 2)
            + " kg BOD/kg MLSS.d"

    End Sub
End Class
```

Example 6.6

Estimate the sludge age for a conventional activated sludge plant from the relationship:

$$\frac{1}{\theta_c} = YU - k_d$$

Where:

θ_c = mean cell residence time (d)

Y = sludge yield coefficient

U = specific utilization rate (per day) = (mass BOD removed)/(mass volatile solids in basin) (kg BOD/ kg MLVSS.d)

Given: Y = 0.8, kd = 0.05/d, MLVSS = 3000 mg/l, retention time = 0.2 days, and incoming BOD5 of 180 mg/l.

Solution

$$\frac{1}{\theta_c} = YU - k_d = \frac{YL_i}{MLVSS.t} - k_d = \frac{YL_i}{MLVSS.t} - k_d = \frac{0.8 \times 180}{3000 \times 0.2} - 0.025$$

Then, θ_c = 5.3 days

Activated Sludge

It involves generation of an activated mass of aerobic microbial culture that enable stabilization of wastes. The process is a continuous one.

The factors that influence the process include:
- Wastewater flow quantity and quality.
- Sludge characteristics.
- Degree of mixing and turbulence.
- Operation and maintenance aspects.

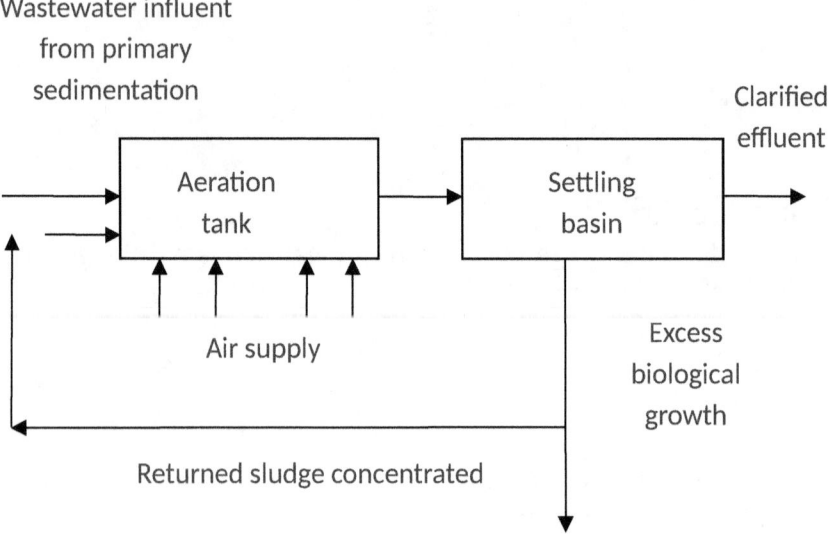

Fig. 6.1 Aeration tank

Table (6.3) Design Criteria for Conventional activated sludge

Parameter	Value
MLSS (mg/L)	1500 - 3000
Volumetric organic loading (g BOD/m^3/d)	500 - 700
Aeration detention time (hr) (based on average daily flow)	4 - 8
F/M ratio (g BOD/g MLSS/d)	0.1 - 0.6
Sludge retention time (d)	5 - 15
Sludge age (d)	3 - 4
Recycling ratio	0.25 - 0.5
Sludge yield index (kg solids/kg BOD removed)	0.7 - 0.9

Settling Properties
Sludge volume index, SVI, Mohlman SVI

It is a measure of the settleability of the activated sludge. The test also monitors the operation of an aeration system. SVI, is the volume (in ml) occupied by 1 g of asettled suspended solids.

SVI = V$_s$*1000/MLSS

Where:

SVI = Sludge volume index, mL/g.

V$_s$ = Settled volume of sludge in a 1 L graduated cylinder in 30 minute (mL/L, or %)

MLSS = Mixed liquor suspended solids, mg/L.

1000 = mg/g

Table (6.4) Sludge Classification According to SVI

SVI Value (mL/g)	Criteria
Less than 40	Excellent settling properties
40 - 75	Good settling properties
76 - 120	Fair settling properties
121 - 200	Poor settling properties
Greater than 200	Bulking sludge

Example 6.7

1) The concentration of the mixed liquor suspended solids, MLSS, in the basin of a step aeration system is 2300 mg/L, and the volume of settled sludge after 30 minutes in a 1 l graduated cylinder is 200 mL. Calculate the SVI Comment on the result.
2) Write a computer program to compute SVI in the basin of a step aeration system given concentration of the mixed liquor suspended solids, and the volume of settled sludge after 30 minutes in a 1 l graduated cylinder.
3) Verify your program by solving example 6.7.1.

Solution

$SVI = V_s*1000/MLSS = 200*1000/2300 = 87$ mL/g.
The obtained value is less than 120, therefore, it lies within the fair range for a settling sludge.

Program 6.7 Listing:

SVI in the basin of a step aeration system

```
'************************
'EXAMPLE 6.7: SVI
'************************
Public Class Form1

    Private Sub Button1_Click(ByVal sender As
        System.Object, ByVal e As
        System.EventArgs) Handles Button1.Click
        Dim MLSS, V, SVI
        MLSS = Val(TextBox1.Text)
        V = Val(TextBox2.Text)
        SVI = V * 1000 / MLSS
        Label3.Text = "SVI = " + SVI.ToString + " mL/g"
        If SVI < 40 Then
            Label3.Text += vbCrLf +
                "Excellent settling properties"
        ElseIf SVI < 76 Then
            Label3.Text += vbCrLf +
                "Good settling properties"
        ElseIf SVI < 121 Then
            Label3.Text += vbCrLf +
                "Fair settling properties"
        ElseIf SVI < 201 Then
            Label3.Text += vbCrLf +
```

```
                    "Poor settling properties"
        Else
            Label3.Text += vbCrLf + "Bulking sludge"
        End If
    End Sub

    Private Sub Form1_Load(ByVal sender As
        System.Object, ByVal e As
        System.EventArgs) Handles MyBase.Load
        Me.Text = "Example 6.7: SVI"
        Me.FormBorderStyle =
            Windows.Forms.FormBorderStyle.FixedSingle
        Me.MaximizeBox = False
        Label1.Text = "MLSS conc. (mg/L)"
        Label2.Text = "Sludge volume (ml)"
        Label3.Text = ""
        Button1.Text = "&Calculate SVI"
    End Sub
End Class
```

Sludge Age (SA), Mean Cell Residence Time (MCRT), Solids Retention Time (SRT), or Cell Age (CA):

For an activated sludge unit, the sludge age (SA) is defined as the total sludge in the biological treatment process divided by the daily waste sludge.

SA = (mass sludge solids in aeration tank (kg))/[mass sludge solids wasted daily (kg/d)]

$SA = V*MLVSS/q_w*SS$

Where:

SA =   Sludge age, d.

V = Volume of aeration tank, m^3.

MLSS = Concentration of mixed liquor suspended solids, mg/L.

q_w = Waste sludge flow, m^3/d.

SS = Suspended solids in waste sludge, mg/L.

Table (6.5) Mean Cell Residence Time for Some Treatment Units[15]

Process	SA (days)
Conventional activated sludge	5 - 15

Step aeration	20 - 30
Contact stabilization	5 - 15
Modified aeration	0.2 - 0.5
Extended aeration	5 - 15
High rate aeration	5 - 10

Example 6.8

In a conventional activated sludge plant the aeration volume is 60 l/capita, while the MLSS is 2400 mg/l and sludge wasting amounted to 0.025 kg/capita.d. Compute the mean cell residence time.

Solution

MCRT = $(60 \times 10^{-3} \text{ m}^3/\text{c} \times 2400 \times 10^{-3} \text{ kg/m}^2)/(0.025 \text{ kg/c.d}) = 5.8$ days

Sludge Density Index (SDI), or Donaldson Index:

Is the reciprocal of the SVI multiplied by 100.

Thus,

SDI = 100/SVI

Where:

SDI = Sludge density index, g/mL.

SVI = Sludge volume index, mL/g.

SDI varies from around 2 for a good sludge to about 0.3 for a poor sludge, i.e. $0.3 < SDI < 2$.

Exercise 6.2

SLR

a) In a conventional activated sludge plant the volume aeration basin is 8000 m³, the daily wastewater flow is 30000 m³ and influent BOD_5 is 175 mg/l. If the MLSS is 2400 mg/l, compute the **sludge loading rate**. (Ans.0.27 kgBOD/kgMLSS.d)

b) A conventional activated sludge basin is 7 m wide, 30 m long and 4 m deep. The influent wastewater flow is 3000 m³/d with a 5-day BOD of 130 mg/l. The MLSS of the aeration basin is 2200 mg/l. Compute the **F/M** ratio, the **volumetric organic loading** and the **retention time.** (Ans.0.21 kgBOD/kgMLSS.d, 0.46 kgBOD/m³.d)

c) A step aeration plant is working under the following conditions:

Influent sewage flow = 20000 m³/d

Aeration volume = 4500 m³

Average influent BOD_5 = 125 mg/l
MLSS concentration = 2200 mg/l
Compute the:
   a) **Detention time.**
   b) Volumetric **organic loading.**
   c) **BOD load.**
   d) **Sludge loading rate.** (Ans.5.4 hr, 0.56 kgBOD/m³.d, 2500 kg/d, 0.25 kgBOD/kgMLSS.d)

d) An aeration basin of volume 200 m3 operates at a MLSS of 2500 mg/l. The daily sewage flow is 1500 m³ with an average BOD of 230 mg/l. The concentration of suspended solids in the effluent amounts to 60 mg/l. If the average waste sludge solid is 75 kg/d, compute the:
   a) **Detention time.**
   b) **F/M ratio.**
   c) **BOD loading.**
   d) BOD **sludge age.**
   e) Suspended solids **sludge age.** (Ans.3 hr, 0.69 kgBOD/kgMLSS.d, 345 kg/d, 1.4, 3 d)

e) Compute the volumetric **organic loading** rate for an aeration basin of volume 7000 m³ serving a population of 25000 whose specific waste production is 60 g BOD/capita.d. Determine also the **sludge loading ratio** for the tank if it is known that the MLSS is 4000 g MLSS/m³. (Ans. 0.21 kgBOD/m³.d, 0.053 kgBOD/kgMLSS.d)

f) Assuming that the average rate of sludge production for the temperature range of 10 – 50° C follows the relationship:

$$\frac{1}{\theta_c} = YU - k_d$$

, estimate the **mean cell residence time** given that: Y = 0.7, kd = 0.1 per day, MLVSS = 3500 mg/l, detention time = 0.2 days, and incoming 5-day BOD is 250 mg/l. (Ans.6.7 d)

g) a) Explain the concept of **"Food-to-microorganisms-ratio"**, outline its importance in biological treatment of wastes. (SQU, 1991).

   b) Settled sewage at an average flow rate of 2.7 Ml/d and with a 5-day BOD of 200 mg/l is treated in an activated sludge plant in a simple aeration tank. The aeration unit is of square cross-section of side length 15m and it is 3m in depth. The mixed liquor volatile suspended solids concentration is 2200 mg/l.

315

i] Calculate the **detention time** [in hours].

ii] Find the **volumetric organic loading rate** in kg/m³.d, and

iii] Determine **food-to-microorganism-ratio**. (SQU, 1991). (Ans. 6 hr, 0.8 kgBOD/m³.d, 0.36 kgBOD/kgMLSS.d)

SVI & SDI

h) In a conventional activated sludge basin the operating MLSS is 2400 mg/l and the settled volume of sludge in 30 minutes in a 1 l graduated flask is 230 ml. Compute the **SVI**. (Ans. 96 mL/g).

i) Determine the **SVI** for an aeration tank used to treat a dairy wastewater, knowing that the MLSS in the aeration basin is 2100 mg/l while the settled volume in the SVI test is 210 ml. (Ans. 100 mL/g).

j) Calculate the **SVI** for an activated sludge settling tank with a MLSS is 2400 mg/l if it is given that the settled volume of sludge in 30 minutes is 32%. (Ans. 133 mL/g).

k) The aeration volume for an extended aeration unit is 250 l/capita, while the MLSS concentration is 2500 mg/l. If the sludge wasting is 0.5 kg/kg BOD removed, compute the **sludge age** knowing that the BOD load is 0.05 kg/capita.d. (Ans. 25 d).

l) An activated sludge treatment plant treats a daily flow of 800 m³/d with the following characteristics: MLSS = 3500 ppm, SS = 300 mg/l, solids settled in 30 min = 25%, and detention time in the aeration tank = 10 hr. Compute the:
   a) **Sludge density index**.
   b) **Sludge volume index**.
   c) **Volume of aeration** unit. (Ans. 0.014 g/mL, 71.4 mL/g, 8,000 m³).

m) Discuss the importance **of sludge age** with respect to the design and performance of activated sludge plants.

n) A sample of mixed liquor is taken from the outlet of an aeration tank and it was allowed to settle for 30 minutes in a 1000 ml measuring cylinder whereupon, the sludge volume index amounted to a value of 50. Estimate the **volume** occupied by the solids after 30 minutes settlement if it is known that the MLSS is 2200 mg/l. (Ans. 110 mL).

o) Control analysis on an activated sludge plant indicated the following: MLSS = 4200 mg/l, solids settled in 30 min. = 25%, wastewater treated is 0.1 m³/s, suspended solids = 220 mg/l, and volume of aeration tank is 2500 m³. Determine **SVI, SDI** and **SA**. (Ans. 60 mL/g, 0.02 g/mL, 5.5 d).

p) a] What is meant by the term **mixed liquor**?

   b] An aeration tank is operating with a MLSS concentration of 1800 mg/L. In a SVI test the sludge that settled after 30 minutes occupied a volume of 450 mL in a 1-L graduated cylinder.

   i] Find the **SVI** of the sludge.

   ii] Would you expect this sludge to **settle** satisfactorily in the secondary clarifier?

   C] The tank mentioned in (b) above is 30 m long, 10 m wide, and has a side water depth [SWD] of 4 m. For a flow of 4 ML/d determine the **food-to-microorganism ratio** for the system. Take BOD of waste entering the aeration tank to be 130 mg/L. (SQU, 1991). (Ans. 250 mL/g, 0.24 /d).

q) a] What is meant by **sludge bulking**? How is the SVI related to bulking? How may bulking be controlled?

   b] What is meant by the term **mixed liquor**?

   c] An activated sludge plant has been selected for treating a settled sewage at a daily flow rate of 3000 m³. The 5-day BOD of the waste is 200 mg/L. The plant consists of two aeration tanks each 4 m deep, 5 m wide, and 25 m long. The MLVSS concentration is 2100 mg/L. compute:

   i] The **detention time**.

   ii] **Volumetric organic loading** rate.

   iii] Sludge loading ratio (**F/M**). (SQU, 1993).

r) a) Suppose an industry decided to build a wastewater treatment plant and hired an engineer to design it for them. One of the first steps would be to **sample** the waste and run some analysis to determine what is characteristics are. If you had to specify the teats to be run, what would you choose as the FIVE most important wastewater parameters of interest? Name the five and state why you want to know these values.

   b) Explain the concept of **food to microorganism ratio**, and its importance in biological treatment of wastes.

Settled sewage at an average flow rate of 2 ML/d and with a BOD_5 of 250 mg/L is treated in an activated sludge plant in a simple aeration tank 15 metres square and 4 metres deep. The mixed liquor volatile suspended solids concentration is 2500 mg/L.

   i) Calculate the nominal average **detention time** (in hours).
   ii) Find the **volumetric organic loading rate** (in leg m^3/d).
   iii) Determine the **F/M** ratio. (UAE, 1989).

c) A sample of mixed liquor is taken from the outlet of an aeration tank and allowed to settle for half an hours in a one litre measuring cylinder. The sludge volume index SVI was found to be 40 mL/g. if the volume occupied by the solids, after the indicated settling period, amounts to 0.072 litre, determine the **MLSS** of the sample. Find also the sludge density index (**SDI**) of the sample. (UAE, 1989). (Ans. 1800 mg/L, 2.5 g/mL).

6.4 Trickling Filters

Settled sewage is biologically treated by aid of microorganisms attached to the filter media. The filter bed is composed of a highly permeable media through which wastewater trickles.

The process influential parameters include:

1) Wastewater characteristics.
2) Filter media properties.
3) Hydraulic and organic loading.
4) Filter design and construction.
5) Recycling ratio.

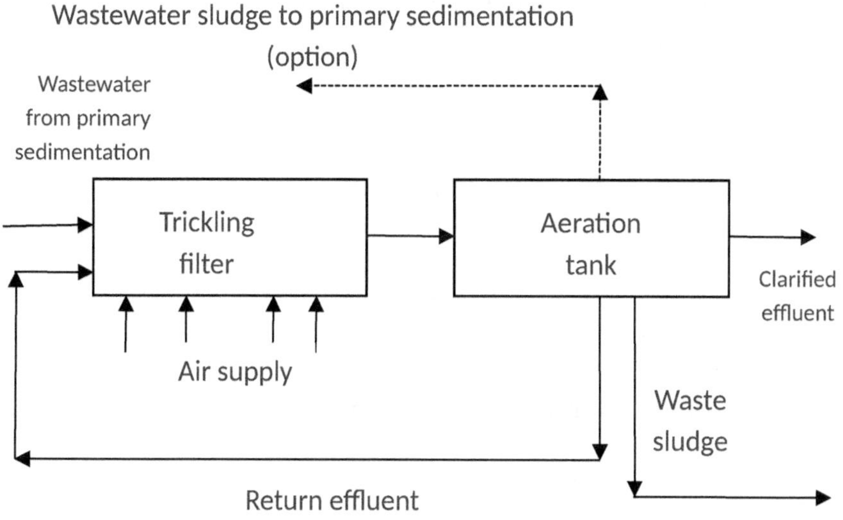

Wastewater sludge to primary sedimentation
(option)

Fig. 6.2 Trickling filter.

Example 6.9

Determine the filter diameter for a high rate filter using the National Research Council (NRC) formula for a sewage given the following information:

Population served = 5000, average daily sewage flow = 340 l/capita, BOD removal efficiency = 80%, recirculation ratio = 4:1, influent BOD_5 of waste = 150 mg/l.

Solution

NRC formula states that:

$$E_1 = \left[\frac{L_e \quad L_i}{L_i} \right]_1 = \frac{100}{1 + 0.44 \sqrt{\dfrac{W_1}{V_1 F_1}}}$$

Where:

E_1 = Efficiency of first-stage trickling filter treatment, %.

W_1 = influent BOD load tof first-stage filter, kg/d.L_i = Influent BOD load to filter, mg/L.

$W_1 = Q*L_i = BOD_{load}$

319

V_1 = Volume of single-stage filter media, m^3.
F_1 = Single-stage recirculation factor.

$$F_1 = \frac{1 + r_1}{(1 + 01 r_1)^2}$$

r_1 = Ratio of recirculated flow to the raw incoming waste flow

$$r = \frac{Q_r}{Q_{sew}}$$

Q_r = recirculated waste flow.
Q_{sew} = Flow of settled sewage, m^3/d.
L_e = Effluent BOD from first-stage, mg/L.
Therefore,
W = 0.34 x 5000 x 150/1000 = 255 kg/d
F = $(1 + 4)/(1 + 0.1 \times 4)^2$ = 2.55
E = 80%
Then,
80 = $(100)/(1 + 0.44(255/V \times 2.55)^{0.5})$
Thus,
V = 309.8 m^3
Use a depth of 3 m then, area A = 309.8/3 = 103.3 m^2 (= $\pi.D^2/4$)
This yields a filter diameter of 11.5 m

Example 6.10
Calculate the effluent BOD concentration from a one stage high rate trickling filter with the following characteristics:
Wastewater flow = 180 m^3/hr
Influent BOD = 280 mg/l
Volume of filter = 800 m^3
Filter depth = 2 m
Recirculation = 100% flow.
Use Velz formula and take k = 0.49.

Solution
Velz formula:

$$L_e = \frac{\left[L_i + r * L_e \right] e^{-kh}}{1 + r}$$

Where:

L_e = Effluent BOD concentration, mg/L.

L_i = Influent BOD concentration, mg/L.

r = Ratio of recirculated flow to the wastewater flow

k = Experimental coefficient.

   = 0.49 for high rate filters .

   = 0.57 for low rate filters. (/m)

h = Filter depth, m.

Thus,

$$L_e = \frac{\left[280 + 1 * L_e\right]e^{-0.49 \times 2}}{1+1}$$

Which yields, L_e = 64.7 mg/l

It is to be noted that Velz formula is valid for BOD removals of 90 percent or less. (In this case removals equals $(290 - 65.7)/280 = 77\%$).

The formula has been modified by Eckenfelder to include the effect of some variables such as flow rate and specific surface.

Rankin formula may be used for a single filter in the form:

$$L_e = \frac{L_i}{3 + 2R}$$

Where:

L_e = BOD of the settled filter effluent, mg/L.

L_i = BOD of the settled sewage, mg/L.

R = Recirculation ratio.

The formula applies to all plants treating sewage by pre-sedimentation, high rate filtration and final settling. This is where BOD included does not exceed 0.7 kg/m^3/d, and where recirculation, if applied, maintains a dosing rate between 93 and 244×10^3 m^3/ha/d (13).

Table (6.6) Typical Design Information for Trickling Filter[15,22]

Parameter	Low Rate Filter	High Rate Filter	Roughing Filter
Organic	0.07 to 0.32	0.32 to 1.0	0.8 to 6.0

Loading (kg/m³/d)			
Hydraulic loading (m³/m²/d)	1 to 4	10 to 40	30 to 200
Depth (m)	1.5 to 3 (2)	1 to 2 (2)	4.5 to 12
Recirculation ratio	0	0.5 to 3	Varies
Filter media	Crushed stone, gravel, slag, etc.	Crushed stone, gravel, rock, slag, synthetic material, etc.	Synthetic materials, red wood, etc.
Power requirement (kW/1000 m³)	2 to 4	6 to 10	10 to 20.
Sloughing	Intermittent	Continuous	Continuous.
Dosing intervals	Not greater than 5 minutes (generally intermittent).	Not greater than 15 seconds (Continuous).	Continuous.

Exercise 6.3

Filter dimensions

1) A high rate trickling filter is designed to treat an average daily flow of 2000 m³. The incoming wastewater has a BOD of 250 mg/l while the effluent wastewater is required to have a BOD_5 of 50 mg/l. Use the NRC equation to find the required filter **dimensions** if there is no recirculation. (Ans. 26 m).

2) A single stage trickling filter is to be used to treat an industrial waste having a BOD5 of 350 mg/l. The desired effluent quality is 40 mg/l. Given that the recirculation ratio is 4:1 and Filter depth of 2 m, use the NRC equation for the single stage filter to compute the needed filter **diameter** for a flow of 0.02 m³/sec. (Ans. 23 m).

3) Find, by the NRC method, the required **volume** of a single stage filter needed to treat 45 l/sec of waste having a 5-day BOD of 180 mg/l if the BOD reduction to be achieved is 75% (Assume that the sewage is not to be recirculated). (Ans. 1220 m³).

4) For the filter mentioned in the previous problem use a depth of 1.8 m and determine:

   a) **Diameter** of filter.
   b) The **BOD loading**.
   c) The **volumetric loading rate**. (Ans. 30 m, 700 kg/d, 0.57 kgBOD/m^3.d).

5) A single stage filter is to treat settled sewage at a rate of 800 m^3/d. Determine the **volume** and **diameter** of filter needed to achieve a BOD reduction of 80%, if the strength of the waste is 400 mg/l and the recirculated flow is 400 m^3/d. (Use NRC formula and take depth of filter to be equal to 1.5 m). What would be the increase in volume if the recirculation ratio is unity. (Ans. 991 m3, 36 %).

6) A waste treatment plant incorporates primary settlement and a single stage percolating filter. Use the NRC equation and the following data to compute required **volume** of filter media:
   Population served = 5000
   Waste
   Waste incoming BOD$_5$ = 300 mg/l
   Effluent BOD from filter = 30 mg/l
   BOD reduction efficiency in primary settlement = 45%
   Recirculation ratio = 2:1 (Ans. 428 m^3).

7) A sewage flow of 20 l/sec with a 5-day BOD of 180 mg/l is applied to a single stage trickling filter 1.8 m in depth. Use the NRC equation to determine:
   a) **Volume** of filter media required to remove 80% of the incoming BOD without recirculation.
   b) The required filter **diameter**.
   c) The **BOD load**. (Ans. 964 m^3, 26 m, 311 kg/d).

8) The sewage from a certain town is to be treated by an Imhoff tank, trickling filters and secondary sedimentation. The design parameters are as follows:
   Design flow = 0.09 m^3/sec
   Population to be served = 45000
   Waste production = 50 g BOD/capia.d
   BOD reduction by Imhoff tank = 30&
   BOD reduction by trickling filter = 90%
   Determine:
   a) **Organic loading** of the trickling filter by using Rumpf equation

$$E = 93 - \frac{0.017\,W}{V} \text{ ı g BOD/d.}$$

b) **Volume** of filter media required.

c) **Number** of filters to be used and **diameter** of filter; take a depth of 3 m. (Ans. 176 gBOD/m^3.d, 8949 m^3, 31 m).

9) Determine the **depth** of a trickling filter treating settled sewage with an ultimate oxygen demand (L$_o$) of 240 mg/l using the following data:

K = 0.57 per m.

Effluent BOD$_5$ = 40 mg/l.

Influent BOD$_5$ = 90% of ultimate oxygen demand

Recirculation = 100% (Ans. 2.5 m).

$$L_e = \frac{\left[L_i + r_1 * L_e \right] e^{-kd}}{1 + r_1}$$

10) Discuss the differences between suspended and fixed growth reactors. Give two examples for each. Which system would you recommend for a rural town in the Sudan? State your reasons. Outline the use of the NRC equation in a trickling filtration system

$$\eta = \frac{L_i - L_e}{L_i} = \frac{1}{1 + 0.44 \sqrt{\dfrac{W}{VF}}}$$

A treatment works incorporates primary sedimentation and trickling filtration to treat wastewater of 8500 inhabitants. The sedimentation unit has an efficiency of 35 percent. The wastewater flows at a rate of 325 L/capita/day with a BOD$_5^{20}$ of 310 mg/l. The trickling filter has the following characteristics: removal efficiency = 82 percent, recirculation ratio = 3:1, and filter depth = 2.5 m.

- Determine the trickling filter **diameter** using the National Research Council formula.
- Estimate the trickling filter **efficiency** for temperatures of 22 and 25 °C.
- Comment on your findings. (OIU, UNESCOC, 1999). (Ans. 22 m).

11) Define trickling filter and WSP. Design a low rate filter to treat 6.0 Mld of sewage of BOD of 210 mg/l. The final effluent should be 30 mg/l and organic loading rate is 320 g/m^3/d. check hydraulic loading rate and organic loading rate. Given that: 30% of BOD load removed in primary sedimentation. (B.Sc. UoD, 2014) (Ans. 48 m, 3.33m^3/d/m^2 , 326.18 g/d/m^3)

$$E_1 = \frac{L_i - L_e}{L_i} = \frac{100}{1 + 0.44\sqrt{\dfrac{W_1}{V_1 F_1}}} \qquad F_1 = \frac{1 + r_1}{\left(1 + 0.1 * r_1\right)^2}$$

$$E_2 = \left(\frac{L_i - L_e}{L_i}\right)_2 = \frac{100}{1 + \dfrac{0.44\sqrt{\dfrac{W_1}{V_1 F_1}}}{1 - E_1}}$$

Effluent concentration

12) Calculate the **effluent BOD concentration** from a two stage percolating filter using the data given below:
Population served = 5000
Daily wastewater discharge = 1 m^{3}l/capita
Influent BOD concentration = 250 mg/l
Volume of each filter used = 800 m^3
Filter depth = 1.8 m
Recirculation ratio for first stage = 1.25
Recirculation ratio for second stage = 1
The NRC formula

$$E_1 = \left[\frac{L_i - L_e}{L_i}\right]_1 = \frac{100}{1 + 0.44\sqrt{\dfrac{W_1}{V_1 F_1}}} \quad \text{first stage}$$

And

$$E_2 = \left(\frac{L_i - L_e}{L_i}\right)_2 = \frac{100}{1 + \dfrac{0.44\sqrt{\dfrac{W_1}{V_1 F_1}}}{1 - E_1}} \quad \text{second stage (Ans. 32 mg/L).}$$

13) a) Define the term "**Sloughing** of a trickling filter".

   b) Using the NRC formula determine the **effluent quality** of a two-stage trickling filter treating 180 m^3/hr settled sewage with a 5-day BOD of 150 mg/L. Filters capacity and conditions of flow are as indicated in the table below (UAE, 1989) (Ans. 14 mg/L).

Item	First filter	Second filter
Volume (m^3)	800	800
Depth (m)	2	2
Recirculation flow (m^3/min)	3.75	3

14) a) How does a trickling filter operate? Outline major trickling filter **operational** problems.

   b) Using the NRC formula compute the **effluent BOD** & **efficiency** of trickling filter 12 m in diameter with 2.1 m depth of media, for a raw wastewater flow of 1500 m^3/d with 130 mg/L BOD & a recirculation flow of 750 m^3/d.

   c) What is the BOD removal **efficiency** of a single-stage rock-filled trickling filter secondary at 16°C, r_1 = 0.5 & BOD loading of 750 g/m^3.d. (UAE, 1991). (Ans. 33 mg/L, 75, 66%).

Influent concentration

15) In a sewage treatment plant, the wastewater of a 5-day **BOD concentration** of L_i is subjected to primary sedimentation, trickling filtration and final settlement. The degree of treatment acquired an effluent BOD concentration of 30 mg/l. If the BOD removal efficiency for primary filtration is 40%, while that of the trickling filter and final sedimentation is 70%, compute the value of L_i. (Ans. 167 mg/L).

16) a) One operational problem with trickling filters is "**ponding**", the excessive growth of slime on the rocks and subsequent clogging of the spaces so that the water no longer flows through the filter. Suggest some cures for the ponding problem.

   a) A trickling filter of diameter 12 m and depth 2 m is used for treating a wastewater effluent of 150 m^3/hr with a 5-day BOD of "Y" mg/L. given that the recirculation flow amounts to 3125 litre/minute, find the value of **Y** to yield a final effluent 5-day BOD of 30 mg/L. (UAE, 1989).

(use NRC equation: $E=\dfrac{100}{1+0.44\left(\dfrac{W}{VF}\right)^{0.5}}$ Ans. 101.5 mg/L).

17) a) Indicate **assumptions** included in the National Research Council, NRC, formula for estimating efficiency of a trickling filter.

   b) A wastewater treatment plant has a primary sedimentation tank of efficiency 35 %, a trickling filter and a final clarifier. Wastewater of a 5-day BOD concentration of X, in mg/L is introduced at a flow rate, Q, of 0.02 m³/s. The trickling filter has a filter diameter, D, of 10 m, filter depth, h, of 2.5 m, recirculated flow, Q_r, of 1.6 m³/min., and a final effluent 5-day BOD, L_e, of 25 mg/L, temperature correction factor, Tc of 1.035. Determine:

   i.   **Influent** 5-day **BOD** concentration, L_i to the trickling filter unit using the National Research Council, NRC, formula.

   ii. **Efficiency** of trickling filter unit as predicted by the NRC equation.

   iii. **Influent** (raw wastewater) 5-day BOD concentration X, in mg/L to the treatment plant.

   iv. **Overall efficiency** of plant (including primary sedimentation, first & second stage filter).

   v.   **Efficiency** of trickling filter unit during winter (15°C average temperature) and summer conditions (25°C average temperature). (B.Sc. UoD, 2013) (Ans. 105 mg/L, 77%, 162 mg/L, 85%, 65%, 91%).

$$E=\left[\dfrac{L_i-L_e}{L_i}\right]=\dfrac{100}{1+0.44\sqrt{\dfrac{W}{VF}}}\quad r=Q_R/Q$$

$$E_T=E_{20}(T_c)^{T-20},\ F=\dfrac{(1+r)}{(1+0.1*r)^2}$$

Filter Efficiency

18) Calculate the **overall efficiency** of a two stage filter having identical filters in series with depths of 2 m and diameter of 16 m.

 Use the following data in design calculations:

 Wastewater flow = 800 m^3/d

 Influent 5-day BOD = 250 mg/l

 Effluent 5-day BOD not exceeding 20 mg/l

 Recirculation ratio for first stage = 1.25

 Recirculation ratio for second stage = 0.5

 Use NRC equation.

 In aforementioned problem the system was replaced by a single stage filter of depth 2m. Determine the needed filter diameter (Assume that the recirculation ratio amounts to 4:1).

 /L (Ans. 63%).

19) a) One operational problem with trickling filters is "**Ponding**", the excessive growth of slime on the rocks and subsequent clogging of the spaces so that the water no longer flows through the filter. Suggest some cures for the ponding problem.

 b) A trickling filter of diameter 12 m and depth 2 m is used for treating a wastewater effluent of 150 m3/hr with a 5-day BOD of "**Y**" mg/l. Given that the recirculated flow amounts to 3125 l/min., find, by using the NRC equation, the value of "Y" to yield a final, effluent BOD5 of 30 mg/l. (SQU, 1991). (Ans. 101 mg/L).

20) a] Differentiate between unit **operation** and unit process in wastewater treatment.

 b] A trickling filter, with a diameter of 26 m and depth of 2 m, receives a wastewater flow of 6500 m^3/d containing 610 kg of BOD. Calculate the BOD in g/m^3.d and the hydraulic loading in m^3/m^2.d. Determine the NCR **efficiency**, assuming that the flow includes a recirculation rate of 0.5. (SQU, 1991).

21) a] Differentiate between unit operations and unit processes in wastewater treatment.

 b] A trickling filter, with a diameter of 26 m receives a wastewater flow of 6500 m^3/d containing 610 kg of BOD. Calculate the **BOD** in g/m^3.d & the **hydraulic loading** in m^3/m^2.d.

 Determine the NRC **efficiency**, assuming that the flow

includes a recirculation ratio of 0.5 & the effluent BOD = 30 mg/L. (SQU, 1991). (Ans. 12 $m^3/m^2/d$, 68%, 1.6 $kg/m^3.d$).

22) A primary effluent is introduced to a trickling filter at a flow rate of 70 $m^3/hour$. The effluent has a BOD of 120 mg/L. The trickling filter has a diameter of 15 m and with a media depth of 3 m. The recirculation flow is 30 $m^3/hour$. Using the NRC formula compute the **effluent BOD** and **efficiency** of the trickling filter unit. (SQU, 1993). (Ans. 23 mg/L, 81%).

6.5 Waste Stabilization Ponds

Waste Stabilization Ponds are biological treatment processes whereby stabilization of organic matter is achieved by the aid of microorganisms in shallow basins. They are regarded as low cost treatment systems especially for developing countries.

The factors that need to be considered in design aspects include:

- Availability of required land area.
- Site selection with respect to topography, climatological conditions and hydrological aspects.
- Wastewater flow pattern and characteristics.
- Loading rates and available nutrients.
- Pond design methodologies.

Facultative waste stabilization ponds
Example 6.11

1) Determine the size of a facultative pond needed to treat wastewater of BOD_5 of 400 mg/l given the following data:

   Effluent BOD = 60 mg/l
   Wastewater flow = 2000 m^3/d
   Depth of pond = 1.5 m
   Lowest temperature = 24° C
   Removal rate constant for pond, $k_p = 0.35/day$ at 20° C.
   $(k_p)_T = (k_p)_{20}.(1.05)^{T-20}$

2) Write a computer program to compute the size of a facultative pond needed to treat wastewater given influent BOD_5, effluent BOD, wastewater flow, depth of pond, lowest temperature and removal rate constant for pond.

3) Verify your program by solving example 6.11.1.

Solution

Assuming complete mixing conditions then,

$$\frac{L_e}{L_i} = \frac{1}{1+k_p t}$$

Where

L_e = Effluent BOD, mg/L.

L_i = Influent BOD, mg/L.

t = Retention time, or mean hydraulic retention time, d.

k_p = Removal rate constant for waste stabilization pond, d^{-1}.

$[k_p]_{24}$ = $[k_p]_{20}\{1.05\}^{[T-20]} = 0.35(1.05)^{[24-20]} = 0.425$ /d

Then,

60/400 = 1/{1 + 0.425*t}.

This yields a detention time of t = 13.3 days = V/Q

The volume, V = t*Q = 13.3*2000 = 26600 m^3.

Surface area = Volume/depth = 26600/1.5 = 17733 m^2.

Program 6.11 Listing:

Size of a facultative pond needed to treat wastewater

```
'***************
'EXAMPLE 6.11
'***************
Public Class Form1

    Private Sub Form1_Load(ByVal sender As
        System.Object, ByVal e As
        System.EventArgs) Handles MyBase.Load
        Me.Text = "Example 6.11"
        Me.MaximizeBox = False
        Me.FormBorderStyle =
            Windows.Forms.FormBorderStyle.FixedSingle
        Label1.Text = "Effluent BOD, Le (mg/L)"
        Label2.Text = "Influent BOD, Li (mg/L)"
        Label3.Text = "Wastewater flow (m3/d)"
        Label4.Text = "Depth of pond (m)"
        Label5.Text = "Removal rate constant at 20C"
        Label6.Text = "Lowest Temp. (C)"
        Label7.Text = ""
        Button1.Text = "&Calculate"
    End Sub

    Private Sub Button1_Click(ByVal sender As
```

```
System.Object, ByVal e As
System.EventArgs) Handles Button1.Click
 Dim Le, Li, Q, d, kp, kp24 As Double
 Dim t, temp, V, SA As Double

 Le = Val(TextBox1.Text)
 Li = Val(TextBox2.Text)
 Q = Val(TextBox3.Text)
 d = Val(TextBox4.Text)
 kp = Val(TextBox5.Text)
 temp = Val(TextBox6.Text)

 kp24 = kp * (1.05 ^ (temp - 20))
 t = ((Li / Le) - 1) / kp24
 V = t * Q
 SA = V / d
 Label7.Text = "Detention time = " +
    FormatNumber(t, 2) + " days"
 Label7.Text += vbCrLf + "Volume = " +
    FormatNumber(V, 2) + " m3"
 Label7.Text += vbCrLf + "Surface area = " +
    FormatNumber(SA, 2) + " m2"
  End Sub
End Class
```

Example 6.12

A facultative type aerated pond is to serve 10000 people and treat sewage of the following properties:

Wastewater flow = 150 l/capita.d

Influent 5-day BOD = 54 g/capita.d

Design the pond to achieve an effluent BOD of 30 mg/l. (Assume winter temperature of lagoon to be 15° C. k_p = 2.2 /day at 20° C.).

Solution

$[k_p]_{16}$ = $2.2(1.05)^{[16-20]}$ = 1.72 /d

Li = $54 \times 10^3/150$ = 360 mg/l

$$\frac{L_e}{L_i} = \frac{1}{1 + k_p t}$$

$$\frac{30}{360} = \frac{1}{1 + 1.72t}$$

From which, t = 6.4 days

$V = t.Q = 6.4 \times 150 \times 10000/1000 = 9600 \ m^3$
Assuming a depth of 3 m then,
Surface area of pond = 9600/3 = 3200 m^2.

Maturation ponds
Example 6.13
- A maturation pond is capable of reducing the number of *E.typhi* by 99.5%. Compute pond retention time and volume needed for treating a daily sewage flow of 250m3. (take k' (bacterial die away rate) = 0.8 /day).
- Write a computer program to compute pond retention time and volume needed for treating a daily sewage flow given its amount, bacterial die away rate, and percent bacterial reduction.
- Verify your program by solving example 6.13.1.

Solution
The rate of die-away of faecal organisms in a single maturation pond is given by:
$$\frac{N_e}{N_i} = \frac{1}{1+k't}$$

Where:

N_e = Effluent bacterial number, number of bacteria/100 mL.
N_i = Influent bacterial number, number of bacteria/100 mL.
k' = Bacterial-die away rate, d^{-1}.
t = Retention time, d.

$$\frac{N_e}{N_i} = \frac{100-99.5}{100t} = 5 \times 10^{-3}$$

Therefore,
$$5 \times 10^{-3} = \frac{1}{1+0.8't}$$
Which yields, t = 249 days = V/Q
$V = 250 \ m^3/d \times 249 \ d = 62250 \ m^3$

Program 6.13 Listing:

Pond retention time and volume needed for treating a daily sewage flow

```
'***************
'EXAMPLE 6.13
'***************
Public Class Form1

    Private Sub Form1_Load(ByVal sender As
        System.Object, ByVal e As
        System.EventArgs) Handles MyBase.Load
        Me.Text = "Example 6.13"
        Me.MaximizeBox = False
        Me.FormBorderStyle =
            Windows.Forms.FormBorderStyle.FixedSingle
        Label1.Text =
            "Percentage reduction of bacteria"
        Label2.Text = "Daily sewage flow (m3)"
        Label3.Text = "Bacterial die away rate (/day)"
        Label4.Text = ""
        Button1.Text = "&Calculate"
    End Sub

    Private Sub Button1_Click(ByVal sender As
        System.Object, ByVal e As
        System.EventArgs) Handles Button1.Click
        Dim k, t, Q, V, NeNi, perc As Double

        perc = Val(TextBox1.Text)
        Q = Val(TextBox2.Text)
        k = Val(TextBox3.Text)
        NeNi = (100 - perc) / 100

        t = ((1 / NeNi) - 1) / k
        V = t * Q
        Label4.Text = "Detention time = " +
            FormatNumber(t, 2) + " days"
        Label4.Text += vbCrLf + "Volume = " +
            FormatNumber(V, 2) + " m3"
    End Sub
End Class
```

Example 6.14

Compute the number and size of maturation ponds needed to treat wastewater flowing at a rate of 300 m^3/d containing 10^6 *E. coli*/100

333

ml. The number of microorganisms emerging with the effluent is 2000/ml. (Take k' = 2 /day for the species and pond retention time of 20 days).

Solution

For multi-celled ponds:

$$\frac{N_e}{N_i} = \frac{1}{(1+k't)^n}$$

Where n is the number of ponds in series.

Then,

$$\frac{2000}{10^6} = \frac{1}{(1+2 \times 20)^n}$$

Therefore, number of ponds = 2

Total volume required = 2 x 20 x 300 = 12000 m³

Exercise 6.4

WSP size

1) Determine **size** of facultative pond which is required to give an effluent 5-day BOD of 60 mg/l for a flow with the following characteristics:

    Influent 5-day BOD = 350 mg/l

    Amount of discharged swage = 1000 m³/d

    The pond is of depth 15 m and liquid temperature is 25° C. Assume rate constant for BOD removal k_p to be 0.3 /day at 20° C. (Ans. 12620 m²).

2) **Design** a flow through aerated lagoon to treat wastewater by using the following data:

    Wastewater flow = 4000 m³/day

    Influent BOD_5 = 220 mg/l

    Effluent soluble BOD_5 = 20 mg/l

    Removal rate constant at 20° C = 2.5 /day

    Mean monthly temperature for the coldest month = 10° C. (Ans. 26144 m²).

3) What is the **volume** of an oxidation pond required to serve a population (P) of 3000 discharging waste with the following characteristics:

    Flow rate (Q) = 250 l/capita.d

    BOD_5 = 200 mg/l

Water temperature (T) = 15° C.

Volume of pond, V = 3.5×10^{-5} PQ.BOD$(1.085)^{35-T}$

Removal rate constant at 20° C = 2.5 /day (Ans. 26838 m³).

4) a) Write briefly about **stabilization ponds** as system for wastewater treatment.

  a) Find the **surface area** of a waste stabilization pond treating wastewater with a 5-days BOD of 185 mg/L and flowing at a rate of 85 m³/hour. Using

$$\frac{L_e}{L_i} = \frac{1}{1+k_p t}$$ thematical relationship for a single completely mixed lagoon, find the surface area of the pond given that:

  a) Removal rate constant for the pond is 0.3 per day at 20° C
  b) Effluent 5-day BOD from pond = 50 mg/L
  c) Minimum temperature = 12°C
  d) Temperature correction factor 1.035 (k_{pT} = $k_{p20}(1.035)^{T-20}$)
  e) Depth of pond = 1.4 m.

  b) What changes will be incurred in (b) if two ponds are connected in series? (U of K, 2002).

5) Write briefly about **combined suspended and attached** growth system for wastewater treatment. Illustrate your discussion with appropriate sketches and examples.

  What are the major factors that influence **operation** and maintenance of a waste stabilization pond?

  Wastewater with a 5-days BOD of 324 mg/L and flowing at a rate of 100 m3/hour is to be treated in a waste stabilization pond. Using

$$\frac{L_e}{L_i} = \frac{1}{1+k_p t}$$

as the mathematical relationship for a single completely mixed pond, determine the surface area of the pond given that:

  • Removal **rate constant** for the pond is 0.2 per day at 20° C
  • **Effluent 5-day BOD** from pond = 22 mg/L
  • **Depth** of pond = 1.25 m.

- What changes will be incurred if **two** ponds are used in series? (U of K, 2001).
6) a) Write briefly about waste stabilization pond as a combined suspended and attached growth system.
   a) What are the major factors that influence operation and maintenance of a pond system?
   b) Wastewater with a 5-days BOD of 200 mg/L and flowing at a rate of 1.6 m³/min is to be treated in a waste stabilization pond. Using

$$\frac{L_e}{L_i} = \frac{1}{1+k_p t}$$

as the mathematical relationship for a single completely mixed pond, determine the **surface area** of the pond given that:

- Removal rate constant for the pond is 0.25 per day at 20° C
- Effluent 5-day BOD from pond = 25 mg/L
- Depth of pond = 1.4 m. (U of K, 2001). (Ans. 46080 m²).

7) Assuming existence of complete mixing within pond, & without re-circulation of solids in facultative ponds, influent BOD mass balance of pool yields the following equation. **Define** parameters incorporated in the equation.

$$\frac{L_e}{L_i} = \frac{1}{1+k_p t} \qquad (k_p)_T = (k_p)_{20} * (T_c)^{T-20}$$

Wastewater with a 5-day BOD of 200 mg/L and flowing at an hourly rate of 40 m³ is to be treated in a waste stabilization pond. Determine the **surface area, length and width** of the pond given the data presented in table (1). (SQU, 1991). (Ans. 17455 m², 94, 187 m).

Pond characteristics

Parameter	Value
Lowest temperature	12 °C
Removal rate constant for the pond at 20°C	0.32 per day
Effluent 5-day BOD from the pond	40 mg/L
Depth of pond	1.1 m
Temperature correction factor	1.06
Pond length	Twice its breadth

8) Wastewater with a 5-day BOD of 180 mg/L & flowing at a rate of 1.2 m³/minute is to be treated in a waste stabilization pond. Determine the surface area of the pond given the following data: (B.Sc. UoD, 2014) (Ans. 43,637 m²)

>Summer lowest temperature = 43 °C,
>
>Winter lowest temperature = 18 °C,
>
>Removal rate constant for the pond = 0.2 per day at 20°C,
>
>Temperature correction factor = 1.05,
>
>Effluent 5-day BOD from the pond = 20 mg/L, &
>
>Depth of pond = 1.1 m.
>
>$[k_p]_T = [k_p]_{20}\{1.05\}^{[T-20]}$, $V = \tau^*Q$, $L_e/L_i = 1/\{1 + k_p^*\tau\}$

WSP detention time & efficiency

9) A pond of depth 1.5 m is to treat a flow of 120 l/capita.d with a 5-day BOD of 50 g/capita.d. Given that the BOD loading rate of the pond is 100 kg/ha.d, determine:
   c) **Detention time**.
   d) Pond **efficiency**. (Assume temperature of 10° C, k_p = 0.3 /day at 20° C). (Ans. 63 d, 92%).

10) A stabilization pond operating at a temperature of 10° C and treating wastewater of 5-day BOD of 250 mg/l provided a BOD removal of 85%. Compute the **detention time** (Assume k_p at 20° C equals 0.4 /day). (Ans. 23 dm).

11) A pond of depth 1.5 m is to treat a flow of 150 L/c/d with a 5-day BOD of 60 g/c/d. given that the BOD loading rate of the [pnd is 120 kg/ha.d; determine the **detention time.** (Take k_p = 0.2 /day). (UAE, 1991).

WSP effluent BOD

12) Describe the process of operation of a waste stabilization pond. Illustrate your answer with appropriate sketches. (UAE, 1991).

A wastewater flow of 5000 m³/d is treated in a facultative oxidation pond that is 2 m deep with a surface area of 20 ha. The wastewater has a soluble BOD_5 of 150 mg/L & a reaction rate coefficient of 0.3 /d. Determine the soluble **BOD** of the **effluent**. (Assume a completely mixed reactor without solids recycles.) (UAE, 1991). (Ans. 6 mg/L).

13) A wastewater flow of 10,000 m^3/d is treated in a facultative oxidation pond that is 1.5 m deep with a surface area of 20 ha. The wastewater has a soluble 5-day BOD of 250 mg/l and a reaction rate coefficient of 0.3 /d. Determine the **soluble BOD** of the **effluent** [assume a completely mixed reactor with-out solids recycle]. (SQU, 1991). (Ans. 25 mg/L).

WSP removal rate constant

14) An oxidation pond of depth 1.5 m is to treat sewage flowing at a daily rate of 150 l/capita.d with a 5-day BOD of 60 g/capita.d. The loading rate of the pond at 120 kg/ha.d, enabled a BOD reduction of 90%. Determine the corresponding removal **rate constant** of the pond (Temperature is 10° C). (Ans. 0.18 /d).

WSP effluent bacteriological quality

15) Compute the **number** of faecal bacteria emerging from a sewage treatment plant incorporating a facultative pond and two maturation ponds in series with the aid of the following data:
Number of *E. coli* initially present in sewage = 6×10^6/100 ml
Facultative pond detention time = 30 days
Maturation pond detention time = 7 days
Bacterial die-away rate, k' = 2 /day (Ans. 437/100 mL).

Table (6.7) Design Information for Waste Stabilization Ponds

	Aerobic	Anaerobic	Facultative
Influent	Sewage with high organic load, high degree of solidssystem or anaerobic pond	Sewage from a sewerage	Sewage from facultative ponds
Treatment	Partial		
Disposal of effluent	to facultative pond	Maturation pond	Agricultural irrigation, fish farming, aquatic birds.
Depth, m	2 to 4	1 to 1.5	1

Detention time, day	8 to 20	20 to 180	5 to 10
Main biological action	Organisms that do not require DO for feeding & reproduction.	Anaerobic &aerobic	Aerobic organisms
Operation	Parallel or series connection ponds in series, parallel useful for large ponds.	At least 3	One or more, series or parallel
Color	Grayish black	Green or brownish green.	Green
Frequency of sludge removal, year	2 to 12	8 to 20	probably never.
Optimum temperature, °C	30	20	20
Oxygen requirement	-	-	0.7 to 1.4 times removed BOD.
Chemicals needed	Nutrients when there is a deficiency, no other chemicals	Nutrients when there is a deficiency, no other chemicals	
pH	6.8 to 7.2	6.5 to 9.0	6.5 to 8.0
Expected Problems	Odors, large land requirements ground water pollution.	Odors when loading is high, groundwater pollution, Reduction in biological activity in cold climates.	Reduction in problems associated with biological activity under cold. weather conditions

6.6 Oxidation Ditches

Oxidation ditches are extended aeration systems. Aeration is usually accompanied by the aid of aeration rotors placed across the ditch to provide aeration and recirculation.

Example 6.15

1) Design an oxidation ditch to serve a population of 5000 people using the data given below:
   Sewage flow = 120 l/capita.d
   Influent 5-day BOD = 54 g/capita.d
   Lowest operating temperature = 15° C
   Sludge loading rate = 0.18 kg BOD/kg MLSS
   MLSS = 4000 mg/l
2) Write a computer program to design an oxidation ditch given population to be served, sewage flow, influent 5-day BOD, lowest operating temperature, sludge loading rate, and MLSS.
3) Verify your program by solving example 6.15.1.

Solution

a) The aeration tank

SLR = $L_i.Q_w$/MLSS.V

Q_w = 120 l/capita.d x5000/1000 = 600 m³/d

L_i = 54 x 5000/1000 = 270 kg/d = 270 kg/d/600 m³/d = 450 mg/l

Therefore, volume of ditch = 450 x 600/0.18 x 4000 = 375 m³

Detention time, t = V/Q = 375/600 = 15 hours

b) The settling basin

Assume overflow rate of 27 m3/m2.d and retention time of 2 hours, and design for a peak overflow of 3xinflow.

Therefore, area of tank = 3 x 600/27 = 67 m²

(Diameter of 9.5 m)

Depth of tank = Q.t/A = 3 x600 x 2/24x67 = 2.2 m

Weir overflow rate = Q/πD = 800/πx9.5 = 20 m³/m.d

c) The oxygen requirement

Take oxygen requirement as twice BOD_5, i.e.

O_f = 2 x 54 x 5000/1000 = 540 kg/d = 23 kg/hr

Convert the rating of the cage rotor to field conditions by using:

$$O_o = \frac{O_f}{\dfrac{C_{sw} - C_i}{C_s}(1.02)^{T-20}.\alpha}$$

Where:

O_o = Oxygen transfer rate at standard conditions, kgO_2/kWh

O_f = Oxygen transfer rate under field conditions, kgO_2/kWh

α = (Oxygen transfer rate in water)/(Oxygen transfer rate in tap water at same temperature) (α range between $0.65 - 0.98$ (0.7 for sewage)

C_s = oxygen saturation concentration of distilled water at $20°$ C.

C_1 = actual concentration of dissolved oxygen likely to exist in aeration tank under operating conditions ($1 - 2$ mg/l).

T = temperature in $°$ C.

C_{sw} = oxygen saturation concentration for water.

From tables, C_s = 10.2 for T = $15°$ C

Take c_{sw} = $0.95C_s$ = 0.95×10.2 = 9.69 mg/l

And c1 = 1.5 mg/l for the aeration tank, α = 0.7

Cage rotor rating at standard conditions = 3 kgO_2/m.hr

Therefore,

$$3 = \frac{O_f}{\dfrac{9.69 - 1.5}{10.2} (1.02)^{15-20} \times 0.7}$$

Then, O_f = 1.53 kgO_2/m length.hr

Required length of rotor = 23 (kg/hr)/1.53(kg/m.hr) = 15 m

Program 6.15 Listing:

Design an oxidation ditch

```
'*****************
'EXAMPLE 6.15
'*****************
Public Class Form1
    Dim Cs(31) As Double

    Private Sub Button1_Click(ByVal sender As
        System.Object, ByVal e As
        System.EventArgs) Handles Button1.Click
        Dim Qw, Li, T, MLSS As Double
        Dim V, P, SLR, A, D, depth As Double
        Dim time, Overflow, _Of As Double

        Qw = Val(TextBox1.Text)
        Li = Val(TextBox2.Text)
        T = Val(TextBox3.Text)
        P = Val(TextBox4.Text)
        SLR = Val(TextBox5.Text)
```

```
D = Val(TextBox6.Text)
MLSS = Val(TextBox7.Text)


Qw = Qw * P / 1000
Li = Li * P / Qw
V = (Li * Qw) / (MLSS * SLR)
TextBox8.Text = "Volume of ditch = "
    + V.ToString + " m3"
TextBox8.Text += vbCrLf
'calculate detention time
time = V / Qw * 24
TextBox8.Text += "Detention time = "
    + time.ToString + " hr"
TextBox8.Text += vbCrLf
'Assume overflow rate of 27 m3/m2.d
'and retention time of 2 hours,
'and design for a peak overflow of 3xinflow
A = 3 * Qw / 27
depth = (3 * Qw * 2) / (A * 24)
Overflow = Qw / (Math.PI * D)
TextBox8.Text += "Tank Area = "
    + FormatNumber(A, 2) + " m2"
TextBox8.Text += vbCrLf
TextBox8.Text += "Tank Depth = "
    + FormatNumber(depth, 2) + " m"
TextBox8.Text += vbCrLf
TextBox8.Text += "Weir overflow rate = "
    + FormatNumber(Overflow, 2) + " m3/m.d"
TextBox8.Text += vbCrLf
'we changed the value of Li, so retrieve
'it again
Li = Val(TextBox2.Text)
'Take oxygen requirement as twice BOD5
_Of = (2 * Li * P) / 1000
_Of /= 24 'convert to kg/hr
TextBox8.Text += "Taking oxygen requirement
                as twice BOD5"
TextBox8.Text += vbCrLf

Dim L, Oo, C1, Csw, alpha, Of2 As Double
'Take csw = 0.95Cs,
'And c1 = 1.5 mg/l for the aeration tank,
' ▯ = 0.7
'Cage rotor rating at standard conditions =
'3 kgO2/m.hr
If T = 0 Then Csw = 0.95 * Cs(0) _
Else Csw = 0.95 * Cs(T)
C1 = 1.5 : alpha = 0.7 : Oo = 3
TextBox8.Text += "Taking Csw = 0.95Cs, C1 =
            1.5mg/l, alpha = 0.7"
```

```
      TextBox8.Text += vbCrLf
      TextBox8.Text += "Cage rotor rating at
            standard conditions = 3 kgO2/m.hr"
      TextBox8.Text += vbCrLf

      Of2 = Oo * (((Csw - C1) / Cs(T)) *
            ((1.02) ^ (T - 20)) * alpha)
      L = _Of / Of2
      TextBox8.Text += "Therefore, Of = " +
         FormatNumber(Of2, 2) + " kgO2/m length.hr"
      TextBox8.Text += vbCrLf
      TextBox8.Text += "Required length of rotor = "
         + FormatNumber(L, 2) + " m"
End Sub

Private Sub Form1_Load(ByVal sender As Object,
    ByVal e As System.EventArgs) Handles Me.Load
      Me.Text = "Example 6.15"
      Me.MaximizeBox = False
      Me.FormBorderStyle =
         Windows.Forms.FormBorderStyle.FixedSingle

      Label1.Text = "Sewage flow (l/capita.d)"
      Label2.Text = "Influent 5-d BOD (g/capita.d)"
      Label3.Text = "Lowest operating temp. (C)"
      Label4.Text = "Population"
      Label5.Text =
         "Sludge Loading Rate (kgBOD/kgMLSS)"
      Label6.Text = "Diameter of tank (m)"
      Label7.Text = "MLSS"
      TextBox8.Text = ""
      TextBox8.Multiline = True
      TextBox8.Height = 60
      TextBox8.ScrollBars = ScrollBars.Vertical
      Button1.Text = "&Calculate"
      'initialize table of oxygen concentration
      'staturation
      Cs(0) = 14.6
      Cs(1) = 14.2
      Cs(2) = 13.8
      Cs(3) = 13.5
      Cs(4) = 13.1
      Cs(5) = 12.8
      Cs(6) = 12.5
      Cs(7) = 12.2
      Cs(8) = 11.9
      Cs(9) = 11.6
      Cs(10) = 11.3
      Cs(11) = 11.1
      Cs(12) = 10.8
```

343

```
        Cs(13) = 10.6
        Cs(14) = 10.4
        Cs(15) = 10.2
        Cs(16) = 10.0
        Cs(17) = 9.7
        Cs(18) = 9.5
        Cs(19) = 9.4
        Cs(20) = 9.2
        Cs(21) = 9.0
        Cs(22) = 8.8
        Cs(23) = 8.7
        Cs(24) = 8.5
        Cs(25) = 8.4
        Cs(26) = 8.2
        Cs(27) = 8.1
        Cs(28) = 7.9
        Cs(29) = 7.8
        Cs(30) = 7.6
    End Sub
End Class
```

Exercise 6.5

1) **Design** an oxidation ditch to serve a population of 2000 inhabitants. The sewage flow has the following characteristics:
   Sewage flow = 80 l/capita.d
   Influent 5-day BOD = 50 g/capita.d
   Lowest operating temperature = 10° C
   Sludge loading rate = 0.2 kg BOD/kg MLSS
   MLSS = 4200 mg/l
   Diameter of cage rotor used = 0.7 m
   Rotors standard rating = 3 kgO$_2$/m length.hour
   Rototrs immersion depth = 15 cm
   Rotational speed of rotors = 70 rpm.

2) **Design** an oxidation ditch that operates at a food-to-microorganisms ratio of 0.05 kg BOD/kg MLSS.d with no pre-settlement. Use the following data in design:
   Number of persons served = 4000
   BOD contribution = 40 g/capita.d
   MLSS = 4500 mg/l (Ans. 711 m^3).

Table (6.8) Design Information for Oxidation Ditches

Parameter	Value
Sludge loading rate (kg BOD/kg MLSS.d)	0.03 – 0.15
Aeration requirement (kgO$_2$/kg BOD$_5$ applied)	1.2 - 2
Excess sludge production (g/capita.d)	5 - 30
Sludge drying beds land area (m^2/capita)	0.15 – 0.3
Nitrogen : BOD removed requirement	1:120
Phosphorous : BOD removed	1:600
MLSS desired (mg/l)	3000 - 5000
Sludge residence time (days)	20 - 30
Recirculation ratio	0.25 – 0.75
Sludge yield index (kg solids/kg BOD removed)	0.3 – 0.6
Aeration channel detention time (days)	1
Channel depth (m)	1.2 – 1.8
Channel geometry	45° or vertical side walls

6.7 Septic Tank

The septic tank incorporates principally the processes of grit separation and primary settlement.

Factors that influence the design and operation of the system include:

1) Wastewater flow pattern and characteristics.
2) Availability of proper disposal options.
3) Socioeconomic factors.
4) Desludging procedure.
5) Location of tank with respect to surrounding features.
6) Availability of needed labor, tools, materials and method of know how.

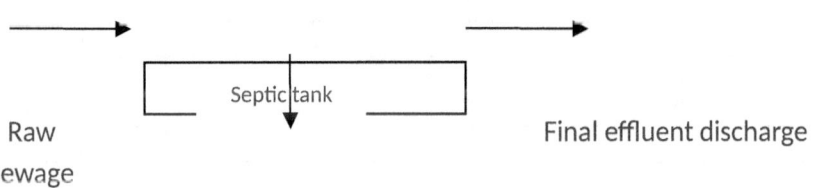

Raw sewage

Final effluent discharge

Septic tank sludge for disposal

Fig. 6.3 Septic tank.

345

Example 6.16

Design a septic tank for a family of 8 persons. The water consumption is 120 l/capita.d. The disposal of effluent is in a drain field by surface irrigation. (Assume rate of sludge accumulation is 0.04 m^3/capita.year)

Solution

Assume that the wastewater flow entering the tank is 75% of water consumption and that the detention time at start-up is 3 days.
Thus, tank volume = 0.75 x 120 x 3/1000 = 2.2 m^3
Take, tank dimensions: 1 m depth, 1 m width and 2.2 m length. The length of the first compartment = 1.5 m from inlet end.
Desludging interval = (wastewater flow, m^3/cpita.d)/(rate of sludge accumulation, m3/capita.year) = 0.75 x 120/0.04 x 1000 = 2.25 years

Table (6.9) Septic Tanks for Individual Dewellings[21]

Tank volume, m^3	Recommended dimensions, m		
	Inside length	Inside width	Liquid depth
1.1	1.5	0.7	1.1
1.5	1.8	0.9	0.9
1.9	1.8	0.9	1.2
2.8	2.3	1.1	1.2
3.8	2.7	1.2	1.2
4.7	3.4	1.2	1.2
5.7	3.2	1.5	1.2
7.6	4.3	1.5	1.2
9.5	4.4	1.8	1.2
11	5.2	1.8	1.2
13	4.9	1.8	1.5
15	4.9	2..2	1.5
19	5.9	2.2	1.5
23	6.2	2.4	1.5
26	7.3	2.4	1.5
30	7.0	2.4	1.8
38	8.5	2.4	1.8

Exercise 6.6

- **Design** a septic tank to serve a family of 10 persons. The water consumption is 120 l/capita.d and wastewater entering the tank is about 0.8 of water consumption.
- **Design** a septic tank to serve a family dwelling given that the estimated daily flow of sewage is 1500 liters.
- **Design** a septic tank to serve a family dwelling of 6 persons by using the formula: capacity of tank (m³) = 180 P + 2000, where P is the number of people served.
- A family of 15 persons dwells in a suburban house. The place is supplied with drinking water from central water distribution system. It is required to design a septic tank to serve the inhabitants. **Design** the tank and the surface irrigation system to be used given the following data:

Percolation test result: time for 1 cm fall = 4 minutes

Specific sewage production = 150 l/capita.d

Detention time in septic tank = 2 days

Persons served	6	20	50
Depth, cm	120	150	200 – 250 (max)

Percolation

Time for 1 cm fall (minutes)	Effective absorption area (m²/capita)
Less than 0.8	3.1
1.2	3.7
1.5	4.3
2.0	5.0
4.0	6.2
6.0	8.0
12.0	11.0
24.0	15.0
Greater than 24	Tile field is not available

Table (6.10) Effectiveness of some wastewater treatment processes

Treatment unit	SS	BOD	Total nitrogen	Oil & grease	Bacteria	Viruses
Pre-sedimentati	VG	G		NE	R	R

on						
Fine screening	R	R	NE	NE	NE	NE
Dissolved air flotation	R	R		R	NE	NE
Convention al activated sludge	E	E	VG	G	VG	G
Oxidation ponds	R	G	R	R	VG	G
Trickling filters	VG	VG	R	R	R	R
Septic tank	G	G	R	R	R	
Anaerobic lagoon	G	G				
Chemical treatment	G	R	R			
Reverse osmosis	VG	VG	G	G	G	R

Key:

E	Effective	R	Reasonable	VG	Very good
NE	Not effective	G	Good		

6.8 Sludge Treatment and Disposal
6.8.A. Sludge Digestion (Anaerobic)

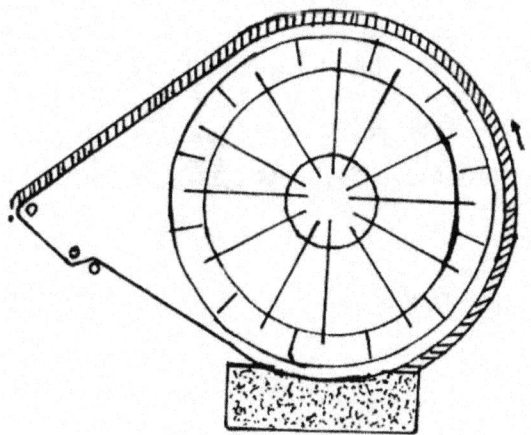

Sludge digestion signifies the controlled degradation of organic substances, normally occurring under anaerobic conditions. Factors that influence the process include: pH, temperature, nutrients, toxic substances e.g. heavy metals, volatile solids, ammonia, type and characteristics of decomposed materials and mixing.

Example 6.17

Design a sludge digester of depth 6 m for a community of 1000 people, given that the per capita capacity is 0.07 m³.

Solution

Capacity of sludge digestion = 0.07 x 1000 = 700 m³
Required area = 700/6 = 117 m²
This gives a diameter of 12.3 m

Example 6.18

1) Compute the volumetric gas production rates from a beef manure given that the influent volatile solids concentration to the digester amounts to 150 kg/m³ and the hydraulic retention time is 10 days. The maximum specific growth rate of microorganisms is 0.13 per day at a temperature of 20° C. take the ultimate methane yield as

equal to 0.25 m³ gas/ kg VS, k_n coefficient of 1.2 and digester volume of 4 m³.

2) Write a computer program to compute the volumetric gas production rates from a waste given influent volatile solids concentration to the digester, hydraulic retention time, maximum specific growth rate of microorganisms, temperature, ultimate methane yield, k_n coefficient and digester volume.

3) Verify your program by solving example 6.18.1.

Solution

The volumetric gas production or specific yield could be determined from the relationship:

$$V_g = \frac{(Y_t * VS)}{t}\left[1 - \frac{k_n}{t\mu_{max} - 1 + k_n}\right]$$

Where:

V_g = Volumetric gas production rate, or the specific yield, m³ gas/m³ digester/d.

Y_t =   Ultimate gas yield, m³ gas/kg VS added.

VS =   Concentration of influent volatile solids, kg/m³.

k_n =   Kinetic coefficient, dimensionless.

t =   Hydraulic detention time, days.

u_{smax} = Maximum specific growth rate of microorganisms, per day

Then,

$$V_g = \frac{(0.25 \, x \, 150)}{10}\left[1 - \frac{1.2}{10 \, x \, 0.12 - 1 + 1.2_n}\right] = 0.75 \quad \text{gas m}^3 / \text{m}^3 \text{ digester}$$

volume.d.

Therefore, daily amount of gas produced = 0.75 x 4 = 3 m³

Program 6.18 Listing:

Volumetric gas production rates from a waste

```
'********************
'EXAMPLE 6.18
'********************
Public Class Form1

    Private Sub Button1_Click(ByVal sender As
        System.Object, ByVal e As
        System.EventArgs) Handles Button1.Click
```

```
          Dim Vg, Yt, VS, kn, t, umax, V, gas As Double
          Yt = Val(TextBox1.Text)
          VS = Val(TextBox2.Text)
          kn = Val(TextBox3.Text)
          t = Val(TextBox4.Text)
          umax = Val(TextBox5.Text)
          V = Val(TextBox6.Text)

          Vg = ((Yt * VS) / t) *
             (1 - (kn / (t * umax - 1 + kn)))
          Label7.Text = "Vg = " + FormatNumber(Vg, 2)
             + " gas m3 /m3 digester volume.d"
          Label7.Text += vbCrLf
          gas = Vg * V
          Label7.Text += "Daily amount of gas produced = "
                + FormatNumber(gas, 2) + " m3"
     End Sub

     Private Sub Form1_Load(ByVal sender As
        System.Object, ByVal e As
        System.EventArgs) Handles MyBase.Load
        Me.Text = "Example 6.18"
        Me.FormBorderStyle =
           Windows.Forms.FormBorderStyle.FixedSingle
        Me.MaximizeBox = False
        Label1.Text =
           "Ultimate gas yield (m3 gas/kg VS)"
        Label2.Text =
           "Concentration of influent VS (kg/m3)"
        Label3.Text = "Kinetic coefficient"
        Label4.Text = "Hydraulic detention time (days)"
        Label5.Text = "Maximum specific growth rate
(/day)"
        Label6.Text = "Digester volume (m3)"
        Label7.Text = ""
        Button1.Text = "&Calculate"
     End Sub
End Class
```

Table (6.11) Design Information for the Conventional Anaerobic Digester

Parameter	Value
Volatile Solids Loading, kg/m^3/d	0.3 - 2
Volatile Solids Destruction, %	40 - 50
Gas Production, m^3 gas/kg VS	0.2 - 1.5
Influent Sludge Solids, kg/m^3/d	2 - 5

Total solids decomposition, %	30 - 40
pH	6.5 - 7.4
Alkalinity concentration, mg/L	2000 - 3500
Solids retention time, day	30 - 90
Digester capacity, m^3/capita	0.1 - 0.17
Gas composition, % 1) Methane 2) Carbon Dioxide 3) Hydrogen Sulfide	 65 – 70 32 – 35 Trace
Temperature, °C	30 - 35

6.8.B. Sludge Dewatering

Sludge dewatering or water removal from sludge is of significance for effective, efficient and ultimate disposal. there are many factors that influence sludge dewaterability, they include: presence of fine particles, sludge solids content, shearing strength, protein content, anaerobic digestion, pH, particle charge, moisture content and filter aids used.

One of the simplest methods used for determining how well a sludge dewaters is the specific resistance to filtration test. The specific resistance is the resistance to filtrate egress caused by a cake of unit weight of dry solids per unit filter area; or, it is the pressure difference required to cause a unit flow rate of filtrate having unit viscosity through a unit weight of filter cake.

Table (6.12) Sludge Dewaterability.

Specific resistance value (m/kg)	Sludge characteristics
10^{11} - 10^{12}	Easily filtered sludge
10^{14} - 10^{15}	Poorly filtered sludge

Example 6.19

The following results were obtained from a specific resistance determination on a sample of digested sludge:

Time (sec) (t)	Volume of filtrate (ml) (V)
120	4.9
180	6.3
240	7.4

352

300	8.4
420	10.3
600	12.75
780	14.7
1080	17.6
1200	18.8

Vacuum pressure = 68.95 kN/m^2
Filtrate temperature = 20° C
Filtrate viscosity = 1.002x10^{-3} Ns/m^2
Solids content = 21.4 kg/m^3
Area of filtration = 38.48x10^{-4} m^2
Volume of sludge used = 100 ml
Plot the values of t/V against V and hence obtain the slope and compute the specific resistance to filtration.

Solution

From the data given, the following table would be obtained:

Time (t) (sec)	Volume of filtrate (V) (ml)	(t/V)x10^6 (s/m^3)
120	4.9	24.49
180	6.3	28.57
240	7.4	32.43
300	8.4	35.71
420	10.3	40.78
600	12.75	47.06
780	14.7	53.06
1080	17.6	61.36
1200	18.8	63.83

From graph of t/V against V, the slope b would equal to 2.84x10^{12} s/m^6
From Carman and Coackley's equation, the specific resistance is:

$$r = \frac{2\,b\,PA^2}{\epsilon\,C}$$

Where:
r = Specific Resistance of Sludge Cake to filtration, m/kg.
b = Slope of the straight line of t/V versus V, s/m^6.
P = Pressure applied, N/m^2.
A = Area of filtration, m^2.
ϵ = Viscosity of filtrate, N*s/m^2.

353

C = Solids content, kg/m³.

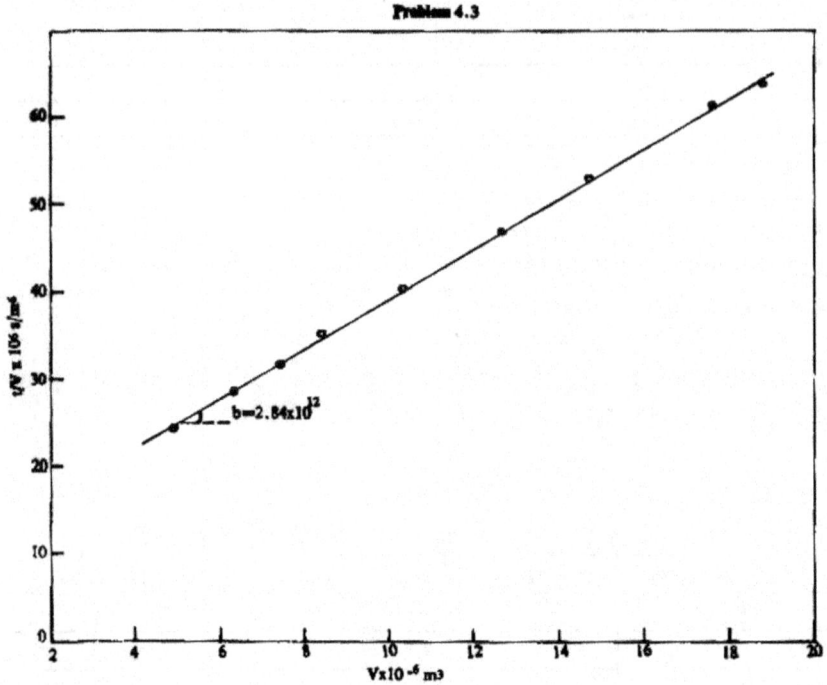

Substituting given values in the above mentioned equation yields:

$$r = \frac{2 \times 2.64 \times 10^{12} \times 38.48 \times 10^3 \times (38.48 \times 10^{-4})^2}{1.002 \times 10^{-3} \times 21.4} = 27 \times 10^{13} \, m/kg$$

Since this value lies between 10^{14} and 10^{15} m/kg, then it indicates a poor filtering sludge.

Example 6.20

For a sample of digested sewage sludge the following specific resistance values were obtained for the corresponding applied pressures:

Pressure applied, kN/m²	Specific resistance, $rx10^{-13}$, m/kg
293.04	52.95
586.075	84.52
117.15	158.85

1758.225	210.78
2344.3	276.97276.97

Use the given data to calculate the compressibility coefficient of the sludge.

Solution

From the given data the following table may be constructed:

Pressure applied, kN/m²	Specific resistance, rx10⁻¹³, m/kg	Log P	Log r
293.04	52.95	2.4669	14.7239
586.075	84.52	2.7680	14.9270
117.15	158.85	3.0690	15.2010
1758.225	210.78	3.2451	15.3238
2344.3	276.97276.97	3.3700	15.4424

The compressibility coefficient could be found from the relationship:

$r = r'*P^{c-}$

Where:
r = Specific resistance to filtration at applied pressure P, m/kg.
r_s'= A constant.
$\bar{c}$ = A constant termed the coefficient of compressibility, (varies between 0 and 1).

The above equation may be put in the form:

$\text{Log} r = \bar{c} \text{Log} P + \log r'$

By plotting the logarithm of the specific resistance (Log r) as a function of the logarithm of the corresponding applied pressure (Log P), then the slope of the line obtained is the coefficient of compressibility, $\bar{c}$. From the graph:
$\bar{c} = 0.75$

355

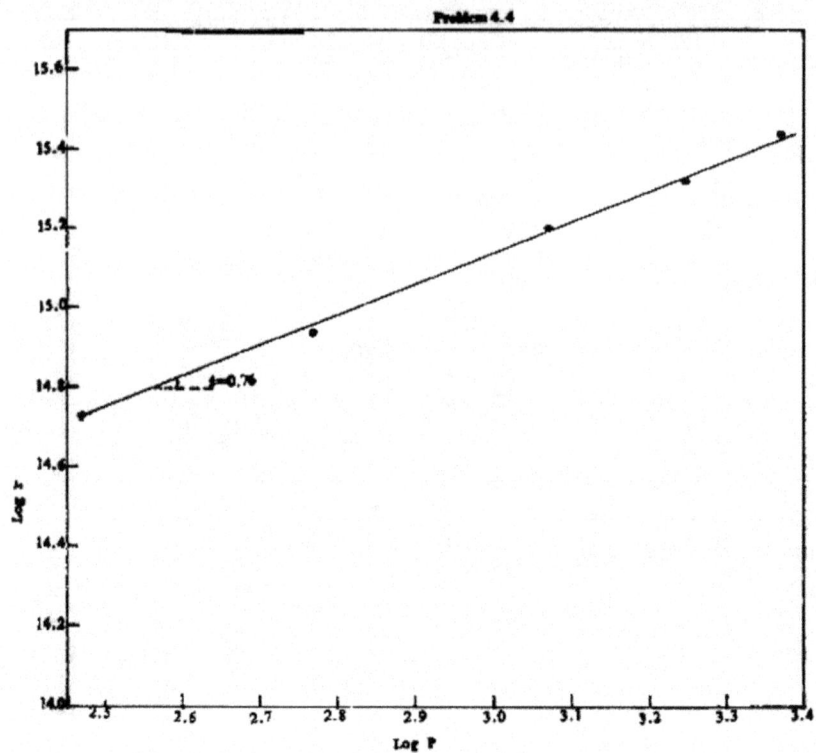

Theoretical Yields of Rotary Filters
Example 6.21

Find the theoretical yield from a vacuum filter given that the pressure difference across the cake is 98 kPa, moisture content 95%, specific resistance at applied pressure is 0.3×10^{13} m/kg, time for one revolution of filter 120 sec, viscosity of filtrate 1.01×10^{-3} Ns/m2. The fraction of filter area used for cake formation 0.26 (Take the cake correction factor as 1.9).

Solution

The yield (Y) of dry suspended solids from a vacuum filter is given by:

$$Y = \frac{F_c \sqrt{[2PC_1 F_f]}}{\epsilon r \theta} \quad (kg/m^2.s)$$

where:

P = pressure difference across cake (N/m²)

C1 = mass dry SS/unit volume of liquid in sludge (kg/m^3)

ε = viscosity of filtrate (Ns/m^2)

F$_f$ = fraction of filter area used for cake formation, area below sludge surface (m^2)

r = specific resistance at pressure P (m/kg)

θ = time for one revolution of filter (sec)

F$_c$ = cake correction factor.

Therefore,

$$Y = \frac{1.9 x C \sqrt{2 \times 98 \times 10^3 \times \dfrac{5 \times 10^{-3}}{95 \times 10^{-6}} \times 0.26}}{1.01 \times 10^{-3} \times 0.3 \times 10^{13} \times 120} = 5.16 \times 10^{-3} \, kg/m^2$$

Exercise 6.7

Specific resistance

1) In a specific resistance test on a mixed sludge sample the following values were recorded:

Time (s)	0	60	120	165	300	900	1200
Filtrate volume (ml)	0	2.7	3.8	4.5	6.1	10.6	12.1

Find the **specific resistance** value for the sludge given:

Filtrate viscosity = 1 x10^{-3} Ns/m^2

Filtrate area = 3.8 x10^{-3} m^2

Suspended solids concentration = 20 kg/m^3

Vacuum pressure = 60 kPa (Ans. $7*10^{14}$ m/kg).

2) The following results were obtained from a specific resistance test carried out in the laboratory for a digested sewage sliudge after being conditioned with 60% kieselguhr (based on the dry solids content of the sample):

Time (s)	2	3	4	5	7	10	13	18	20
Filtrate volume (ml)	12.8	16.6	18	20.2	23.5	28.1	31.8	37.3	39

Moisture content of the mixture of slugdge and kieselguhr = 97%

Manometer pressure reading = 69 kPa

Diameter of Watman number 1 filter paper = 70 cm

Filtrate temperature = 19°C.

Specific resistance of digested sludge prior to conditioning = 28.4x10^{13} m/kg

Filtrate viscosity = 1.027x10^{-3} Ns/m^2 at temperture of 19° C

Find the degree of **dewaterability** of the conditioned sludge. (Ans. $4.9*10^{13}$ m/kg, 83%).

3) a] Outline major methods of sewage sludge dewatering. Discuss the significance of dewatering as a process that precedes final disposal.

   b] In a specific resistance test using the Buchner funnel technique, on a sample of activated sludge, the following values were recorded:

Time, t (min.)	Volume of filtrate, V (ml)		
	Test [1]	Test [2]	Test[3]
1	1.3	1.4	1.5
2	2.3	2.4	2.5
4	4.1	4.2	4.3
8	6.8	6.9	7.0
15	10.3	10.4	10.5

   Find the value of the **specific resistance** of the sludge given:

   Vacuum utilized = 97.5 Kpa

   Filtrate viscosity = $1.011*10^{-3}$ Ns/m²

   Volume of filtrate used = $50*10^{-6}$ m³

   Solids content = 7.5 %

   Filtrate temperature = 20°C

   Diameter of Whatman No.1 filter paper = 7.5 cm. (SQU, 1991). (Ans. $2.4*10^{14}$ m/kg).

4) Compute the **specific resistance** of a sample of activated sludge given the following data obtained from a small laboratory test filter:

Time (s)	120	180	420	600	780	900
Filtrate volume (ml)	2.5	3.5	6.9	8.9	10.8	11.9

   Vacuum applied = 90 kPa

   Solids content = 4%

   Filtrate viscosity = $1.027x10^{-3}$ Ns/m²

   Volume of filtrate used = $50x10^{-6}$ m³

   Area of filtration = $3.8x10^{-3}$ m² (Ans. $18.8*10^{13}$ m/kg).

5) In a specific resistance test the following data were obtained:

Volume of sample, $Vx10^{-6}$ (m^3)	22	23	25	28	30	32	34	37
Time/volume, $(t/V)x10^6 = s/m^3$	20.2	21	22.6	24.9	26.5	28	30.4	31.9

Compute the value of the **specific resistance** given: $\varepsilon = 1.01x10^{-3}$ Ns/m², P = 96 kN/m², A = 4x10⁻³ m², moiture content = 93% (Ans. $3.2*10^{13}$ m/kg).

$$r=\frac{2\,bPA^2}{\mu\,C}=\frac{2*0.78*10^{12}*96*13^3*\left(4*10^{-3}\right)^2}{75.27*1.01*10^{-3}}=3.15*10^{13}\,m/kg$$

6) Discuss the problem of sludge dewaterability. Use the following data to compute the **specific resistance** of the sludge:

Time (s)	30	60	90	120	150	180	210	240	270	300	330	360	390
volume of filtratex10⁻⁶ (m³)	8	14	19	23	27	31	34	37	40	43	45	48	51

Applied vacuum pressure = 60 kPa
Viscosity of filtrate = 1.005x10⁻³ Ns/m²
Area of filtration = 3.86x10⁻³ m²
Solids content = 6.541 kg/m³ (Ans. $2.6*10^{13}$ m/kg).

7) The results of a pressure filtration test on a conditioned sewage sludge are as tabulated below:

Time of filtration (sec)	30	90	180	300	480	840
Filtrate volumex10⁻⁶ (m³)	6.5	11.5	16	21	26.5	35

Determine the **specific resistance** to filtration given that:
Pressure applied = 60 kPa
Solids content = 40 kg/m³
Filtrate viscosity = 1.01x10⁻³ Ns/m²
Filter area = 4x10⁻³ m² (Ans. $3.3*10^{13}$ m/kg).

8) What are the major factors that influence filterability of sewage sludge? Sludge dewatering data were collected from a Buchner funnel test as shown below:

Time (s)	Volume of filtrate (mL)
14.5	66
29.5	92
45.0	112

59.0	129
70.0	134
89.0	156
102.0	167
120.0	180

The specific conditions of the test were:

$A = 104.6 \text{ cm}^2$ $P = 15$ in Hg

$C = 0.056$ g/mL   Temperature of filtrate = 20°C

Determine the **specific resistance** of the sludge. (UAE, 1991). (Ans. $7.6*10^{11}$ m/kg).

9) Outline appropriate minimum evaluation procedure to improve the planning, functioning and utilization of future sanitation project in an urban residential and industrial area.

Discuss factors that govern the **dewaterability** of sewage sludges.

In a specific resistance test vacuum pressure used is 69 kPa, absolute viscosity = 1.1×10^{-3} Ns/m^2, solids content 0.075 g/mL, area of Whattman number one filter paper = 44.2 cm^2, and plot of t/V versus V yielded a slope of 8.25×10^{12} s/m^6. Using Carman-Coackley equation:

$$r = \frac{2bPA^2}{\mu C}$$ ind the **specific resistance** of this sludge. (U of K, 2002).

10) Outline appropriate minimum evaluation to improve the planning, functioning and utilization of future water supply and or sanitation projects.

Discuss methods of sludge dewatering.

In a specific resistance test vacuum pressure used amounted to 68.95 kPa, absolute viscosity = 1.139×10^{-3} Ns/m^2, solids content 79 kg/m^3, diameter of Whattman number one filter paper = 7 cm, and plot of t/V versus V yielded a slope of 1.387×10^{12} s/m^6.

Using Carman - Coackley equation:

$$r = \frac{2bPA^2}{\mu C}$$ find the **specific resistance** of this sludge. What will be the value of the specific resistance for a solids content of 4 percent? (U of K, 2001).

11) "Evaluation is a systematic way of learning from experience and of using the lessons learned both to improve the planning of future projects and also to take corrective action to improve the functioning, utilization and impact of existing projects". Discuss this statement for minimum evaluation procedure for water supply (**or** sanitation) projects.

Outline objectives of sludge dewatering as a physical unit operation utilized in rendering moisture content of sludges. In a specific resistance test vacuum pressure used = 60 kPa, viscosity = 1.139×10^{-3} Ns/m^2, solids content = 0.084 g/mL, area of filtration 38.48×10^{-4} m^2, and plot of t/V versus V yielded a slope of 1.387×10^{12} s/m^6. Using Carman - Coackley equation:

$$r = \frac{2bPA^2}{\mu C}$$ find the **specific resistance** of this sludge. (U of K, 2001). (Ans. 2.6×10^{16} m/kg).

12) Define Integrated Water resources Management (IWRM) concept and process. How can you implement the concept to the conditions of your area?

13) a. Discuss the problem of sludge dewatering.

b. Compute the specific resistance of a sample of activated sludge given the following data obtained from a small laboratory test filter:

Time (sec)	120	180	420	600	780	900
Volume of filtrate(ml)	2.5	3.5	6.9	8.9	10.8	11.9

Vacuum applied = 90 kPa
Solids concentration = 4%
Filtrate viscosity = 1.027×10^{-3} Ns/m^2
Area of filtration = 3.8×10^{-3} m^2 (U of K, 1986). (Ans. 18.8×10^{13} m/kg).

14) c) a) Why are wastewater sludges often difficult to dewater? (U of K, 1985).

b) The following results were obtained from a specific resistance test carried out in the laboratory for digested sewage sludge after being conditioned with 60% (of the dry solids content of the sample) kieselguhr:

Time(min.)	2	3	4	5	7	10	13	18	20
Filtrate volume, (ml)	12.8	15.6	18	20.2	23.5	28.1	31.8	37.3	39

Moisture content of the mixture of sludge and kieselguhr = 97%

Manometer pressure reading = 69kPa

Diameter of Whattman No.1 filter paper = 70mm

Filtrate temperature = 19°C

Specific resistance of digested sludge prior to conditioning = 28.4 x 10^{13} m/kg

Filtrate viscosity = 1.027 x 10^{-3} Ns/m^2 at temperature of 19°C

Find how much is the degree of **dewaterability** of the conditioned sludge. (U of K, 1985). (Ans. 5.4*10^{13} m/kg, 86%).

15) Outline objectives of sludge dewatering as a physical unit operation utilized in rendering moisture content of sludges.

The following results were obtained from a specific resistance test on a sample of digested sewage sludge:

Sample: original digested sewage sludge Volume of sample used = 100×10^{-6} m^3 Solids content = 0.0214×10^3 kg/m^3		Vacuum applied = 68.95 kN/m^2 Area of filtration = 38.48×10^{-4} m^2 Filtrate temperature = 20 °C Filtrate viscosity = 1.002×10^{-3} Ns/m^2	
	Volume of filtrate x 10^{-6} m^3		
Time (sec)	Test (1)	Test (2)	Test (3)
120	4.90	4.95	4.85
180	6.30	6.40	6.20
240	7.45	7.35	7.40
300	8.35	8.40	8.45
420	10.35	10.30	10.25
600	12.75	12.70	12.80
780	14.70	14.75	14.65
1080	17.60	17.55	17.65
1200	18.75	18.85	18.80

1. Plot the values of t/V versus V.

2. Using Carman - Coackley equation: $r = \dfrac{2bPA^2}{\mu C}$

find the **specific resistance** to filtration.

3. Comment on your result. (OIU, UNESCOC, 2004). (Ans. $27*10^{14}$ m/kg).

16) a) What is sludge **dewatering**? Briefly outline sludge dewatering methods. (SQU, 1992).

  b) In a specific resistance test using the Buchner funnel technique on a sample of sludge, the following values were recorded:

Time, minutes	1	2	3	4	5
t/V min/mL	0.77	0.87	0.97	1	1.09
Volume of filtrate, mL	1.3	2.3	3.1	4.0	4.6

  i) Find the value of the **specific resistance** of the sludge given:
  Vacuum applied = 69 kPa
  Filtrate viscosity = $1.1*10^{-3}$ Ns/m^2
  Volume of filtrate used = 50 mL
  Solids concentration = 0.075 g/mL,
  Area of Whatman No.1 filter paper used = 44.2 cm^2.

  ii) Is this sludge amenable to dewatering by vacuum filtration? (SQU, 1992). (Ans. $1.9*10^{14}$ m/kg, No,).

17) a) Comment on the methods of dewatering and on the significance of dewatering as a process preceding the final sludge disposal. b) The following results were obtained from specific resistance determination on a sample of digested sewage sludge:-

Time (t), (sec)	Volume of filtrate (v), (ml)
0	0.0
120	4.9
180	6.3
240	7.4
300	8.4
420	10.3
600	12.75
780	14.7
1080	17.6
1200	18.8

Vacuum pressure	68.95 KN/m²
Filtrate temperature	20°C.
Filtrate viscosity	1.002 X 10⁻³ N S/m²
Solids content	21.4 kg/m³
Area of filtration	38.48 X 10⁻⁴ m²
Volume of sludge used	100 x 10⁻⁶ m³

18) Plot the values of t/v against V and hence obtain the slope and calculate the **specific resistance**. (U of K, 1983). (Ans. $27*10^{13}$ m/kg).

19) a) What is the significance of **dewaterability** of sewage sludges.
b) Write briefly about methods used to dewater sewage sludges. Which method would you recommend to Sudan and why?
c) In a specific resistance test using the Buchner funnel technique, on a sample of activated sewage sludge, the following values were recorded:-

Time (t), (min)	Volume of Filtrate (V), (ml)		
	Test (1)	Test (2)	Test (3)
0	0.0	0.0	0.0
1	1.3	1.4	1.5
2	2.3	2.4	2.5
4	4.1	4.2	4.3
8	6.8	6.9	7.0
15	10.3	10.4	10.5

Find the value of the **specific resistance** of the sludge, given:
Vacuum utilized   = 97.5 kPa
Filtrate viscosity   = 1.011 X 10⁻³ Ns/m²
Volume of filtrate used   = 50 X 10⁻⁶ m³
Solids content   = 7.5 percent
Diameter of whatmen No.1 filter paper = 7.5 cm. (U of K, 1984). (Ans. $2.4*10^{14}$ m/kg).

20) a] Name major factors that influence sludge dewterability.
b] Find the **specific resistance** of a sample of activated sludge given the following data obtained from a small laboratory test filter.

Time, s	V	t/V
120	2.5	48
180	3.5	51.4

420	6.9	60.9
600	8.9	67.4
780	10.8	72.2
900	11.9	75.6

Vacuum applied   90 kPa
Solids concentration   4%
Filtrate viscosity $1.027*10^{-3}$ Ns/m^2
Area of filtration $3.8*10^{-3}$ m^2
Comment on the results you obtained. (SQU, 1991). (Ans. $18.8*10^{13}$ m/kg).

Compressibilty

21) Two sewage sludges (I and II) have been tested experimentally. The specific resistance tests revealed the following results:

sludge	Applied pressure, kN/m^2	specific resistance, $r x 10^{-13}$ m/kg
I	90	1.5
II	196	2.1

The compressibilty coefficient of sludge I amounts to 0.65. Compare the **dewaterability** of the two sludges.

22) The specific resistance of a sludge is $0.2x10^{13}$ m/kg at an applied pressure of 49 kPa. Find the **specific resistance** of the sludge at a pressure of 69 kPa if the compressibility coefficient of the sludge is 0.7. (Ans. $0.25*10^{13}$ m/kg).

23) Determination of specific resistance at different pressures gives a slope of 0.75 for a plot of log r versus log P for sludge (A). The value of the specific resistance at 90 kPa is $2(10)^{13}$ m/kg for sludge (A). a value quoted in the literature is (in s^2/g) $4(10)^9$ at 196 kPa for sludge (B). Compare the **dewaterability** of the two sludges. (UAE, 1991). (Ans. $3.6*10^{13}$, $3.9*10^{13}$ m/kg).

24) c. The specific resistance of a sludge is $0.2x10^{13}$ m/kg at an applied pressure of 49 kPa. Find the **specific resistance** of the sludge at a pressure of 69 kPa if the compressibility coefficient of the sludge is 0.7. (U of K, 1986). (Ans. $0.25*10^{13}$ m/kg).

25) Determination of specific resistance at different pressures give a slope of 0.65 for a plot of logr versus logp for sludge A. The value of the specific resistance at 90 kPa is $1.5 x 10^{13}$ m/kg for sludge A. A value quoted in the literature is $2.1 x 10^9$ s^2/g at 196

kPa for sludge B. Compare the **dewaterability** of the two sludges. (U of K, 1985). (Ans. $2.5*10^{13}$ m/kg).

Filter yield

26) A specific resistance determination on a sample of sewage sludge of a moisture content of 96% gave the value of $0.2x10^{13}$ m/kg. Find the **yield** of the sludge solids expected from dewatering such a sludge by a vacuum filter, given: $F_c = 2$, $P = 100$ kN/m^2, $\varepsilon = 1x10^{-3}$ Ns/m^2, $F_f = 0.27$, $\theta = 2$ min. (Ans. $6.1*10^{-3}$ kg/m^2.s).

27) A vacuum filter of diameter 2.5 m with a cycle time of 120 sec/revolution is to dewater sewage sludge of solids content 30 kg/m^3. The pressure diffrenvce developed across the filter amounted to 69 kN/m^2. If the specific resistance of the sludge is $3x10^{12}$ m/kg, find the filter **yield**. (Assume $\varepsilon = 10^{-5}$ Ns/m^2, $F_f = 0.17$, $F_c = 1.4$). (Ans. $6.1*10^{-3}$ kg/m^2.s).

28) For a conditioned sample of sewage sludge the following results have been determined:
Sludge solids content = 2%
specific resistance = $6x10^{11}$ m/kg
Filtrate viscosity = 10^{-3} Ns/m^2
Vacuum filter diameter = 2 m
Filter dip in sludge = 0.12 m
Solids content of dewatered sludge = 15%
Cycle time = 110 sec
Compute filter **yield**. (Ans. $8.5*10^{-3}$ kg/m^2.s).

6.9 Disposal into Receiving Bodies of Water

The effects of rwaw discharges into bodies of water include:
- Change in characteristics of receiving water.
- Endangering of aquatic life.
- Creation of malodour and nuisance.
- Production of taste and odour problems.
- Build-up of concentration of toxic materials.
- Generation of public health hazards.
- Deceleration of the process of self-purification.

Example 6.22

The dissolved oxygen (DO) of a stream is (p) mg/l. A city in the vicinity discharges its wastewater into the stream. The waste discharged has a DO content of zero. The ratio of stream flow to that of the wastewater is 8:1. If the DO of the mixture is 9 mg/l, find the value of (p). (SQU, 1991).

Solution

$$DO_m \, Q_m = Q_s . DO_s + Q_w . DO_w$$

$$DO_m \left(1 + \frac{Q_{sr}}{Q_w}\right) = DO_r \frac{Q_r}{Q_w} + DO_w$$

$9(1 + 8) = P \times 8 + 0$

$P = 9 \times 9 \times 8 = 81 \div 8 = 10$

Example 6.23

1) A factory produces 21600 m³/d of wastewater which finds its way into a neighbouring stream that flows at a rate of 10 m³/s. The velocity of the stream is about 2 km/hr. The temperature of the stream was found to be about 20° C, while that of the discharged wastewater is 30° C. The laboratory analysis indicated that the 5-day BOD at 20° C of the wastewater is 400 mg/l while that of the stream is 2.5 mg/l. The amount of dissolved oxygen contained in the stream is about 25 percent of the satutration value, while the wastewater is anaerobic. From the given data compute the:

   1. Temperture of the mixrture of wastewater discharged and that of the stream.
   2. Amount of the 5-day BOD of the mixture.

3. Critical oxygen deficie.

4. Location of the critical oxygen deficit.
(Assume k_1 and k'_2 at $20°$ C to be 0.3 and 0.7 /day respectively.).

2) Write a computer program to compute temperture of the mixrture of wastewater discharged from a factory and that of the stream, amount of the 5-day BOD of the mixture, critical oxygen deficiet, and location of the critical oxygen deficit given amount of wastewater finding its way into the neighbouring stream, stream flow rate, velocity of the stream, temperature of the stream, temperature of the discharged wastewater, the 5-day BOD at $20°$ C of the wastewater,BOD of the stream, amount of dissolved oxygen contained in the stream, and DO of wastewater and k_1 and k'_2.

3) Verify your program by solving example 6.23.1.

Solution

Wastewater discharge flow, Q_w = 216000/24x60x60 = 2.5 m³/s
Stream discahrge Q_s = 10 m³/s
Velocity of stream = 2x24 = 48 km/d
Temperture of stream, T_s = $20°$ C
Temperture of wastewater, T_w = $30°$ C
BODs = 2.5 mg/l
BODw = 400
From tables, saturation conentration of dissolved oxygen at $20°$ C = 9.2 mg/l. Therefore, Dos = 0.85x9.2 = 7.82 mg/l

a) Temperture of mixture = $(T_w.Q_w + T_s.Q_s)/(Q_w + Q_s)$ = (30x2.5 + 2.5x10)/(2.5 + 10) = $22°$ C

b) BOD$_5$ of the mixture =$(BOD_w.Q_w + BOD_s.Q_s)/(Q_w + Q_s)$ = (400x2.5 + 2.5x10)/(2.5 + 10) = 82 mg/l

c) The Streeter-Phelps oxygen sag equation may be used to determine the critical oxygen deficit

$$D_c = \frac{k'}{k''} L_o e^{-k't_c}$$

Where:
D_c= Critical dissolved oxygen deficit, mg/L.
k' = First order reaction rate constant, /d.
k'_2 = Re-aeration constant, /d.

368

L_0 = Ultimate oxygen demand, mg/L.

t_c = Critical time required to reach the critical point.

K_1 and k'_2 should be corrected for the temperture of 22o C, assuming

$(k_1)_T = (k_1)_{20}.(1.135)^{T-20}$

Then, $(k_1)_{22} = 0.3x(1.135)^{22-20} = 0.386$ /day

For k'2:

$(k'_2)_T = (k'_2)_{20}.(1.024)^{T-20}$

Therefore,

$(k'_2)_{22} = 0.7x(1.024)^{22-20} = 0.724$ /day

Ultimate BOD = L_0xBOD$_5$mixture/(1 – 10^{-k1t}) = 82/(1 – $10^{-0.3x5}$) = 105.55 mg/l

The time required to reach the critical point, t_c is given by:

$$t_c = \frac{1}{k''-k'} \ln\left[\frac{k''}{k'}\left[1-\frac{D_o(k''-k')}{k'L_o}\right]\right]$$

Where:

t_c =   Critical time, d.

D_o =   Initial Oxygen deficit at the point of wastewater discharge, at time t = 0 = DO saturation concentration – DO concentration of the mixture

k" =   Reaeration constant, /d.

k' =   First order reaction rate constant, /d.

L_o =   Ultimate BOD at point of discharge, mg/L.

DO concentration of the mixture = =$(DO_w.Q_w + DO_s.Q_s)/(Q_w + Q_s)$ = (0x2.5 + 7.82x10)/(2.5 + 10) = 6.26 mg/l

Therefore, D_o = 9.2 – 6.26 = 2.94 mg/l

$$t_c = \frac{1}{(0.734-0.386)} \ln\left[\frac{0.734''}{0.386}\left[1-\frac{2.94(0.734-0.386')}{0.386 \, x \, 105.55}\right]\right]$$

= 1.77 days

Thus,

$$D_c = \frac{k'}{k''}L_o e^{-k'tc} = \frac{0.386'}{0.734} x \, 105.55 \, x \, e^{-0.386 x \, 1.77} = 28 \, mg/l$$

d) Location of the critial oxygen deficit, x_c = v.t_c

V = velocity of flow in the stream = 48 km/d

x_c = 48x1.77 = 85 km

369

Table (6.13) Classification of Rivers According to Concentration of Dissolved Oxygen

Classification Scheme BOD_5^{20} value	DO, as a % mg/L) of saturation
Good	Greater than 90
Fair	75 - 90
Doubtful	50 - 75
Polluted	Less than 50

Program 6.23 Listing:
Concentration of pollutant from factory wastewater discharged to a stream

```
'*******************
'EXAMPLE 6.23
'*******************
Public Class Form1
    Dim Cs(31) As Double

    Private Sub Button1_Click(ByVal sender As
        System.Object, ByVal e As
        System.EventArgs) Handles Button1.Click
        Dim Qw, Qs, V, Tw, Ts, BODs, BODw As Double
        Dim Dow, Dos, k1, k2 As Double
        Dim Tm, BODm As Double

        Qw = Val(TextBox1.Text)
        Qs = Val(TextBox2.Text)
        V = Val(TextBox3.Text)
        Ts = Val(TextBox4.Text)
        Tw = Val(TextBox5.Text)
        BODs = Val(TextBox6.Text)
        BODw = Val(TextBox7.Text)
        Dow = Val(TextBox8.Text)
        k1 = Val(TextBox9.Text)
        k2 = Val(TextBox10.Text)

        Qw /= (24 * 60 * 60)
        V *= 24
        Dos = 0.85 * (Cs(20))
        Tm = ((Tw * Qw) + (Ts * Qs)) / (Qw + Qs)
        TextBox11.Text = "Temp. of mixture = "
                + FormatNumber(Tm, 2) + "C"
        TextBox11.Text += vbCrLf
```

370

```vb
      BODm = ((BODw * Qw) + (BODs * Qs)) / (Qw + Qs)
      TextBox11.Text += "BOD of mixture = "
            + FormatNumber(BODm, 2) + " mg/l"
      TextBox11.Text += vbCrLf
      'Use The Streeter-Phelps equation to
      'determine the critical oxygen deficit
      'correct k1 and k2 to mixture temp. Tm
      Dim xc, dc, tc, k1T, k2T, BODu, _Do As Double
      k1T = k1 * (1.135 ^ (Tm - 20))
      k2T = k2 * (1.024 ^ (Tm - 20))
      'calculate ultimate BOD
      BODu = (1.25 * BODm) / (1 - (10 ^ (-k1 * 5)))
      TextBox11.Text += "Ultimate BOD = "
            + FormatNumber(BODu, 2) + " mg/l"
      TextBox11.Text += vbCrLf

      _Do = (Dow * Qw + Dos * Qs) / (Qw + Qs)
      _Do = Cs(20) - _Do
      tc = (1 / (k2T - k1T)) * Math.Log((k2T / k1T) *
            (1 - ((_Do * (k2T - k1T)) / (k1T * BODu))))
      TextBox11.Text += "Time to reach critical
            point = " + FormatNumber(tc, 2) + " days"
      TextBox11.Text += vbCrLf
      dc = (k1T / k2T) * BODu * (Math.E ^ (-k1T * tc))
      xc = V * tc
      TextBox11.Text += "Dc = " + FormatNumber(dc, 2)
            + " mg/l"
      TextBox11.Text += vbCrLf
      TextBox11.Text += "Location of critical oxygen
            deficit = " + FormatNumber(xc, 2) + " km"
End Sub

Private Sub Form1_Load(ByVal sender As
    System.Object, ByVal e As
    System.EventArgs) Handles MyBase.Load
    Me.Text = "Example 6.23"
    Me.FormBorderStyle =
        Windows.Forms.FormBorderStyle.FixedSingle
    Me.MaximizeBox = False
    Label1.Text = "Wastewater discharge flow,
        Qw (m3/d)"
    Label2.Text = "Stream discharge, Qs (m3/s)"
    Label3.Text = "Stream velocity (km/hr)"
    Label4.Text = "Stream temp., Ts (C)"
    Label5.Text = "Wastewater temp., Tw (C)"
    Label6.Text = "BOD5 of stream, BODs (mg/l)"
    Label7.Text = "BOD5 of wastewater, BODw (mg/l)"
    Label8.Text = "Dissolved oxygen in wastewater"
    Label9.Text =
        "First order reaction rate const, k1 (/d)"
```

371

```
        Label10.Text = "Reaeration constant, k2 (/d)"
        TextBox11.Multiline = True
        TextBox11.ScrollBars = ScrollBars.Vertical
        TextBox11.Height = 65
        Me.Height += 65
        Button1.Text = "&Calculate"
        'initialize table of oxygen concentration
        'staturation
        Cs(0) = 14.6
        Cs(1) = 14.2
        Cs(2) = 13.8
        Cs(3) = 13.5
        Cs(4) = 13.1
        Cs(5) = 12.8
        Cs(6) = 12.5
        Cs(7) = 12.2
        Cs(8) = 11.9
        Cs(9) = 11.6
        Cs(10) = 11.3
        Cs(11) = 11.1
        Cs(12) = 10.8
        Cs(13) = 10.6
        Cs(14) = 10.4
        Cs(15) = 10.2
        Cs(16) = 10.0
        Cs(17) = 9.7
        Cs(18) = 9.5
        Cs(19) = 9.4
        Cs(20) = 9.2
        Cs(21) = 9.0
        Cs(22) = 8.8
        Cs(23) = 8.7
        Cs(24) = 8.5
        Cs(25) = 8.4
        Cs(26) = 8.2
        Cs(27) = 8.1
        Cs(28) = 7.9
        Cs(29) = 7.8
        Cs(30) = 7.6
    End Sub
End Class
```

Exercise 6.8

Quantity of stream flow

1) a] Briefly discuss the sources of groundwater **pollution** (draw neat sketches).

    b] Sketch the thermal **stratification** and overturn of a lake of reservoir.

    c] The BOD_5 of an effluent from poorly operating sewage treatment plant is 100 mg/L and the discharge is 1.5 ML/d. the receiving stream has a BOD_5 of 3 mg/L. What minimum stream **flow** is needed for a dilution such that the combined BOD_5 of sewage and stream water is not greater than 10 mg/L. (SQU, 1991).

Sag curve

2) A stream has a normal flow of 3.12 m³/s, its 5-day BOD is 2 mg/l and it is saturated with dissolved oxygen. A sewage effluent is discharged to the stream at a daily rate of 6480 m³, with a 5-day BOD of 30 ppm. Compute the dissolved oxygen **concentration** and dissolved oxygen **deficit** over the next five days and draw the oxygen sag curve. (Assume temperatures are 20° C throughout, saturation of oxygen at this temperature is 9.2 mg/l, take $k_1 = 0.1$ and $k'_2 = 0.4$ per day. Assume velocity of stream to be 2 km/hr). (Ans. 0.65 mg/L).

3) A river flows at 0.5 m³/s with a 5-day BOD of 1 mg/l and it is saturated with oxygen. Sewage effluent is discharged to the stream at the rate of 0.2 m³/s with a 5-day BOD of 25 mg/l and it is 10% saturated with dissolved oxygen. Calculate:

    a) The dissolved oxygen **deficit** after 2 days.

    b) The **critical** oxygen deficit.

    (Assume temperature to be constant at 20° C and $k_1 = 0.1$ and $k'_2 = 0.4$ per day). (Ans. 2.5, 2.48 mg/L).

4) a) A simple river oxygenation model is described by the equation of Streeter and Pheleps as indicated below:

$$D_t = \frac{k_1 L_o}{k_2 - k_1}\left[e^{-k_1 t} - e^{-k_2 t}\right] + D_o e^{-k_2 t}$$

Dentine the parameter indicated in the equation

5) A river flows at 1800 m³/hr with a **5-day BOD** of Y mg/L and it

373

is saturated with oxygen. If the critical oxygen deficit amounts to 2.5 mg/l, compute:

1) The value of **Y** in mg/l.
2) The dissolved oxygen **deficit** at tar two days.

[Assume temperature to be constant at 20°C, $K_1 = 0.1$ and $K_2' = 0.4$ per day]. (K of U, 1988). (Ans. 25, 2.48 mg/L).

Solution

River	Sewage
Q = 1800 ÷ (60×60) = m³/s	qw =1260 = 0.2 m³/s
$BOD_5 = 1$	BOD = Y
$D_O = C_s = 9.2$ from table for t = 20°C	DO = $0.1C_s = 0.92$

$D_c = 2.5$ mg/L

$$BOD_m = \frac{Q_r BOD_r + Q_w BOD_w}{Q_r + Q_w} = \frac{0.5 \times 1 + 0.2Y}{0.5 + 0.2} = \frac{0.5 + 0.2Y}{0.7} = L$$

$$L_o = \frac{L}{1 - 10^{-k't}} = \frac{0.5 + 0.2Y}{0.7(1 - 10^{-0.1 \times 5})} = 1.04 + 0.418Y$$

$$D_o = C_s - DO_m = 9.2 - \left[\frac{9.2 \times 0.5 + 0.92 \times 0.2}{0.5 + 0.2}\right] = 2.37$$

$$t_c = \frac{1}{k_2 - k_1} \ln\left[\frac{k_2}{k_1}\left[1 - \frac{D_o(k_1 - k_1)}{L_o}\frac{}{k_1}\right]\right]$$

$$= \frac{1}{(0.4 - 0.1)} \ln\left[\frac{0.4}{0.1}\left[1 - \frac{2.37}{(1.04 + 0.418Y)}\frac{(0.4 - 0.1)}{0.1}\right]\right]$$

$$= \frac{1}{0.3} \ln\left[4\left[1 - \frac{7.11}{1.04 + 0.418Y}\right]\right]$$

$$D_c = \frac{k_1}{k_2} L_o e^{-k_1 t_c}$$

$$2.5 = \frac{0.1}{0.4}(1.04 + 0.418Y)e^{-0.1 \times \frac{1}{0.3} \ln\left[4\left[1 - \frac{7.11}{1.04 + 0.418Y}\right]\right]}$$

Assumed Y	L.H.S
10	∞
20	2.37
25	**2.5**
30	2.74

Thus, Y = 25 mg/L

2)

$$L_o = 1.04 + 0.418 \times 25 = 11.5$$

$$D_t = k_1 L_o \frac{e^{-k_1 t} - e^{-k_2 t}}{k_2 - k_1} + D_o e^{-k_2 t}$$

$$D_{2days} = \frac{0.1 \times 11.5}{0.4 - 0.1} \left[e^{-0.1 \times 2} - e^{-0.4 \times 2} \right] + 2.37 e^{-0.4 \times 2} = 2.48 \, mg/L$$

6) A simple river oxygenation model is described by the equation of Streeter and Phelps as indicated below:

$$D_t = \frac{k_1 L_o}{k_2 - k_1} \left[e^{-k_1 t} - e^{-k_2 t} \right] + D_o e^{-k_2 t}$$

    i.   Dentine the parameter indicated in the equation
    ii.   Illustrate the limitations of this model
    iii.   A river flows at 0.5 m³/s with a 5-day BOD of 1 mg/L and it is saturated with oxygen. Sewage effluent is discharged to the stream at the rate of 0.2 m³/s with a 5-day BOD of 25 mg/L and is 10% saturated with dissolved oxygen. Calculate:
        a.   The dissolved oxygen **deficit** after 2 days.
        b.   The **critical** oxygen deficit.
        [Assume temperatures to be constant at 20°C, and k' = 0.4 per day. Oxygen saturation concentration at temperature of 20°C amounts to 9.2 mg/L]. (U of K, 1986).

7) Comment about the effects of raw waste disposal into natural bodies of water. A stream has a normal flow of 3.12 m³/s, its 5-day BOD is 2 mg/L and it is saturated with dissolved oxygen. A sewage effluent from a certain factory is discharged to the stream at a daily rate of 6480 m³, with a 5-day BOD of 30 ppm. Compute the dissolved oxygen **concentration** and dissolved oxygen **deficit** over the next five days. (Assume temperatures are 20°c throughout. Take k_1 = 0.1 and k_2' = 0.4 per day). (U of K, 1988).

8) A factory produces 216000 m³/day of wastewater which finds its way into neighboring stream that flows at a rate of 10 m³/s. The velocity of the stream is about 2 km/hr, the temperature of the stream was found to be 20°C while that of the discharged

wastewater is 30°C. The laboratory analysis indicated that the BOD_5^{20} of the wastewater is 400 mg/L while that of the stream is 2.5 mg/L. The wastewater is anaerobic, but the stream contains dissolved oxygen in an amount of about 85% of the saturation value. From the given data compute:

a) The **temperature** of the **mixture** of wastewater discharged and the stream water.

b) The amount of the **five-day biochemical oxygen demand** of the **mixture** of wastewater and stream water.

c) The **critical** oxygen deficit.

d) The **location** of the critical oxygen **deficit**. (Take $k_1' = 0.3$ and $k_2' = 0.7$ per day). (U of K, 1984). (Ans. 22oC, 82 mg/L, 28 mg/L, 85 km).

9) A river flows at 0.5 m³/s with a 5-day BOD of 1 mg/L and it is saturated with oxygen. Sewage effluent is discharged to the stream at the rate of 0.2 m³/s with a 5-day BOD of 25 mg/L and it is 10% saturated with dissolved oxygen. Calculate:

a) The 5-day **BOD** of the **mixture**.

b) The **dissolved oxygen** of the **mixture**.

c) The **critical** oxygen deficit

Assume temperature to be constant at 20°C (oxygen saturation concentration at this temperature is 9.2 mg/L and $k_1 = 0.1$, $k_2' = 0.4$ per day) (U of K, 1988).

Influent BOD

10) Indicate a simple river oxygenation model and elaborate on its justifications. How much **BOD** can be **discharged** into a stream of flow 8 m³/s in order to establish an oxygen concentration not less than 4 mg/L within a distance of 100 km downstream from the point of discharge? (assume that the velocity of stream flow is 1 km/hr, $k' = 0.1$ /d & $k_2' = 0.4$ /d, temperature have a constant value of 20°c). (U of K, 1987). (Ans. 184 mg/L).

Several factors control and govern the operational methods used in Sanitary Land-filling at any given site. Enumerate these factors and discuss their role and significance. (U of K, 1987).

11) a) Discuss effects of raw wastewater discharges into natural bodies of water.

b) A river flows at 95 m^3/min. With a 5-day BOD of 1 mg/L and it is saturated with oxygen. Sewage effluent is discharged to the stream at the rate of 15 m^3/min with a 5-day **BOD** of "x" and is ten percent saturated with dissolved oxygen. If the dissolved oxygen deficit of the mixture after two days amounts to 2-6 mg/L: Find the value of "x" and the critical oxygen deficit
(Assume temperature to be constant at 20°C, $K_1 = 0.1$ and $K_2' = 0.4$ per day) (UAE, 1989)

12) a) Outline the major effects of raw waste disposal into natural bodies of water.

b) A river flows at 30 m^3/minute with a 5-day BOD of 1 ppm and is saturated with oxygen. Sewage effluent with 50day BOD of 25 mg/L is discharged to the stream at the rate of 720 m^3/hour. If the maximum oxygen deficit is not to exceed 2.5 mg/L, find the **dissolved oxygen** of the waste in terms of percent saturation. (Assume temperature to be constant at 20°C and k' = 0.1, $k_2' = 0.4$ per day). (Ans. 10% saturation).

13) "Self-purification is a natural process in water courses receiving domestic and industrial pollutional loads". Discuss the validity of this statement.

What are the limitations of the Streeter-Phelps model?

A river flows at rate of 1800 m^3/hr with a 5-day BOD of 1 mg/L and it is saturated with oxygen. Sewage effluent is discharged to the river at the rate of 12 m^3/minute with a 5-day BOD of "Y" and it is 10 percent saturated with dissolved oxygen. If the critical oxygen deficit amounts to 2.5 mg/L.

• Compute the value of "**Y**" in mg/L.

• Determine the dissolved oxygen **deficit** after three days.
(Assume temperatures to be constant at 20°C, and take k' = 0.1, k" = 0.4 per day). (OIU, UNESCOC, 1999). (Ans. 25, 2.4 mg/L).

14) The treated sewage of 20000 people is discharged to a nearby stream which is flowing at a rate of 0.15 m^3/s with a BOD of 2 mg/l. The water consumption by the inhabitant's amounts to 150 liter/capita.day, and the BOD contribution is 65 g/capita.day. If the BOD in the stream below the outfall is not to surpass 4 mg/l, find the required **efficiency** of the treatment plant needed to cope with this condition and the maximum effluent BOD

377

permissible. (U of K, 1986).

Effluent BOD

15) A small river with a flow of 1.2 m^3/s, 5-day BOD of 2 mg/l and saturated with oxygen is suggested to receive a sewage effluent at the rate of 0.15 m^3/s. The allowed maximum dissolved oxygen deficit is 3 mg/l downstream. The stream temperature is 20° C (Cs = 9.2 mg/l). Find the maximum **effluent BOD**. (Assume the dissolved oxygen concentration of the effluent to be 100% and the temperature remains at 20° C and k_1 = 0.1 and k'_2 = 0.4 /d). (Ans. 16 mg/L).

16) A town of 10000 inhabitants is to discharge its treated domestic sewage effluent into a stream with a minimum flow of 0.13 m^3/sec and 5-day BOD of 2 mg/l. The sewage is produced at a rate of 135 l/capita.d and BOD contribution per capita is 0.054 kg/d. If BOD downstream of the discharge point is not to exceed 4 mg/l, determine the maximum permissible **effluent BOD** and **efficiency** of treatment plant expected to achieve that permissible value. (Ans. 21 mg/L, 95%).

17) The concentration of a certain pollutant upstream a sewage treatment plant is 1 mg/L and the stream flow is 30 m^3/s. The discharge of effluent from the sewage treatment plant is at the rate of 5 m/s. The pollutant concentration downstream of the sewage treatment plant is 3 mg/L. Compute the pollutant **concentration** of the waste **effluent** stream. (SQU, 1992). (Ans. 15 mg/L).

18) The concentration of a certain pollutant upstream a. sewage treatment plant is 20 ppm and the stream flow is 30 m^3/s. The discharge of effluent from the sewage treatment plant is 5 m^3/s. The pollutant concentration downstream of the sewage treatment plant is 30 ppm. Compute the **pollutant concentration** at the waste effluent stream. (SQU, 1991).

19) Drive from first principles the dilution law for a pollutant in a mixture of effluent and a receiving body of water. Comment about limitations of this method of disposal

A sewage treatment works discharges its wastewater at the rate of 5 m^3/s with a pollutant concentration of (y) mg/L, to a neighboring stream. The concentration of that pollutant upstream the sewage treatment plant is 2 mg/L and the river flows at a rate of 40 m^3/s. Given that the pollutant concentration downstream the sewage treatment plant amounts to 10 mg/L, determine the **pollutant concentration** in effluent (y). (U of K, 2004).

Write briefly about cons and pros of separate sewerage system.

20) What are the factors that affect dilution law for a pollutant in a mixture of treated wastewater effluent and a receiving body of water?

A sewage treatment plant discharges its effluent at the rate of 10 m^3/s with a pollutant concentration of (P$_w$) mg/L, to a receiving stream. The concentration of that pollutant upstream the sewage treatment plant is 5 mg/L and the river flows at a rate of 80 m^3/s. Given that the pollutant concentration downstream the sewage treatment plant amounts to 20 mg/L, determine the **pollutant concentration** in effluent (y). (U of K, 2005).

Write briefly about advantages and disadvantages of a combined sewerage system.

21) A town of 15000 inhabitants is to discharge treated domestic sewage to the neighboring stream which has a minimum flow of 0.21 m^3/s and a 5-day BOD of 2 mg/l. Water consumption in the town is 140 l/capita.d, and the BOD contribution is 0.06 kg/person.d. If the BOD in the stream below sewage outfall is not to exceed 4 mg/l, determine the maximum **effluent BOD** permissible. (Ans. 21 mg/L).

BOD & DO of mixture

22) The 5-day BOD of a stream is 1 ppm. A city in the neighborhood discharges its wastewater into the stream. The waste discharged has a 5-day BOD content of 262 mg/l. The stream, flow is eight times that of the wastewater. Find the 5-day **BOD** of the **mixture**. (SQU, 1995).

23) Why is the dissolved oxygen concentration in a stream likely to have a diurnal variation?

A wastewater effluent of 500 L/s with a BOD = 45 mg/L, DO = 0.4 mg/L and temperature of 17.5°C enters a stream where the flow is 4.5 m^3/s with a BOD = 6 mg/L, DO = 8.4 mg/L and temperature of 22.5°C. from laboratory BOD testing, the k$_1$ of the wastewater diluted with stream water is 0.1 per day at 20°C (base 10), the k_2 in the stream below the sewer outfall is 0.53 per day at 20°C (base e) based on tracer studies.

i)   Find the **BOD** and **DO** of the **mixture**.

ii) Calculate the value and **location** of **critical** dissolved oxygen level using the oxygen sag equations.

iii) How much would the **critical oxygen level** increase if the wastewater was aerated to a DO of 8.4 mg/L prior to discharge? (UAE, 1990). (Ans. 10, 7.6, 3.75, 3.54 mg/L).

24) a) Write an equation to determine the concentration of a pollutant in the mixture of effluent and river, C_m (mg/L), given: concentration, C_w (mg/L), and rate of flow, Q_w (m^3/s), of pollutant introduced to the river and the river flow, Q_r (m^3/s), and amount of pollutant initially found in it, C_r (mg/L).

b) A sewage treatment works discharges its wastewater at the rate of 8 m^3/s with a pollutant concentration of 15 mg/L, to a neighboring stream. The concentration of a particular pollutant upstream the sewage treatment plant is 2 mg/L and the river flows at a rate of 0.2 m^3/minute. Find the **pollutant concentration** downstream the sewage treatment plant. (U of K, 2004).

c) Write briefly about methods of wastewater collection in an urban area.

25) What parameters you need to address in order to allow a certain factory to discharge its wastes in the neighbouring stream. The BOD$_5$ and DO of a stream are 1 and 9 mg/L respectively. A city in the vicinity discharges its own wastewater into the stream. The waste discharged has a BOD$_5$ and a DO content of 280 and zero ppm respectively. The ratio of stream flow to that of the wastewater is 8:1. Compute **BOD$_5$** of the mixture and **DO** of the **mixture** in mg/L. If the rate constant is 0.3 per day, determine the **ultimate BOD$_5$** of the mixture of wastewater of stream at outfall. (U of K, 1987, U of K, 1988, SQU, 1993). (Ans. 32, 8, 41 mg/L).

26) What are the significances of the **BOD test**?

b)A BOD test is carried out on domestic sewage which has a rate constant (k_1) value at 20°c of 0.16 day^{-1}. Calculate the **BOD$_5$** (the BOD exerted) as a fraction of the ultimate BOD. (U of K, 1985). (Ans. 84 %).

27) Three factories discharge their wastewater into a nearby sewer. The amount and characteristics of their waste are as tabulated below:

Factory	Amount of waste discharged (m³/s)	BOD_5^{20} of the waste (mg/l)
I	0.1	400
II	0.2	150
III	0.3	275

Compute the **BOD₅** of the **mixture** being discharged to the sewer.

28) The simple river oxygentation model is described by the equation of Streeter and Phelps as indicated below:

$$D_t = \frac{k_1 L_o}{k_2 - k_1}\left[e^{-k_1 t} - e^{-k_2 t}\right] + D_o e^{-k_2 t}$$

   a)   Define the parameter indicated in the equation.
   b)   Illustrate the limitations of this model.
   c)   Sketch a characteristic oxygen sag curve as obtained by the equation.

A sewer carrying 0.3 m³/s foul sewage with a BOD of 250 mg/L joins two other sewers carrying flows of 0.1 m³/s and 0.2 m³/s with BOD₅ of 379 mg/L and 157 mg/L respectively. Calculate the **BOD₅²⁰** of the **mixture**. (U of K, 1985). (Ans. 241 mg/L).

Disposal into estuary

29) a) Discuss the effects of tidal action on estuarine channels receiving waste discharges.

b) A city uses the nearby estuary for its water supply. The city is located 30 km upstream from the ocean. The estuary has a uniform cross sectional area of 450 m², and coefficient of Eddy diffusion (E) has been measured to be 289.35 m²/s. The chloride concentration in the ocean is 19000 ppm. At a fresh water flow of 1800 m³/min, the chloride concentration at the city is 18 mg/L. A dam is to be built 30km upstream of the city to prevent loss of fresh water to the ocean. If the resultant fresh water outflow is reduced to 180 m³/min, what will the **chloride concentration** be. Based upon the answer you obtained, suggest an appropriate solution to this condition. (UAE, 1989) (Ans. 9518 mg/L).

Chapter Seven

General Questions

7.1 Multiple choice questios
MCQs. (B.Sc. UoD, 2014)

1) Population of a city in 2013 is 450,000. If the population is growing at the rate of 0.2 per decade, forecast the population of the city in 2043.
   a) 720,000
   b) 777,600
   c) 815,113
   d) 540,000

2) The term sewage refers to:
   1) Domestic wastewater from within the municipal area.
   2) Industrial wastewater from within the municipal area.
   3) Storm water from within the municipal area.
   4) All the above put together.

3) Partially separate system carries:
   a) Domestic, commercial and institutional wastewaters.
   b) Domestic, commercial, institutional and industrial wastewaters.
   c) Sewage and storm water from the premises of the residential buildings.
   d) Sewage and the urban storm water excluding that from the residential buildings premises.

4) Which of the following is an example for the attached growth biological treatment process?
   a) Aerated lagoon.
   b) Oxidation pond.
   c) Membrane bioreactor.

d) None of these.
5) Which of the following is an example for the suspended growth biological treatment process?
   a) Moving bed bioreactor (MBBR).
   b) Rotating biological contactor (RBC).
   c) Membrane bioreactor (MBR)
   d) None of these.
6) Hardy Cross method is used for:
   a) Analysis of sewerage system.
   b) Analysis of rural water supply system.
   c) Analysis of urban water supply network.
   d) Analysis of water treatment plants.
7) Groundwater usually requires the following prior to use as drinking water:
   a) Coagulation-flocculation.
   b) Chlorination.
   c) Filtration.
   d) Clarification.
8) A conventional sewage treatment plant removes mainly the following from the sewage:
   1) TSS
   2) Nutrients
   3) Pathogens
   4) TDS
9) The per capita consumption of a locality is affected by
   1. Climatic conditions
   2. Quality of water
   3. Distribution pressure
   The correct answer is
   1. Only (i)
   2. Both (i) and (ii)
   3. Both (i) and (iii)
   4. All (i), (ii) and (iii)
10) As compared to geometrical increase method of forecasting population, arithematical increase method gives
   a) Lesser value
   b) Higher value
   c) Same value
   d) Accurate value

MCQs, Circle the most appropriate answer. (B.Sc., UoD, 2013)

a) Jaundice, dysentery and typhoid affect human population due to
   1) **Water pollution**
   2) Air pollution
   3) Noise pollution
   4) Soil pollution

b) Biological oxygen demand of _____ is the least.
   a) sewage
   b) sea water
   c) **pure water**
   d) polluted water

c) _____ is the first step of sewage treatment.
   a) Precipitation
   b) Chlorination
   c) **Sedimentation**
   d) Aeration

d) We and our surroundings together are called...
   a) **environment**
   b) atmosphere
   c) lithosphere
   d) hydrosphere

e) Which of the following is not an environmental problem ?
   i. Wastage of water
   ii. **Conservation of water**
   iii. Deforestation
   iv. Land erosion

f) BOD is _____ in polluted water and _____ in potable water.
   a) **more, less**
   b) less, medium
   c) medium, more
   d) less, more

g) If a customer complains about the drinking water characteristics, the operator should record the complaint and
   a) **investigate immediately**
   b) investigate only if more complains are received
   c) inform the customer that the water should be boiled
   d) inform the customer that the water is safe

h) Under the requirements of the Safe Drinking Water Act, it is the duty of the water purveyor to deliver potable water of proper quantity only as far as the
   1. entry point of the distribution system
   2. customer's curb box and service connection
   3. customer's tab inside the home
   4. furthest water main blow-off or sampling point

i) What is a commonly used indicator of possible health problems found in plants, soil, water and the intestines of humans and warm-blooded animals?
   - Viruses
   - **Coliform bacteria**
   - Intestinal parasites
   - Pathogenic organisms

j) What are disease producing bacteria called?
   1) Parasites
   2) New strain
   3) Sour type
   4) Pathogenic

k) What are the two main causes of hardness in water?
   a) Gold and silver
   b) Calcium and magnesium
   c) Phosphate and nitrate
   d) Oxygen and methane

l) Which source of water has the greatest natural protection from bacterial contamination?
   a) Shallow well
   b) Deep well in gravel
   c) Surface water
   d) Spring

m) Which of the following causes taste problems and has a rotten egg odor?
   a) Chlorine
   b) Benzene
   c) Nitrate
   d) Hydrogen sulfide

n) Water with a pH of 7.9 is considered to be
   a) acidic

    b) basic or alkaline
    c) neutral
    d) undrinkable

o) The hydrologic cycle is composed of all the following terms, except
    a) condensation
    b) evaporation
    c) precipitation
    d) percolation
    e) adsorption

p) Pick up the correct statement from the following:
    a) Excess quantities of iron and manganese in water, cause discoloration of clothes
    b) Lead and barium salts have toxic effect
    c) Arsenic and selenium are poisonous to human health
    d) Higher copper content affects the lungs
    e) All the above.

q) The population growth curve is
    a) S-shaped curve
    b) parabolic curve
    c) circular curve
    d) straight line
    e) none of these.

r) By boiling water, hardness can be removed if it is due to
    1) calcium sulphate
    2) magnesium sulphate
    3) calcium nitrate
    4) calcium bicarbonate
    5) none of these.

s) Aeration of water is done to remove
    a) odour
    b) colour
    c) bacterias
    d) hardness
    e) turbidity.

f) Anaerobic decomposition of wastewater produces
    a) Ammonium chloride
    b) Hydrogen sulfide
    c) Carbon mono oxide

d) Copper sulphate

7.2 Fill-In-The Blanks

Fill-In-The Blanks (B.Sc. UoD, 2014)

Word Bank: Every Word Can Only be Used Once!

Anaerobic. Back washing. Chlorination. Cholera. Malaria. Scrapping.

1. Cleaning of slow sand filter by **scrapping**, and cleaning of rapid sand filter by **back washing**.
2. **Malaria** is not water borne disease.
3. **Cholera** is water borne disease.
4. Pathogens can be killed by **chlorination**.
5. Most of the bacteria in sewage are **anaerobic**.

7.3 True-False statements

True/False statements (B.Sc. UoD, 2014)

1) Bacterial aerobic oxidation of polluted water in biological oxidation ponds is done to purify it. Presence of bacteria helps in coagulation and flocculation of colloids, oxidation of carbonaceous matter to CO_2 and nitrification or oxidation of ammonia derived from breakdown of nitrogenous organic matter to the nitrite and eventually to the nitrate. (**T**)
2) Biological oxygen demand (B.O.D.) value of a sewerage sample is always lower than its chemical oxygen demand (C.O.D.) value. (**T**)
3) Presence of calcium hardness is responsible for the temporary hardness in water. (**T**)
4) A shallow pond in which the sewage is retained and biologically treated is called maturation (oxidation) pond. (**T**)
5) Sedimentation (clarification) is the most practical and economical method for removal of suspended solid matter from polluted water. (**T**)

7.4 Overall view questions

1) The following table represents the label attached to a 0.6 litres Nova drinking water bottle:

387

Valid for 1 year from production date	
Average composition, ppm	
Anions	
Bicarbonates	20.04
Sulphates	35
Chlorides	17
Fluorides	0.80
Nitrates	3.06
Cations	
Calcium	10
Magnesium	4.45
Sodium	16.79
Potassium	1.05
TDS	120
pH	7.0
Total hardness	43.54
Source of water	Nufoud Al Wasse'e

Using information you acquired from the sanitary engineering related lectures, comment on the label. (B.Sc., UoD, 2013)

2) Being the responsible site construction engineer, your advice is required to take a vital decision regarding which wastewater disposal technology to be adopted for Dammam city districts:

Use of a large wastewater treatment plant for the whole city? Or,
Use of wastewater treatment plant for each individual district of the city? Or,
Continue using septic tank systems in the town?

State your reasons and motives for your selection. (B.Sc., UoD, 2013)

3) Discuss the importance of availability of certain engineering maps such as: demographical patterns projections, hardness map, dissolved solids iso-concentration map, groundwater distribution and seasonal surface runoff in Saudi Arabia. Outline how to estimate the water footprints for KSA? (B.Sc., UoD, 2013)

4) How can you use the scientific knowledge you gained from the sanitary engineering course to safeguard the health of an engineering project such as: constructing a hyper mall in Dhahran area? Illustrate your answer with relevant sketches, diagrams and useful data. (B.Sc., UoD, 2013)

5) Being the district engineer in a AlKhobar town you are needed to look after appropriate use of freely distributed pesticides, insecticides and herbicides to farmers and interested personnel. State recommended approaches to avoid contamination and any potential environmental pollution and incurred health hazards. Defend your plan and strategy. (B.Sc., UoD, 2013)

6) As a construction and sanitary engineer in a KSA municipality, what role could you play to positively contribute towards achieving goals of the country's MDG's as per schedule and plan? State your reasons and explanations. (B.Sc., UoD, 2013)

7) As the responsible municipality engineer you observed that the greater portion and share of total residential water use goes to landscape outdoor use and indoor leaks and wastages. What remedial measures would you take? Whose opinion would you seek? State your reasons. (B.Sc., UoD, 2013)

8) Population estimates are needed for the design and operation of water supply and wastewater units. How are population estimates being carried out in KSA? Comment on methods used, frequency, availability of data and significance. What recommendations would you offer to improve procedures and transparency? (B.Sc., UoD, 2013)

7.5 Short questions

Provide answers to any **FOUR** of the following questions related to *water and wastewater quality, and water supply and treatment.* (B.Sc., UoD, 2013)

Describe the purpose of using an **indicator microorganism** is assessing the microbial quality of treated drinking water or the quality of sewage treatment plant effluent.

A wastewater contains 200 mg/L of phenol (C_6H_5OH). Once the stopper is placed in a **BOD** bottle, the system is closed. Does the total amount of organic matter (microorganisms and non-living organic matter) remain constant? Explain your answer.

In cases where water resources are limited, it may be necessary to move from a "demand-driven" water-supply environment to a **"supply-driven"** system, i.e. where the city water supplier limits water usage. For the case where a water supply system already has water metering installed, give 2 ways in which the city could

389

cause the residents to use less water before they are forced to, due to an actual water shortage situation. Do NOT propose actually shutting off the water, which can have serious negative consequences.

List three (3) benefits and three (3) disadvantages or problems associated with using **groundwater** as a drinking water source.

Provide a brief explanation of the differences between *each pair of treatment methods as they relate to drinking water treatment or wastewater treatment.*

  i.   Slow sand filtration and rapid sand filtration
  ii. Sedimentation and air floatation
  iii. Anaerobic and aerobic treatment
  iv. *E. coli* and viruses as related detection of presence of pollution.

Explain the key processes associated with **filtration** as applied in drinking water treatment. In your answer, include a brief discussion of operation and maintenance issues.

7.6 Rational Discussion Questions

You are requested to attempt solving any **FOUR** of the following topics: (B.Sc., UoD, 2013)

   a) Saudi **standards** as compared to WHO guidelines for **drinking water** quality focusing on microbial, chemical, radiological and acceptability aspects.

   b) Many industrial and domestic **water** users are concerned about the **hardness** of their water. As a consultant engineer of the municipality state your reasons for urging concerned authorities to produce a hardness map for the Kingdom of Saudi Arabia

   c) More than 900 million international journeys are undertaken every year. Global travel on this scale exposes many people to a range of health risks. Many of these risks can be minimized by precautions taken before, during and after travel. Outline **health risks** for travellers and illustrate reasons to use an International Travel and Health Interactive map. Would you recommend formation of a water disease map for KSA? State your reasons

d) As a consultant you are assigned forming a capacity building for acquiring **pesticides** regulatory and control tasks (both for public health and agriculture) in KSA. Outline your suggestions for fulfilling duties in evaluating the current situation, plan for development, prepare for acquiring related tasks and conduct regulatory and control tasks.

e) Provide a general definition of integrated water resource management, **IWRM** and briefly explain how this definition could be applied for the protection of surface drinking water supplies in a **KSA** Basin. In your explanation include two important guiding principles that would foster the long term sustainability of surface water supplies. Discuss importance of fostering the principle in KSA development plan.

Solution

IWRM is defined as a process that promotes the coordinated development and management of water, land and related resources, in order to maximize the resultant economic and social welfare in an equitable manner without compromising the sustainability of vital ecosystems.

f) The 8 Millennium Development Goals, MDGs, supplemented by 18 targets and 48 indicators designed to measure progress, constitute a blueprint for development. Illustrate your views concerning the participation of economic, social, and political institutions and the media, as well as the general public, in local and national efforts to achieve the **KSA MDGs.**

g) It is estimated that about 90% of the total water demand in the kingdom was consumed by agriculture. Outline your views in exploiting **groundwater resources** for irrigation purposes and significance of rain water harvesting in **KSA.**

h) With the current water shortage in KSA, there is a need as well as opportunities to look for **alternative sources** of **water** for use in **concrete** production. As a consultant engineer your advice is sought regarding this issue in line with American or KSA standards for concrete water quality. Draw your comments.

i) Discuss problems associated with **water demand** in **KSA**. Outline authorities' efforts to develop conventional water resources (surface and groundwater) and non-conventional ones (desalination of sea water and treated wastewater).

References and bibliography for further reading

1) Abdel-Magid, I.M., Mohammed, A.H., and Rowe, D.R., Modeling Methods for Environmental Engineers, CRC Press/Lewis Publ., Boca Raton, Fl., 1996.
2) Abdel-Magid, I.M., Selected problems in water supply, Khartoum University Press (KUP), P.O.Box 321, Khartoum, Sudan, 1986.
3) Abdel-Magid, I. M., Problem Solving in Environmetal Engineering, Dammm University Press, Dammam, 2013..
4) Davis, M., Water and Wastewater Engineering, McGraw-Hill Professional; 1 edi., 2010.
5) American Water Works Association and American Society of Civil Engineers, Water Treatment Plant Design 5/E, McGraw-Hill Professional; 5 edi., 2012.
6) AWWA (American Water Works Association), Water Environment Federation and E.W. Rice, R.B. Baird A.D. Eaton (Editors), L. S. Clesceri (Editor), Standard Methods for the Examination of Water and Wastewater by American Public Health Association, 2012.
7) Katebi, R., Johnson, M. A. and Wilkie, J., Control and Instrumentation For Wastewater Treatment Plants (Advances in Industrial Control), Springer, 2013.
8) Gray, N., Water Technology, taylor & francis; 3 edi., 2010.

9) Drinan, J. E. and Spellman, F., Water and Wastewater Treatment: A Guide for the Nonengineering Professional, CRC Press; 2 edi., 2012.

10) Urban Drainage and Flood Control District and Elisa Hindman, Urban Storm Drainage Criteria Manual; Volume 3, Stormwater Best Management Practices, Water Resources Pubns; 3rd edi., 2011.

11) Skinner, B., Small-Scale Water Supply: A Review of Technologies, Practical Action, 2003.

12) Mays, L. W., Water Resources Engineering, Wiley; 2 edi., 2010.

13) Sarai, D. S., Water Treatment Made Simple: For Operators, Wiley; 1 edi., 2005.

14)

15) Water Environment Federation, Operation of Municipal Wastewater Treatment Plants (3-Volume Set), McGraw-Hill Professional; 6 edi., 2007.

16) Spellman, F. R., Handbook of Water and Wastewater Treatment Plant Operations, CRC Press; 2 edi., 2008.

17) Degremont, Water treatment handbook, Lavoisier; 7th edi., 2007.

18) Development Information Centre, US Agency for International Development, National Demonstration Water Project, Institute for Rural Water and National Environmental Health Association, Water for the World Series: Designing septic tanks, Technical Note No., SAN.2.D.3.

19) Kerri, K. D., Dendy, B. B., Brady, J. and Crooks, W., Industrial Waste Treatment - A Field Study Training Program, Office of Water Programs, Vol1, 3rd Edi., 2005.

20) Shammas, N. K. and Wang, L. K., Fair, Geyer, and Okun's, Water and Wastewater Engineering: Water Supply and Wastewater Removal, Wiley; 3 edi., 2010.

21) Feachem, M.J., McGarry, M and Mara, D., Water, wastes and health in hot climates, John Wiely, Chichester, 1978.

22) Gerardi, M.H., Settleability Problems and Loss of Solids in the Activated Sludge Process (Wastewater Microbiology, Wiley-Interscience;, 2002.

23) Gunnerson, C. G., and Stuckey, D. C., Anaerobic Digestion Principles and Practice for Biogas Systems, World Bank, Technical Paper Number 49, World Bank, Washington, DC, USA, 1986.

24) Hammer, M. J., Water and Wastewater Technology, Prentice Hall; 2011.

25) Berner, E. K. and Berner, R. A., Global Environment: Water, Air, and Geochemical Cycles, Princeton University Press; 2 edi., 2012.

26) Kawamura, S., Integrated Design and Operation of Water Treatment Facilities, Wiley; 2 edi., 2000.

27) Tchobanoglous, G., Burton, F. L. and Stensel, H. D., Wastewater Engineering: Treatment and Reuse, McGraw-Hill Science/Engineering/Math; 4th edi., 2002.

28) McCabe, W., Smith, J., and Harriott, P., Unit Operations of Chemical Engineering, McGraw Hill Chemical Engineering Series, 2004.

29) Williamson, K., Civil Engineering: Water & Wastewater Treatment Review, Kaplan Publishing, 2007.

30) Inc. Metcalf & Eddy, Tchobanoglous, G., Stensel, H. D., Tsuchihashi, R., and Burton, F., Wastewater Engineering: Treatment and Resource Recovery, McGraw-Hill Science/Engineering/Math; 5 edi., 2013.

31) Munson, B.R., Rothmayer, A.P., Okiishi, T.H., and Huebsch, W.W., Fundamentals of Fluid Mechanics, Wiley, 2012.

32) Nathanson, J.A., Basic Environmental Technology: Water Supply, Waste Management & Pollution Control, Prentice Hall, 2007.

33) Nemerow, N. L., Agardy, F.J. and Salvato, J. A., Environmental Engineering, Wiley, 2009.

34) Popel, H. J., Aeration and gas transfer, Delft University of Technology, Herdruk, 1976.

35) McGraw-Hill, McGraw-Hill Concise Encyclopedia of Engineering, McGraw-Hill Professional; 1 edi., 2005.

36) Rowe, D.R. and Abdel-Magid, I.M., Handbook of Wastewater Reclamation and Reuse CRC Press/Lewis Publ., Boca Raton, FL, 1995.

37) Sawyer, C. N. and McCarty, P. L., Chemistry for environmental engineering and Science, McGraw-Hill Science/Engineering/Math, 2002.

38) Tchobanoglous, G., Burton, F.L., and Stensel, H.D., Wastewater Engineering: Treatment and Reuse, McGraw-Hill Science/Engineering/Math; 2002.

39) Kerry J. Howe, David W. Hand, John C. Crittenden and R. Rhodes Trussell, Principles of Water Treatment, Wiley; 1 edi., 2012.

40) Mihelcic, J. R. and Zimmerman, J. B., Environmental Engineering: Fundamentals, Sustainability, Design, Wiley; 1 edi., 2009.

41) Spellman, F. R. Handbook of Water and Wastewater Treatment Plant Operations, CRC Press; 2 edi., 2008.

42) Viessman, W., Hammer, M. J., Perez, E.M., and Chadik, P.A., Water Supply and Pollution Control, Prentice Hall; 8 edi., 2008.

43) Lin, S. and Lee, C., Water and Wastewater Calculations Manual, McGraw-Hill Professional; 2 edi., 2007.

44) Whipple, G.C. and Whipple, M.C., Solubility of oxygen in sea water, JACS, 33, 1911, 362.

45) WHO Scientific Group, Techniques for the collection and reporting of data on community water supply, WHO Tech. Report Series No. 490, Geneva, 1972.

46) WHO, Guidelines for Drinking-Water Quality, World Health Organization, 2004.

Appendices

 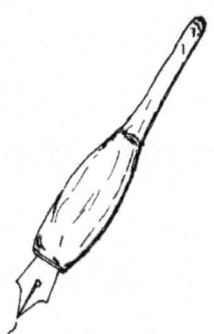

Appendix (A)
Some Physical Properties of Water at Various Temperatures (1)

Temperature, ^{o} C	Density, kg/m^3	Dynamic viscosity μx10^3 Ns/m^2	Kinematic viscosity vx10^6 m^2/s
0	999.8	1.793	1.792
5	1000	1.519	1.519
10	999.7	1.308	1.308
15	999.1	1.140	1.141
20	998.2	1.005	1.007
25	997.1	0.894	0.897
30	995.7	0.801	0.804
35	994.1	0.723	0.727
40	992.2	0.656	0.661
45	990.2	0.599	0.605
50	988.1	0.549	0.556
55	985.7	0.506	0.513
60	983.2	0.469	0.477
65	980.6	0.436	0.444
70	977.8	0.406	0.415
75	974.9	0.380	0.390
80	971.8	0.357	0.367

85	968.6	0.336	0.347
90	965.3	0.317	0.328
95	961.9	0.299	0.311
100	958.4	0.284	0.296

Appendix (B)
Vapor Pressure of Water (2)

Temperature, ° C	Vapor pressure	
	kPa	mm Hg
0	0.611	4.58
5	0.872	6.54
10	1.23	9.21
15	1.71	12.8
20	2.33	17.5
25	3.17	23.8
30	4.24	31.8

Appendix (C)
Distribution Coefficient of Oxygen in Water, k_D (1)

Temperature, ° C	K_D
0	0.0493
10	0.0398
20	0.0337
30	0.0296

Appendix (D)
Solubility of Oxygen in Water Exposed to Water– saturated Air at a Total Pressure of 76 cm Hg (1013 mbar) (44)

Temperature, °C	Chloride concentration in water, g/m³					Difference per 100 mg/L chloride
	0	5000	10000	15000	20000	
	Dissolved Oxygen, g/m³					
0	14.6	13.8	13.0	12.1	11.3	0.017
1	14.2	13.4	12.6	11.8	11.0	0.016
2	13.8	13.1	12.3	11.5	10.8	0.015
3	13.5	12.7	12.0	11.2	10.5	0.015
4	13.1	12.4	11.7	11.0	10.3	0.014
5	12.8	12.1	11.4	10.7	10.0	0.014
6	12.5	11.8	11.1	10.5	9.8	0.014
7	12.2	11.5	10.9	10.2	9.6	0.013
8	11.9	11.2	10.6	10.0	9.4	0.013
9	11.6	11.0	10.4	9.8	9.2	0.012
10	11.3	10.7	10.1	9.6	9.0	0.012
11	11.1	10.5	9.9	9.4	8.8	0.011
12	10.8	10.3	9.7	9.2	8.6	0.011
13	10.6	10.1	9.5	9.0	8.5	0.011
14	10.4	9.9	9.3	8.8	8.3	0.010
15	10.2	9.7	9.1	8.6	8.1	0.010
16	10.0	9.5	9.0	8.5	8.0	0.010
17	9.7	9.3	8.8	8.3	7.8	0.010
18	9.5	9.1	8.6	8.2	7.7	0.009
19	9.4	8.9	8.5	8.0	7.6	0.009
20	9.2	8.7	8.3	7.9	7.4	0.009
21	9.0	8.6	8.1	7.7	7.3	0.009
22	8.8	8.4	8.0	7.6	7.1	0.008
23	8.7	8.3	7.9	7.4	7.0	0.008
24	8.5	8.1	7.7	7.3	6.9	0.008
25	8.4	8.0	7.6	7.2	6.7	0.008
26	8.2	7.8	7.4	7.0	6.6	0.008

27	8.1	7.7	7.3	6.9	6.5	0.008
28	7.9	7.5	7.1	6.8	6.4	0.008
29	7.8	7.4	7.0	6.6	6.3	0.008
30	7.6	7.3	6.9	6.5	6.1	0.008

Appendix (E)
Some Atomic Weights of Relevant Substances

	Symbol	Atomic weight
Aluminum	Al	27
Barium	Ba	137
Calcium	Ca	40
Carbon	C	12
Chloride	Cl	35.5
Copper	Cu	63.5
Hydrogen	H	1
Iodine	I	126.9
Iron	Fe	56
Magnesium	Mg	24.3
Nitrogen	N	14
Oxygen	O	16
Phosphorous	P	31
Potassium	K	39
Silver	Ag	108
Sodium	Na	23
Sulfur	S	32
Zinc	Zn	65

Appendix (F) (1)

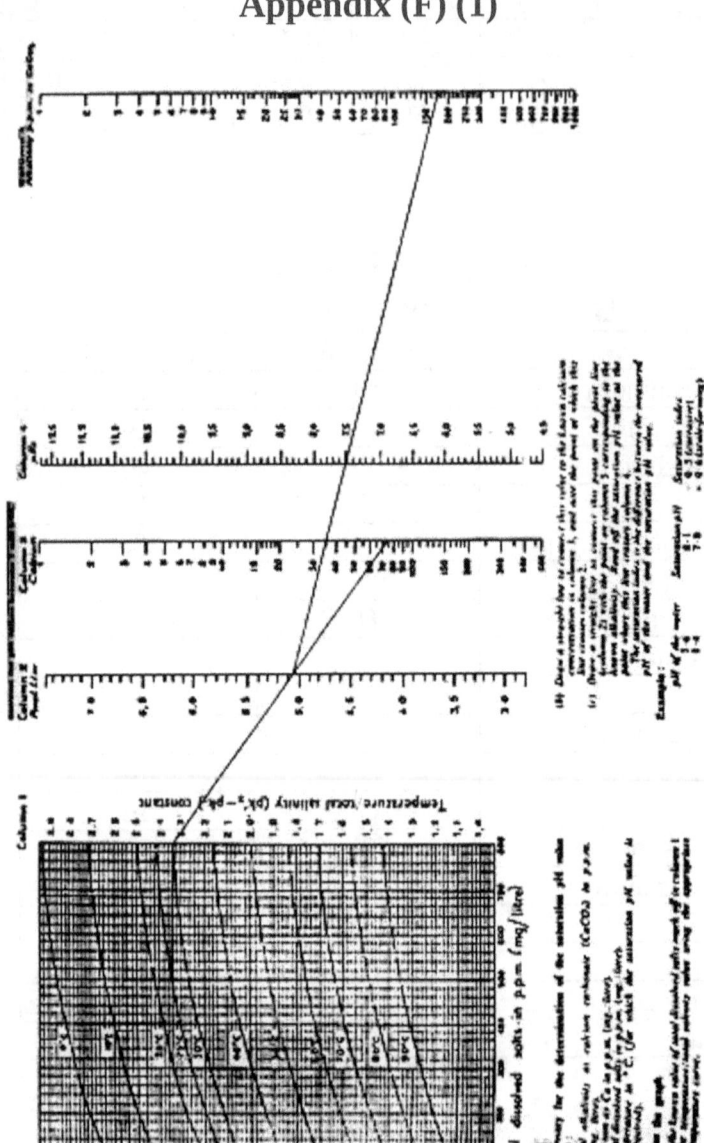

Appendix (G)
Tillman's Curve (1)

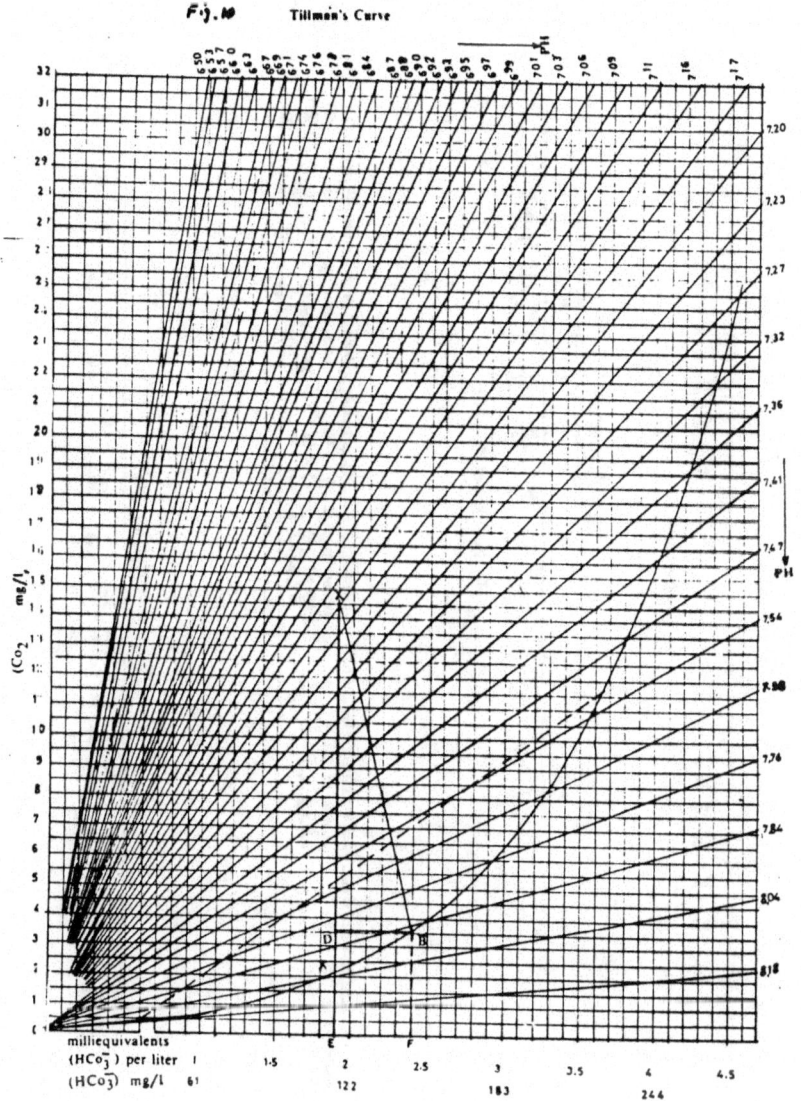

Fig. 10   Tillman's Curve

Appendix (H)
Diagram for solutions of Manning's monograph for circular pipes flowing full (n = 0.013)

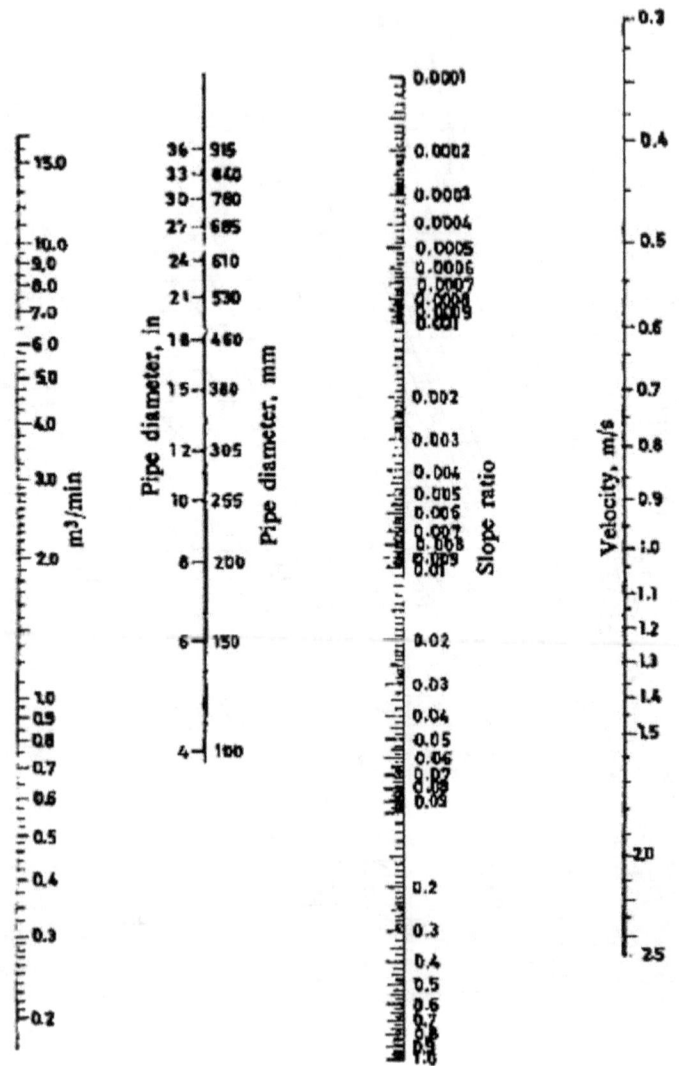

Appendix (H)
Diagram for solutions of Manning's monograph for circular pipes flowing full

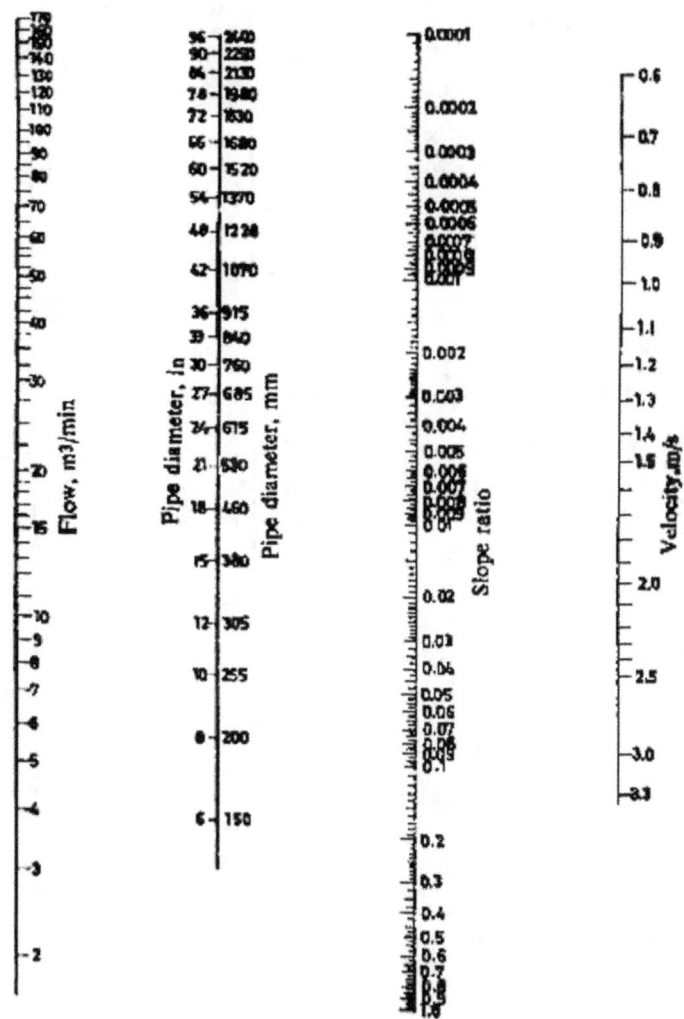

Appendix (H)
Manning's formula pipe flow chart

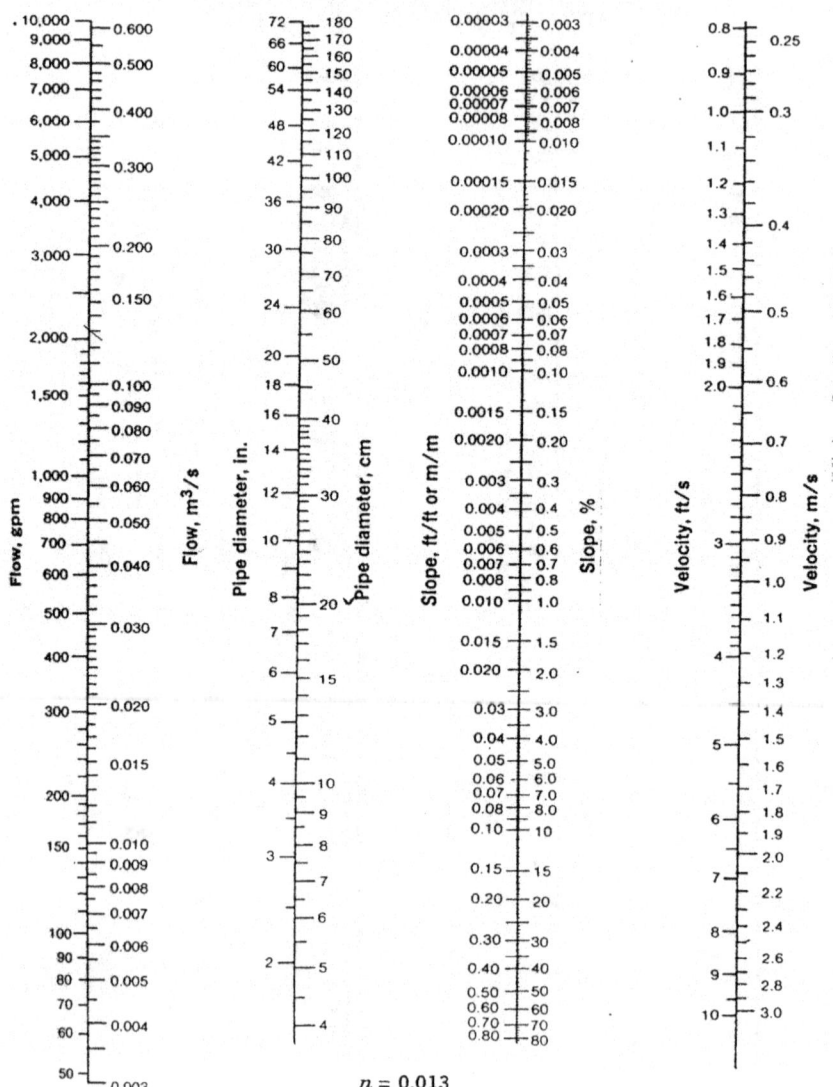

$n = 0.013$

Appendix (I)
The periodic table (Source:
http://volcano.oregonstate.edu/vwdocs/vwlessons/rocks_pics/a13.7.g
if)

Key:

12	← Atomic number
Mg	← Chemical symbol
24.31	← Atomic weight

1a	IIa		IIIb	IVb	Vb	VIb	VIIb	VIII			Ib	IIb	IIIa	IVa	Va	VIa	VIIa	0
1 H 1.008																		2 He 4.00
3 Li 6.94	4 Be 9.01												5 B 10.81	6 C 12.01	7 N 14.00	8 O 15.99	9 F 18.99	10 Ne 20.18
11 Na 22.99	12 Mg 24.31												13 Al 26.98	14 Si 28.09	15 P 30.97	16 S 32.06	17 Cl 35.45	18 Ar 39.95
19 K 39.10	20 Ca 40.08		21 Sc 44.6	22 Ti 47.90	23 V 50.94	24 Cr 51.99	25 Mn 54.94	26 Fe 55.85	27 Co 58.93	28 Ni 58.71	29 Cu 63.54	30 Zn 65.37	31 Ga 69.72	32 Ge 72.59	33 As 74.92	34 Se 78.96	35 Br 79.91	36 Kr 83.80
37 Rb 85.47	38 Sr 87.62		39 Y 88.91	40 Zr 91.22	41 Nb 92.91	42 Mo 95.94	43 Tc 99	44 Ru 101.97	45 Rh 102.91	46 Pd 106.4	47 Ag 107.87	48 Cd 112.40	49 In 114.82	50 Sn 118.69	51 Sb 121.75	52 Te 127.60	53 I 126.90	54 Xe 131.30
55 Cs 132.91	56 Ba 137.34	57-71 below	72 Hf 178.49	73 Ta 180.95	74 W 183.85	75 Re 186.2	76 Os 190.2	77 Ir 192.2	78 Pt 195.09	79 Au 196.97	80 Hg 200.59	81 Tl 204.37	82 Pb 207.19	83 Bi 208.98	84 Po 210	85 At 210	86 Rn 222
87 Fr 223	88 Ra 226	89-103 below	104 Rf 261	105 Ha 260	106 Sg 263													

57 La 138.91	58 Ce 140.12	59 Pr 140.91	60 Nd 144.24	61 Pm 147	62 Sm 150.35	63 Eu 151.96	64 Gd 157.24	65 Tb 158.92	66 Dy 162.50	67 Ho 164.93	68 Er 167.26	69 Tm 168.93	70 Yb 173.04	71 Lu 174.97
89 Ac 227	90 Th 232.04	91 Pa 231	92 U 238.03	93 Np 237	94 Pu 242	95 Am 243	96 Cm 247	97 Bk 247	98 Cf 251	99 Es 254	100 Fm 253	101 Md 256	102 No 254	103 Lw 257

Appendix (J)
Conversion Table

Multiply	by	to obtain
Area		
acre	43560	ft^2
acre	4047	m^2
cm^2	0.155	in^2
ft^2	0.0929	m^2
hecatre, ha	2.471	acre
in^2	6.452	cm^2
km^2	0.3861	$mile^2$
m^2	10.76	ft^2
mm^2	0.00155	in^2
Concentration		
mg/l	8.345	lb/million USA gal
ppm	1	mg/L
Density		
g/cm^3	1000	kg/m^3
g/cm^3	1	kg/L
g/cm^3	62.43	lb/ft^3
g/cm^3	10.022	lb/gal (Br.)
g/cm^3	8.345	lb/gal (USA)
kg/m^3	0.001	g/cm^3
kg/m^3	0.001	kg/L
kg/m^3	0.6242	lb/ft^3
Flowrate		
ft^3/s	448.8	gal/min
ft^3/s	28.32	L/s
ft^3/s	0.6462	M gal/d
M gal/d	1.547	ft^3/s
gal/min	0.00223	ft^3/s
gal/min	0.0631	L/s
L/s	15.85	gal/min
m^3/hr	4.4	gal/min
m^3/s	35.31	ft^3/s
Length		
ft	30.48	cm
in	2.54	cm
km	0.06214	mile
km	3280.8	ft
m	39.37	in
m	3.281	ft
m	1.094	yard
mile	5280	ft
mile	1.6093	km

mm	0.03937	in
Mass		
g	$2.205*10^{-3}$	lb
lb	16	ounce
kg	2.205	lb
tonne, t	1.102	ton (2000 lb)
Power		
Btu	252	cal
Btu	778	ft-lb
Btu	$3.93*10^{-4}$	HP-hr
Btu	$2.93*10^{-4}$	kW-hr
HP	0.7457	kW
Pressure		
atm	760	mm Hg
atm	29.92	in Hg
atm	33.93	ft water
atm	10.33	m water
atm	$1.033*10^4$	kg/m^2
atm	$1.013*10^5$	N/m^2
in water	1.8665	mm Hg
in Hg	0.49116	lb/in^2
in Hg	25.4	mm Hg
k Pa	0.145	lb/in^2, psi
mm Hg	0.01934	lb/in^2
mm Hg	13.595	kg/m^2
lb/in^2	6895	N/m^2
lb/in^2	0.0703	kg/cm^2
Temperature		
Fahrenheit, F	$5*(F-32)/9$	Centigrade, C
Centigrade, C	$(9C/5)+32$	F
C	$C+273.16$	Kelvin, K
Rankine	$F+459.69$	F
Velocity		
ft/s	30.48	cm/s
ft/s	1.097	km/hr
ft/min	0.508	cm/s
cm/s	0.03281	ft/s
cm/s	0.6	m/min
m/s	3.281	ft/s
m/s	196.8	ft/min
Viscosity		
centipoise	0.01	g/cm.s
centistoke	0.01	cm^2/s
Volume		
ft^3	0.02832	m^3
ft^3	6.229	gal (Br.)

ft^3	7.481	gal (USA)
ft^3	28.316	L
gal	0.1337	ft^3
gal (USA)	0.833	gal (Br.)
gal	3.785	L
L	0.001	m^3
L	0.03532	ft^3
L	0.22	gal (Br.)
L	0.2642	gal (USA)
m^3	35.314	ft^3
m^3	1000	L

Appendix (K)
Useful Formulae

	Volume	
Circular Cone l, h, r	$\pi r^2 h/3$	**Area of curved surface = πrl**
Sector of a circle r, l, ϕ included angle	$4\pi r^2\phi/360$	**Length of chord = $\pi r\phi/180$**
Sphere	$\pi r^3/3$	**Surface ara = $4\pi r^2$**
Trapezium h, b, a		**$(a + b)*h/2$**
Triangle A, B, C, a, b, c, ϕ	$\sqrt{s(s-a)(s-b)(s-c)}$ Where $2s = a + b + c$ $(a*b*\sin C)/2 =$	

$a^2 = b^2 + c^2 - 2bc\cos\phi$ $a/\sin A = b/\sin B = c/\sin C$	$(a*c*\sin B)/2 =$ $(b*c*\sin A)/2$	
$\sin(\theta \pm \phi) = \sin\theta\cos\phi \pm \cos\theta\sin\phi$ $\cos(\theta \pm \phi) = \cos\theta\cos\phi \mp \sin\theta\sin\phi$ $\tan(\theta \pm \phi) = (\tan\theta \pm \tan\phi)/(1 \mp \tan\theta\tan\phi)$ $\sin 2\phi = 2\sin\phi\cos\phi$ $\cos 2\phi = \cos^2\phi - \sin^2\phi = 2\cos^2\phi - 1 = 1 - 2\sin^2\phi$		

Appendix (L)
Prefixes of the Metric System

Factor	Prefix	Symbol
$1\ 000\ 000\ 000\ 000 = 10^{12}$	Tetra	T
$1\ 000\ 000\ 000\ = 10^{9}$	Giga	G
$1\ 000\ 000 = 10^{12}$	Mega	M
$1\ 000\ 000\ 000\ 000 = 10^{6}$	Kilo	k
$1\ 000 = 10^{3}$	Hecto	h
$1\ 0 = 10^{2}$	Deca	da
$1\ 0 = 10^{1}$	Deci	d
$0.1 = 10^{-1}$	Centi	c
$0.01 = 10^{-2}$	Milli	m
$0.001 = 10^{-3}$	Micro	μ
$0.000\ 000 = 10^{-6}$	Nano	n
$0.000\ 000\ 000 = 10^{-9}$	Pico	p
$0.000\ 000\ 000\ 000 = 10^{-12}$	Femto	f
$0.\ 000\ 000\ 000\ 000\ 000 = 10^{-15}$	Atto	a

Appendix (M)
Screenshots of the example programs

Program 2.1 – Form1.vb:

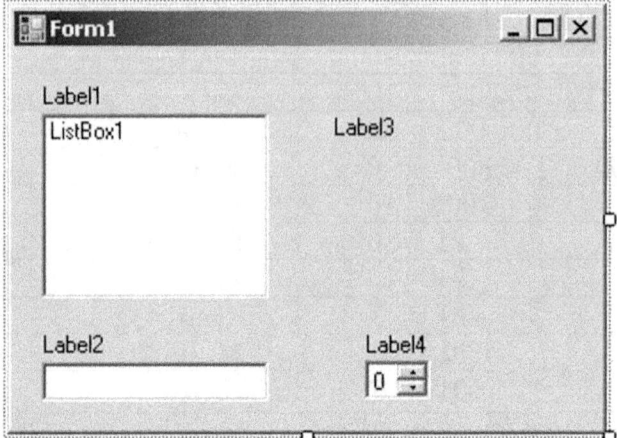

Program 2.2 – Form1.vb:

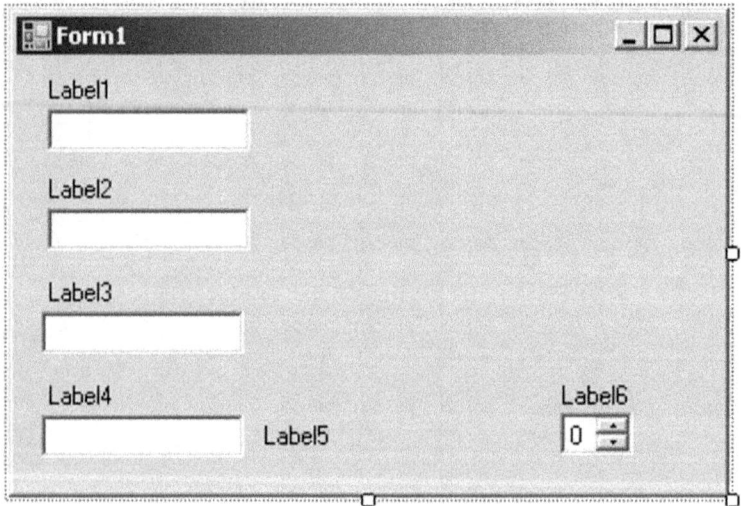

Program 2.3 – Form1.vb:

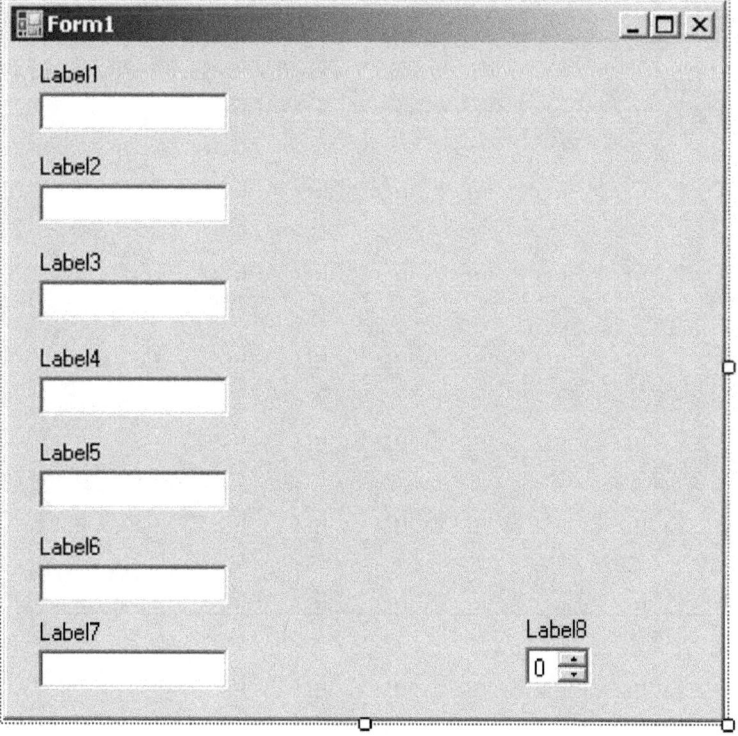

Program 2.4 – Form1.vb:

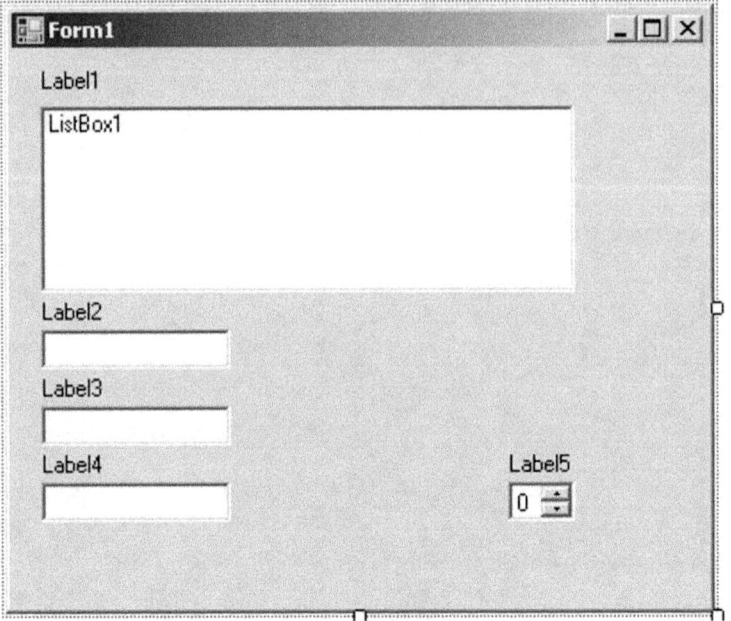

Program 2.6 – Form1.vb:

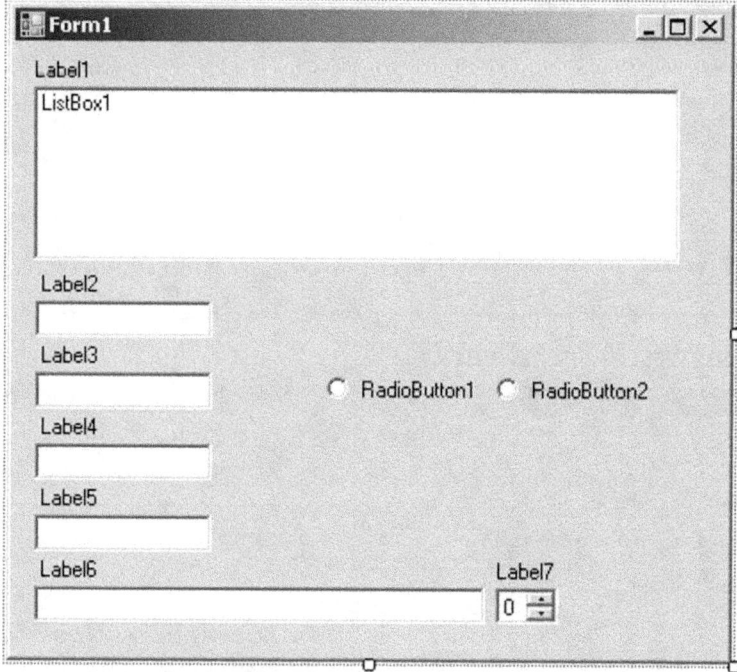

Program 2.7 – Form1.vb:

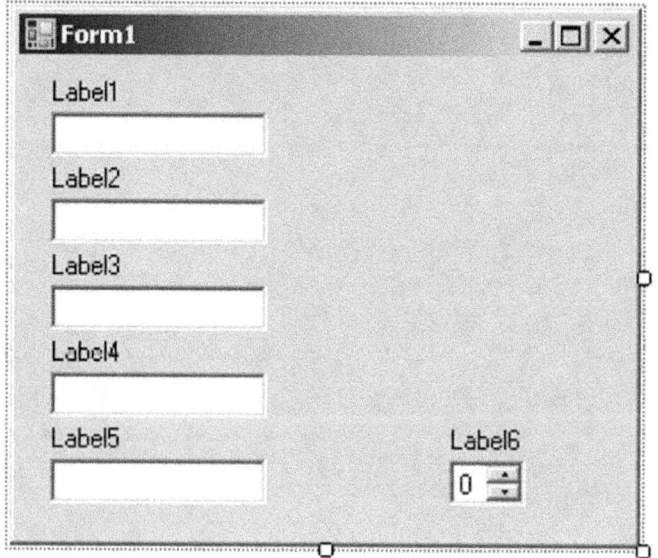

Program 2.9 – Form1.vb:

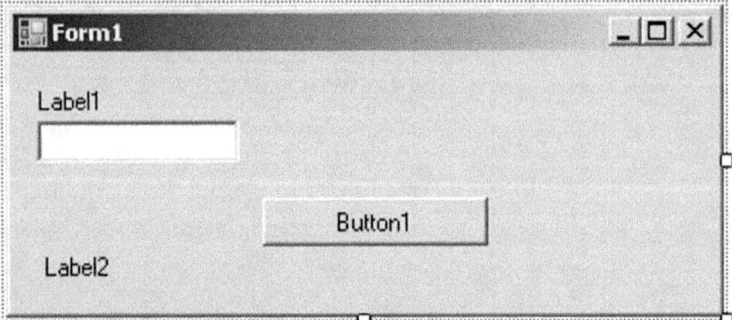

Program 2.11 – Form1.vb:

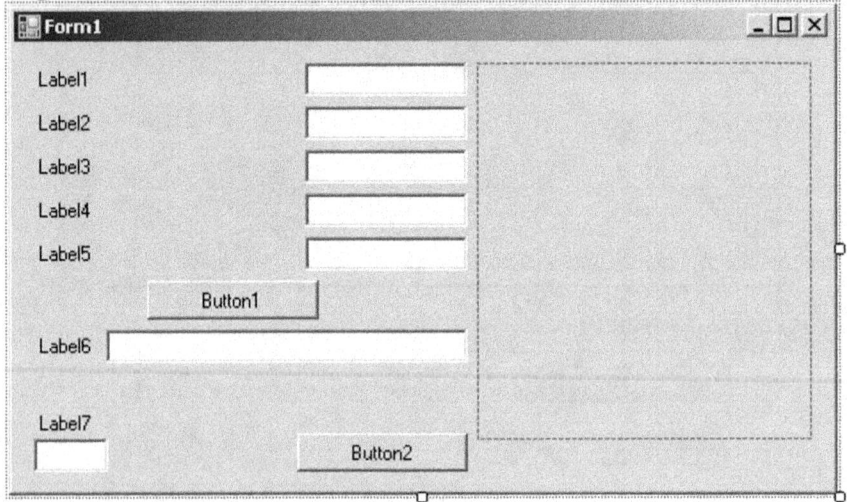

Program 2.13 – Form1.vb:

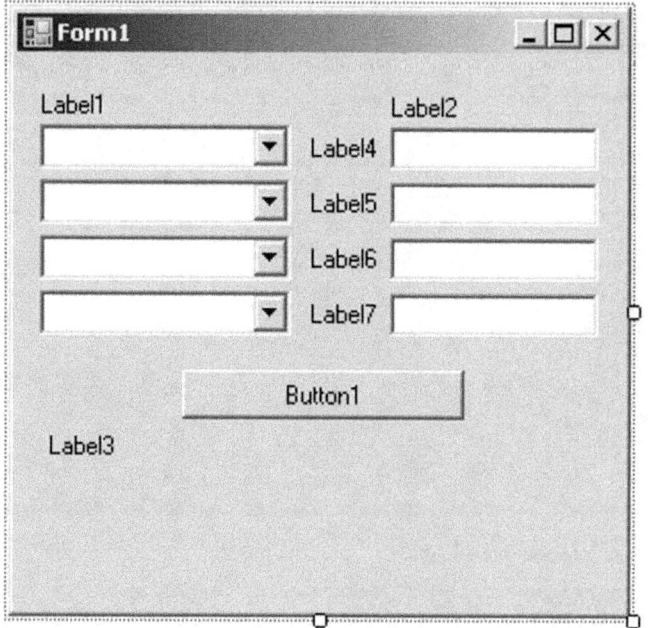

Program 2.15 – Form1.vb:

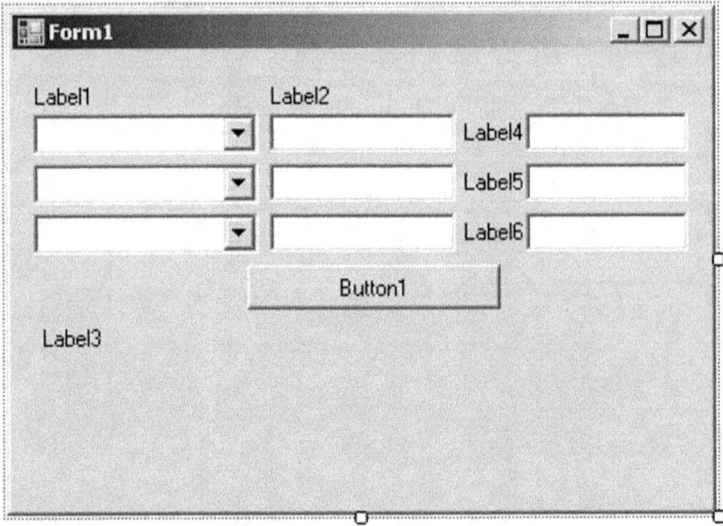

Program 2.17 – Form1.vb:

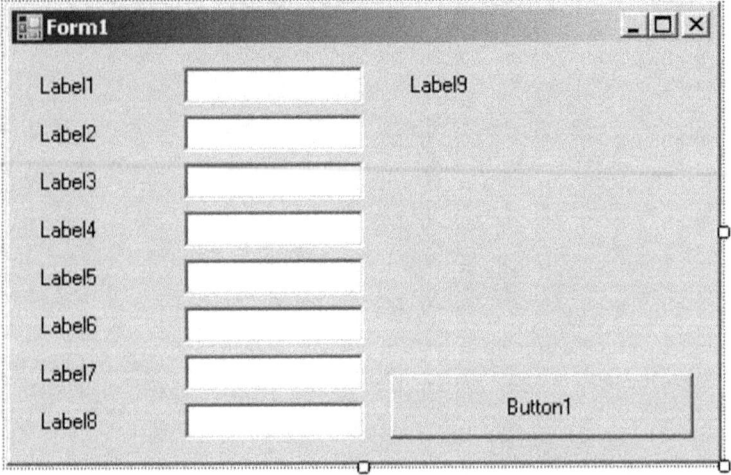

Program 2.18 – Form1.vb:

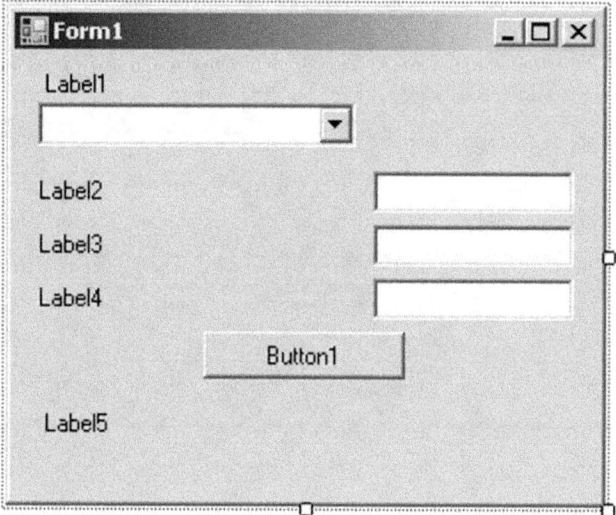

Program 2.19 – Form1.vb:

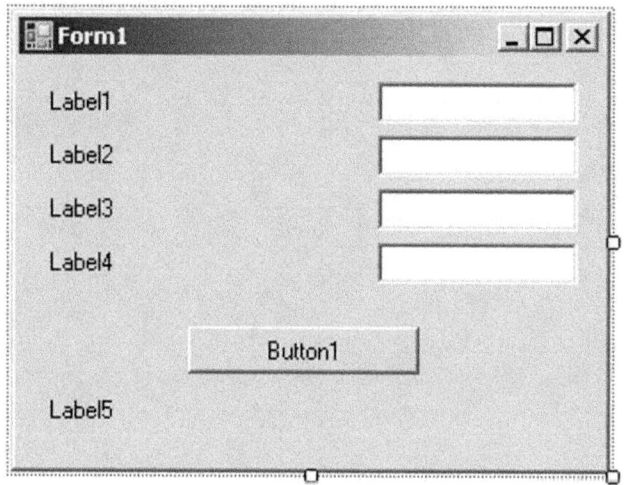

Program 3.1 – Form1.vb:

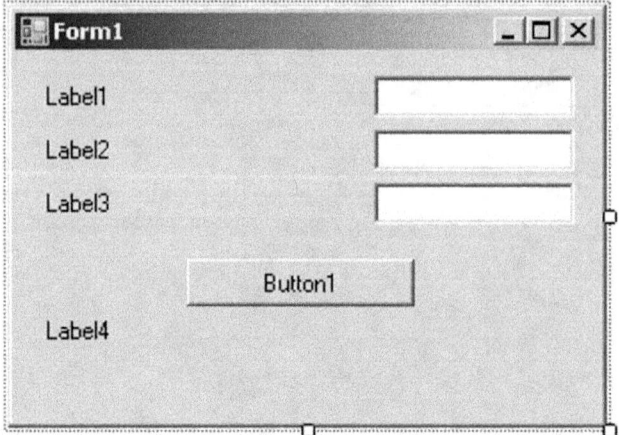

Program 3.6 – Form1.vb:

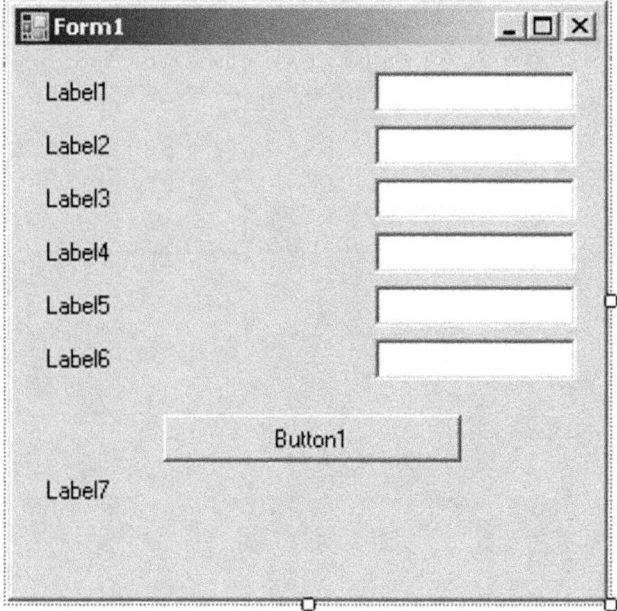

Program 3.7 – Form1.vb:

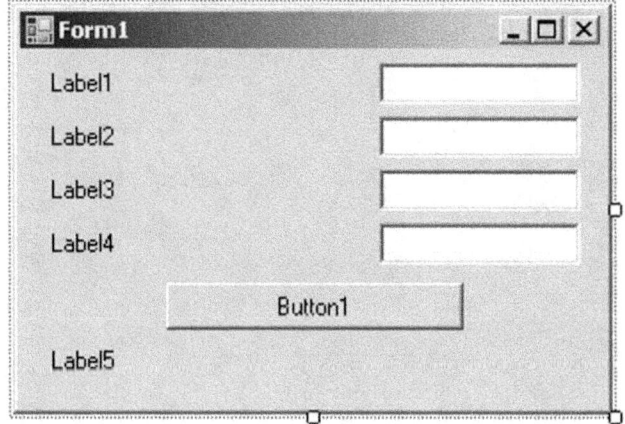

Program 3.8 – Form1.vb:

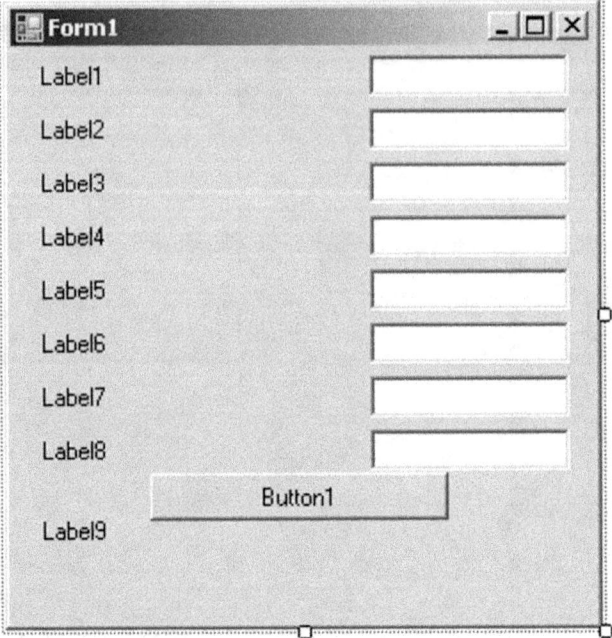

Program 3.10 – Form1.vb:

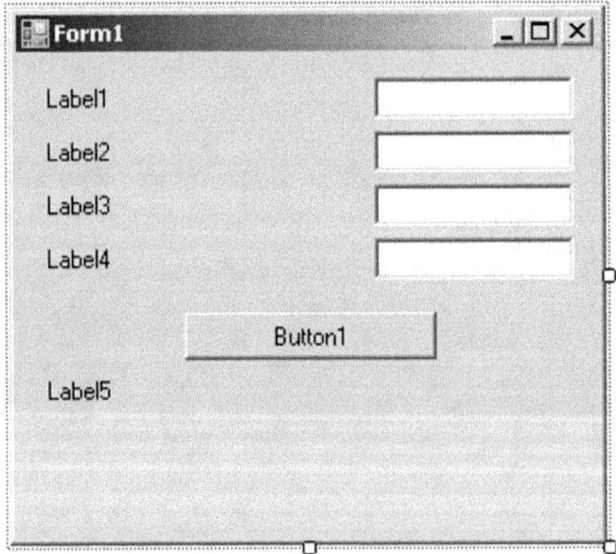

Program 3.12 – Form1.vb:

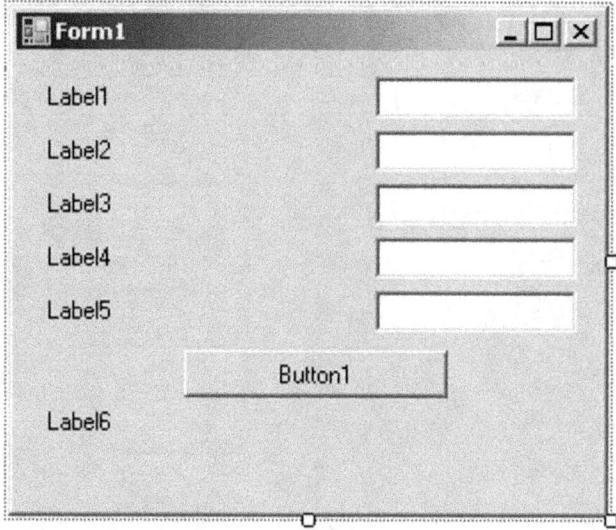

Program 3.13 – Form1.vb:

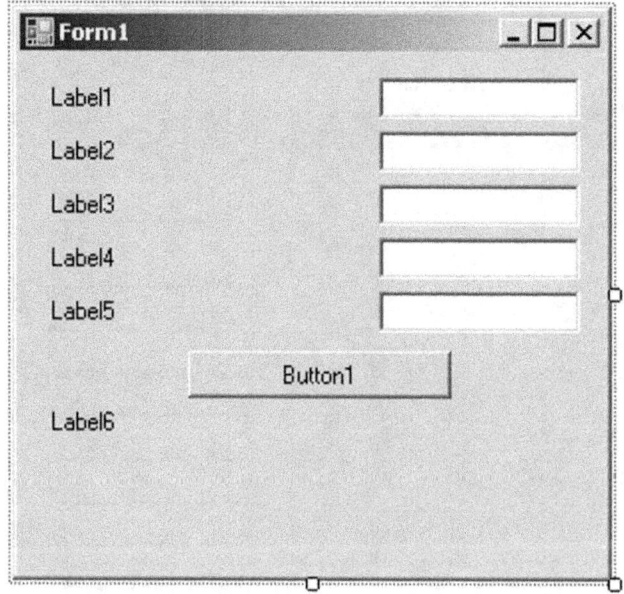

Program 3.14 – Form1.vb:

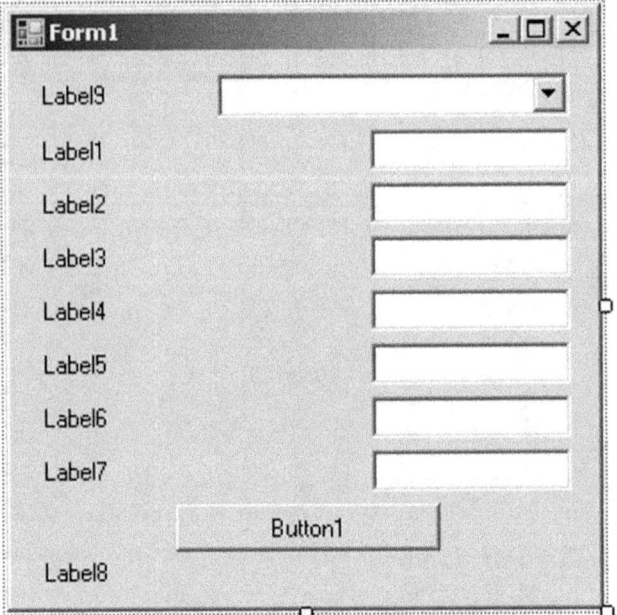

Program 3.16 – Form1.vb:

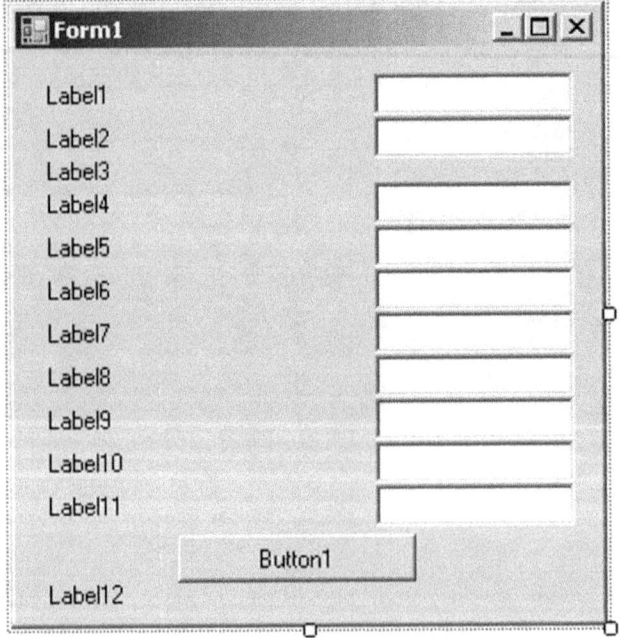

Program 3.18 – Form1.vb:

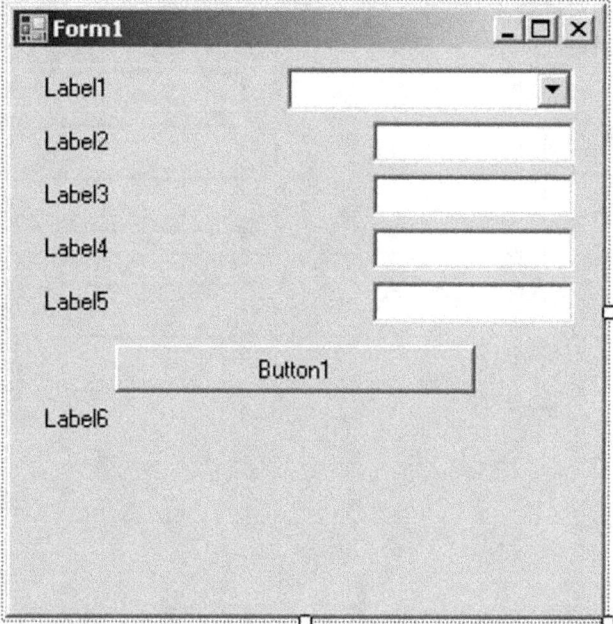

Program 3.21 – Form1.vb:

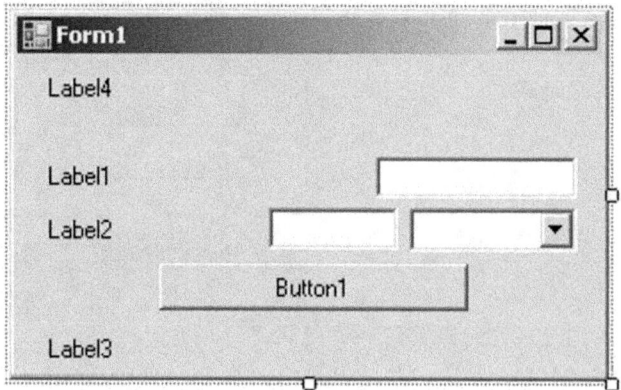

Program 4.1 – Form1.vb:

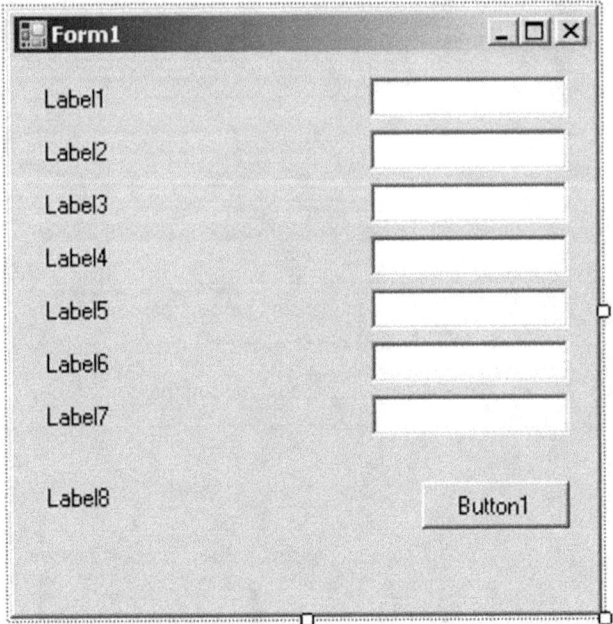

Program 5.1 – Form1.vb:

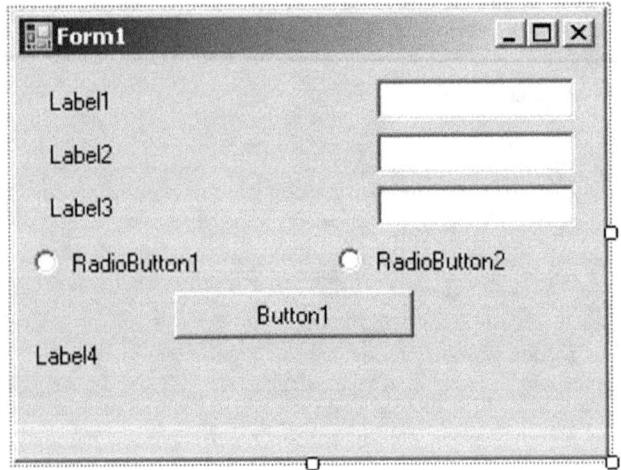

Program 5.2.A.1 – Form1.vb:

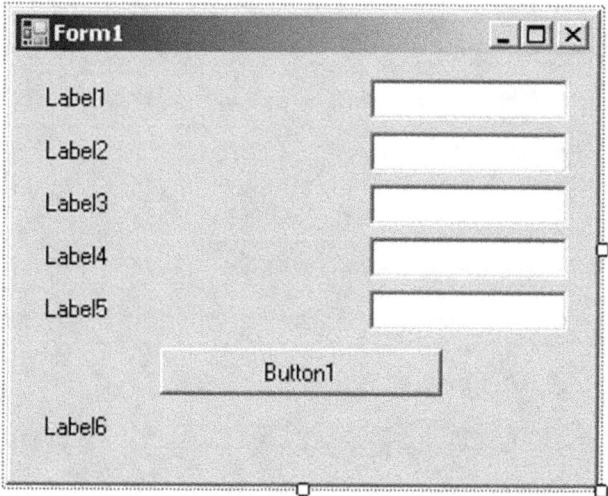

Program 5.2.A.2 – Form1.vb:

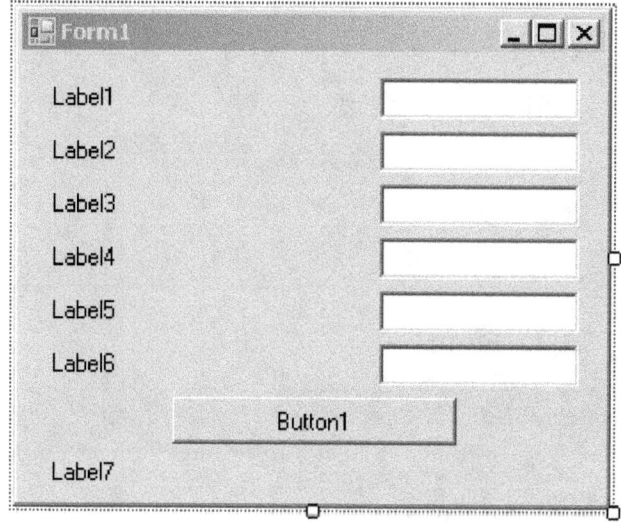

Program 5.2 – Form1.vb:

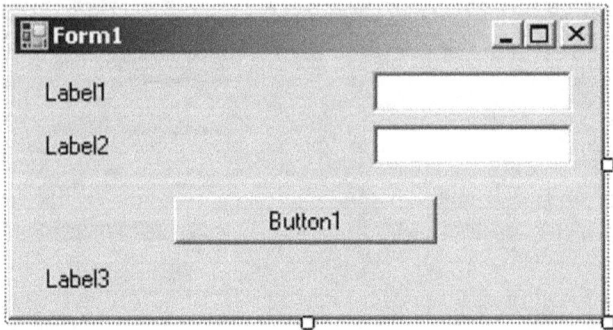

Program 5.3.1.4 – Form1.vb:

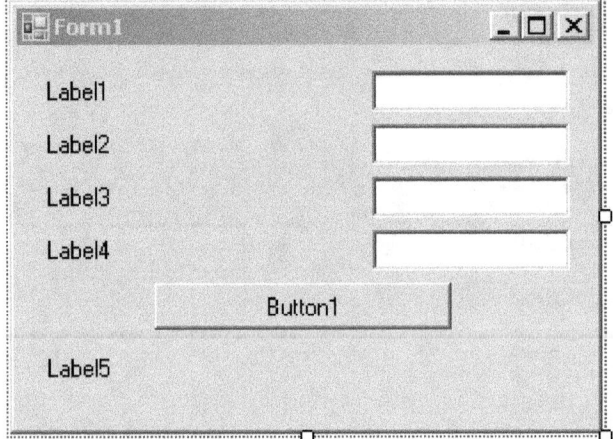

Program 5.3.1.7 – Form1.vb:

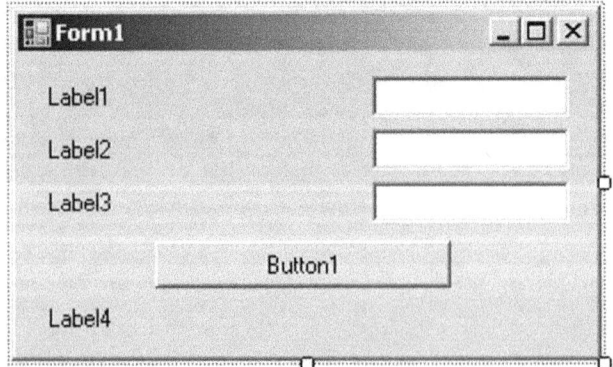

Program 5.3 – Form1.vb:

Program 5.4 – Form1.vb:

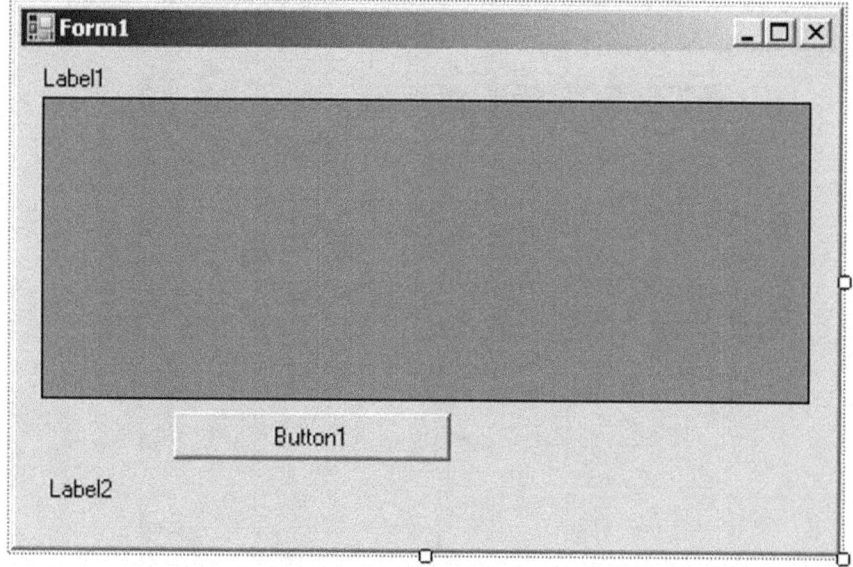

Program 5.5 – Form1.vb:

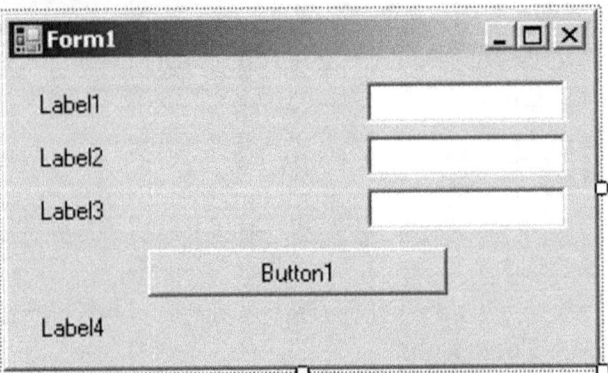

Program 5.6 – Form1.vb:

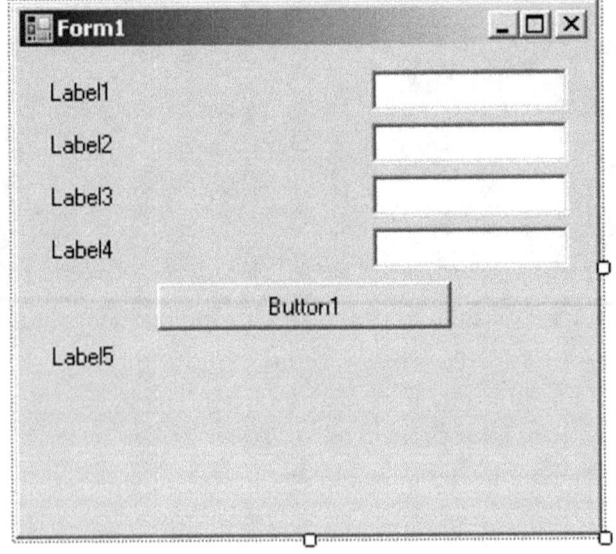

Program 5.8 – Form1.vb:

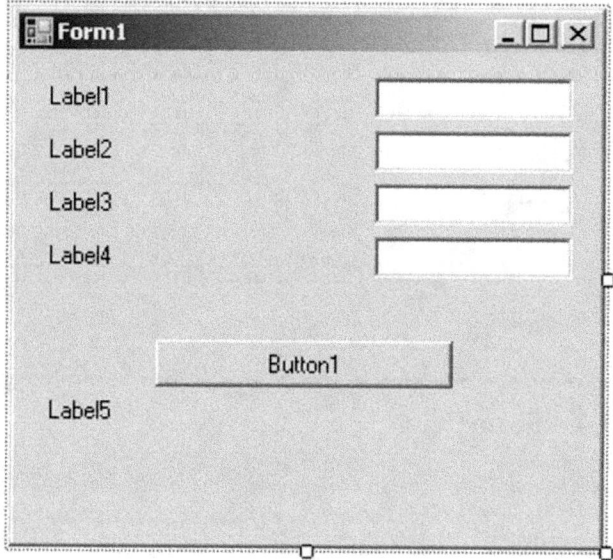

Program 6.1 – Form1.vb:

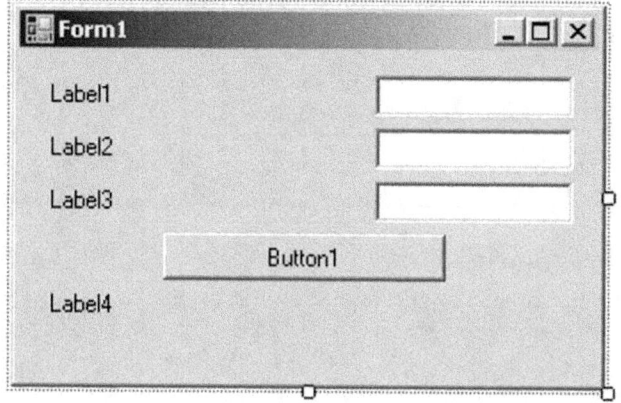

Program 6.2 – Form1.vb:

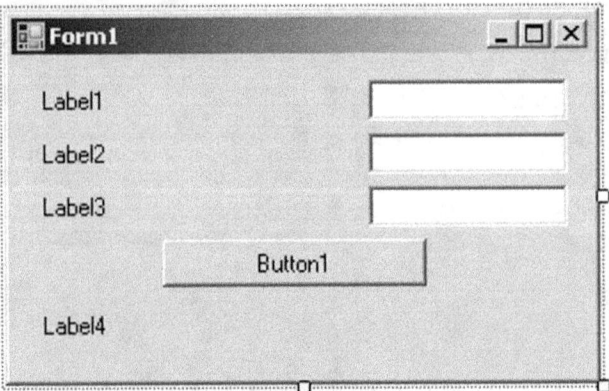

Program 6.4 – Form1.vb:

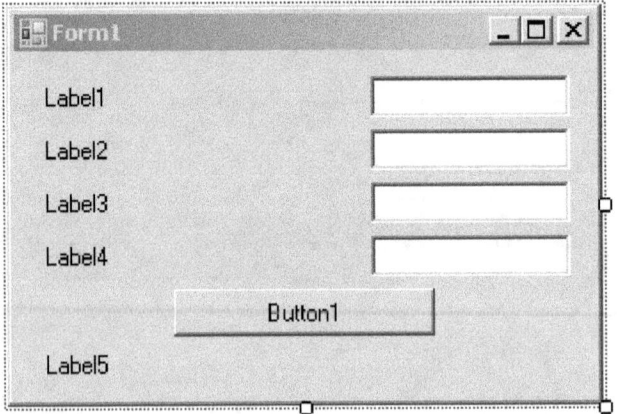

Program 6.5 – Form1.vb:

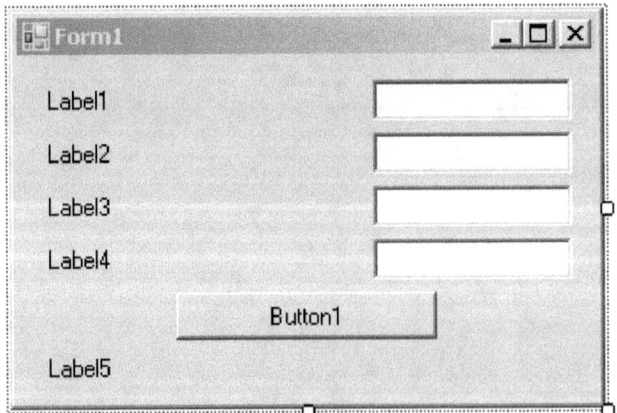

Program 6.7 – Form1.vb:

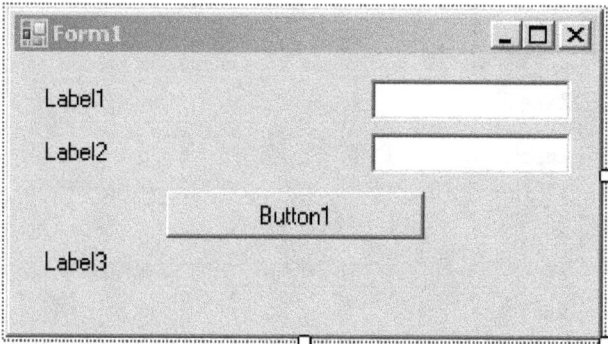

Program 6.11 – Form1.vb:

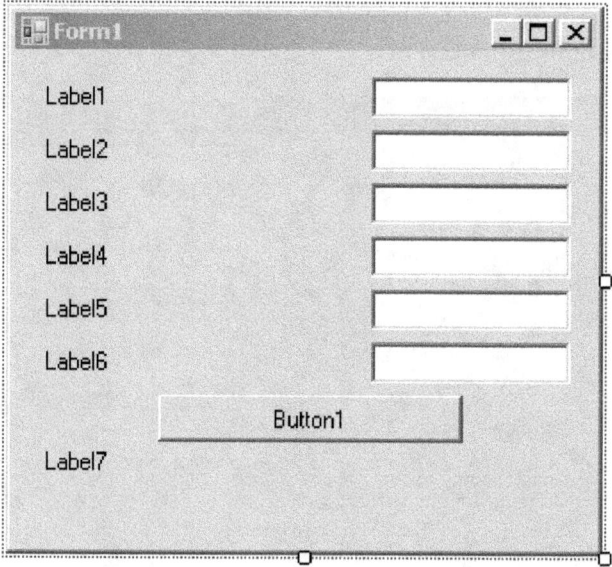

Program 6.13 – Form1.vb:

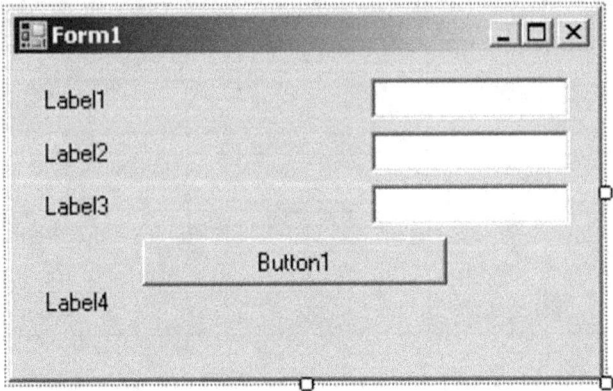

Program 6.15 – Form1.vb:

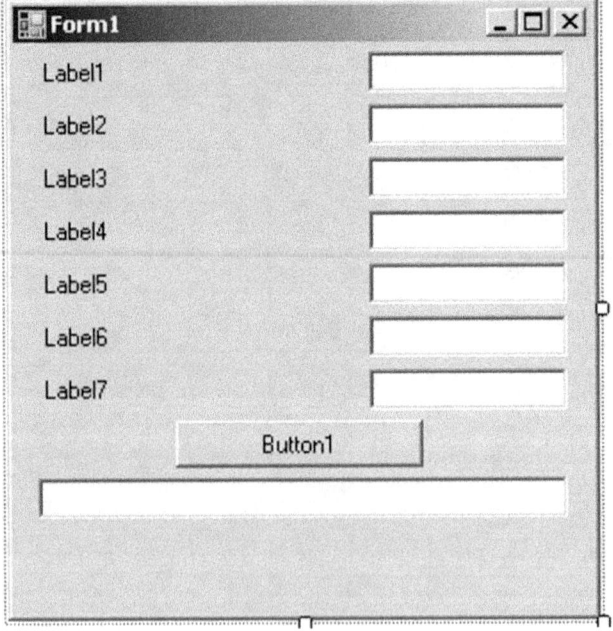

Program 6.18 – Form1.vb:

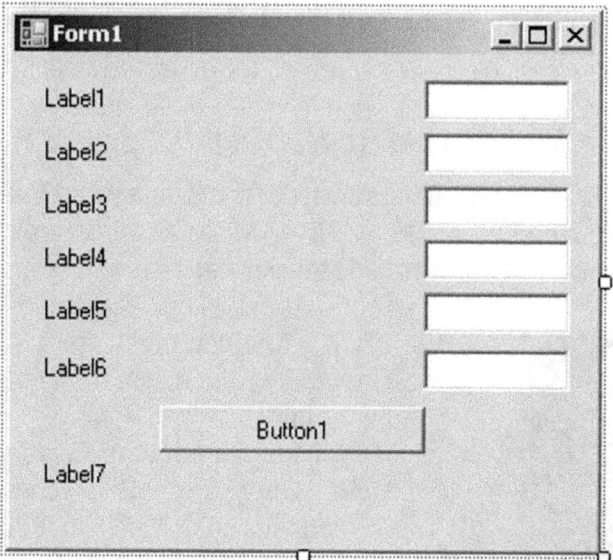

Program 6.23 – Form1.vb:

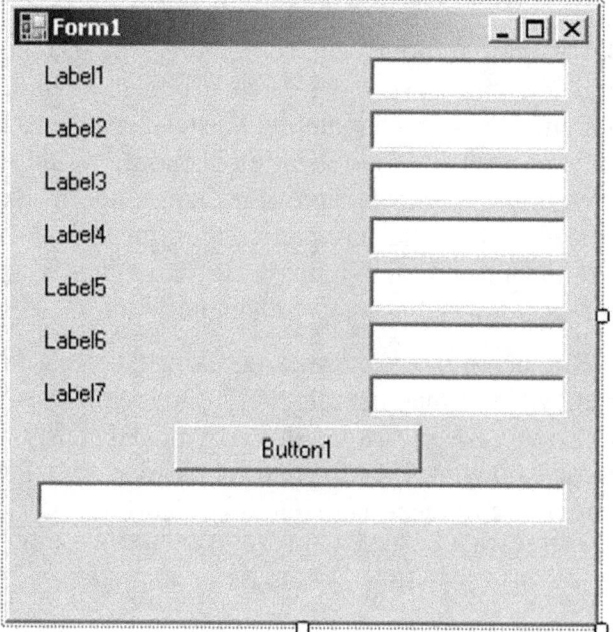

Authors of second edition at a glance

Prof. Isam Mohammed Abdel-Magid:

Prof. Dr. Eng. **Isam Mohammed Abdel-Magid Ahmed**, B.Sc., PDH, DSE, Ph.D., FSES, CSEC, MSECS. Professor of Water Resources and Environmental Engineering.

 Prof. Abdel-Magid received his B.Sc. Civil Engineering from Khartoum University, Sudan; Diploma in Hydrology from Padova University, Italy; M.Sc. in Sanitary Engineering from Delft University of Technology, the Netherlands and Ph.D. in Public Health Engineering from Strathclyde University, U. K in 1977, 1978, 1979 and 1982 respectively.

Prof. Abdel-Magid authored or co-authored over many papers, publications, scientific text and reference books, technical reports, lecture notes in areas of water supply; wastewater disposal, reuse & reclamation and reuse; solid waste disposal; water resources, industrial wastes; and slow sand filtration both in English and Arabic. Prof. Abdel-Magid has participated in several workshops, symposia, seminars and conferences. He has edited and co-edited many conference proceedings and college bulletins.

The Sudan Engineering Society awarded him the prize for the Best Project in Civil Engineering in 1977. In 1986 the Khartoum University Press (KUP) awarded him an Honourly Scarf for Enrichment of Knowledge. MoI Prize for Second Best Performance in 4th year Civil Engineering, Honourly Scarf for Enrichment of Knowledge, UoK Press. Best book of the year - MoIC Sudanese Press Council, ALECSO prize for a book in engineering.

Prof. Abdel-Magid has taught environmental engineering subjects and supervised research projects for graduate and post-graduate

Students at Khartoum University (Sudan), United Arab Emirates University (United Arab Emirates), Sultan Qaboos University (Oman), Sudan University for Science and Technology (Sudan), Juba University (Sudan), Industrial Research and Consultancy Centre (Sudan), Sudan Acdemy for Sciences (Sudan), King Faisal University (KSA), and University of Dammam (KSA). Prof. Abdel-Magid acted as external examiner to different institutions, founder of some institutions, centers & refereed journals.

Current address: Head proofreading and Revision Dept., Center of Scientific Publications and Member of its Council, Building 800, Room 28, Environmental Engineering Department, College of Engineering, University of Dammam, Box 1982, Dammam 31451, KSA. Fax: +96638584331, Phone: 966530310018+, +96633331686, E-mail: iahmed@ud.edu.sa and isam abdelmagid@yahoo.com, Web site: http://www/sites.google.com/site/isamabdelmagid/

Dr. Mohammed Isam Mohammed Abdel-Magid

Dr. Mohammed Isam Mohammed Abdel-Magid (MBBS, BLS, ALS, MRCP-UK Part I&II Written) is a graduate of the College of Medicine, University of Khartoum, Sudan, 2008. He completed basic training with the Ministry of Health, Sudan, then worked as a physician in the department of Internal Medicine, Ribat University hospital, Sudan, and the Ministry of Health, Kingdom of Saudi Arabia.

He is completing his higher training with the membership of the Royal Colleges of Physicians of the United Kingdom (MRCP-UK) of which he completed two parts.

He tutored in problem-based learning teaching sessions in the department of Internal Medicine, Sudan International University, Sudan.

He is a registered practicing physician with the Sudan Medical Council, the Health Authority of Abu-Dhabi (HAAD), and the Saudi Commission of Health Specialties (SCHS). He is a full member of the Society of Acute Medicine of UK (SAM), the European Society for Emergency Medicine (EuSEM), and the European Respiratory Society (ERS).

He is a peer reviewer with the Science Journal of Medicine & Clinical Trial and the Pan-African Journal of Medical Sciences.